Handbook of Animal Models in Transplantation Research

Editors

Donald V. Cramer, D.V.M., Ph.D.
Luis G. Podesta, M.D.
Leonard Makowka, M.D., Ph.D.
Cedars-Sinai Medical Center
Los Angeles, California

Library of Congress Cataloging-in-Publication Data

Handbook of animal models in transplantation research / editors, Donald V. Cramer, Luis G. Podesta, Leonard Makowka.

p. cm.

Includes bibliographical references and index.

ISBN 0-8493-3629-5 (acid-free paper)

1. Transplantation of organs, tissues, etc.—Research—Technique. 2. Surgery, Experimental. 3. Animal models in research. I. Cramer, Donald V. II. Podesta, Luis G. III. Makowka, Leonard, 1953–.

RD120.7.H36 1994

617.9′5′0072—dc20 93-5638

CIP

Direct all inquiries to CRC Press, Inc., 2000 Corporate Blvd., N.W., Boca Raton, Florida 33431.

International Standard Book Number 0-8493-3629-5

Library of Congress Card Number 93-5638

Printed in the United States of America 2 3 4 5 6 7 8 9 0

Printed on acid-free paper

ACKNOWLEDGMENT

The editors wish to express their sincere gratitude and appreciation for the invaluable assistance of Ms. Rhonda McIver in the preparation and editing of this book. Ms. McIver's attention to detail and her tireless efforts allowed for a smooth and efficient completion of this work.

PREFACE

The significant advances in clinical transplantation over the past few decades have allowed for the evolution of organ grafting from an experimental surgical procedure to an effective therapy for the treatment of end-stage organ failure. Most transplant centers now report one year recipient and graft survival rates that exceed 80% for hepatic, renal, and cardiac grafts. Several factors have contributed to the current success of clinical transplantation, including the development of effective immunosuppressive therapies, the perfection of surgical techniques, the establishment of tissue-typing protocols, and an improved understanding of the immunological mechanisms involved in graft rejection. These medical advances have made extensive use of experimental surgical techniques in a variety of animal models of organ transplantation. Reproducible and reliable data have been generated from these experiments and this information has been a key ingredient in the development of protocols for subsequent clinical trials.

As early as 1905 the French surgeon Alexis Carrel developed the vascular techniques that allowed for kidney and heart transplantation in dogs. He was awarded the Nobel Prize for the development of these techniques in 1912 and they now form the basis for modern transplantation surgery. Subsequently, several unsuccessful attempts at clinical organ transplantation were made during 1933–1939, followed by a resurgence of interest in these techniques in the early 1960s. The renewal of interest in transplantation was the result of the development of pharmacologic agents (6-mercaptopurine) capable of suppressing the immune-mediated rejection of allografts. The subsequent development of sophisticated surgical models of organ transplantation in several animal models allowed for a variety of experiments to be performed. These have included the testing of new immunosuppressive drugs and preservation solutions, the examination of the immunological mechanism(s) responsible for graft rejection, the development of new sources of donor organs (xenografts), and the establishment of improved immunologic monitoring techniques posttransplantation. These models have also been utilized in surgical training programs to allow members of transplant surgical teams to acquire the skills necessary for use of these techniques in the operating room.

The *Handbook of Animal Models in Transplantation Research* provides detailed information on the use of different surgical models in experimental organ transplantation. Each chapter includes a historical perspective on the development of a specific transplantation procedure, a review of the anatomy of the appropriate organ systems, a detailed description of the surgical procedures, and a short summary of the potential experimental application of the transplant model. As discussed in the Handbook, the selection of an appropriate model for a specific experimental procedure depends on: (1) the nature of the experimental study, (2) the type of organ system to be evaluated, and (3) the financial resources available. We have compiled a series of 22 chapters which have been prepared by a group of authors with a wide and varied experience in clinical and experimental transplantation. The surgical models for seven different organ systems (heart, heart-lung, kidney, liver, small bowel, pancreas, and multivisceral) have been included with individual chapters that describe the experimental applications in large and small animals. The use of different organ systems in the same experimental setting has proved valuable for the identification of differences in the sensitivity of individual organs to the rejection process. Methods for cellular transplantation, such as isolated hepatocyte and pancreatic islet transplantation, new modes of immunosuppression, and a description of experimental models for xenotransplantation have been included to provide a broad framework for the design of the most appropriate studies.

We intend that this Handbook will provide the clinician and basic scientist with a broad perspective on the current experimental models utilized for organ transplantation and stimulate additional areas of research in this important field. The continued use, development, and refinement of animal models for basic research is essential for the future development and application of organ transplantation as an effective therapeutic modality.

THE EDITORS

Donald V. Cramer, D.V.M., Ph.D., is Director of the Transplantation Biology Research Laboratory in the Department of Surgery at Cedars-Sinai Medical Center in Los Angeles. He obtained his training at the University of California, Berkeley and Davis campuses, including a B.S. degree in 1964 and D.V.M. degree in 1966. He received a Ph.D. from Harvard University Medical School in 1978. He served as an Instructor in Pathology at the University of Pittsburgh, School of Medicine from 1971 to 1972, a Clinical Assistant to the Medical Staff at the Presbyterian-University Hospital (Pittsburgh, PA) from 1971 to 1989, an Associate Professor of Pathology (with tenure), University of Pittsburgh, School of Medicine from 1978 to 1989, and was a member of the Pittsburgh Cancer Institute from 1986 to 1989. In 1989 he assumed his present position.

Dr. Cramer is a member of the American College of Veterinary Pathologists, American Association of Immunologists, International Transplantation Society, American Society of Transplant Physicians, International Society for Heart and Lung Transplantation, the New York Academy of Sciences, Transplantation Society, American Society for Histocompatibility and Immunogenetics, American Association for the Advancement of Science, American Association of Pathologists, Inc., the American Society of Investigative Pathology, and American Society of Transplant Surgeons. He has served on a variety of professional review committees, including Comparative Medicine Review Committee, Division of Research Grants, NIH, Ad hoc reviewer of the National Institutes of Health Study Sections (Aging Review Committee; Surgery and Bioengineering; Surgery, Anesthesiology and Trauma); National Science Foundation and the Veterans Administration, and the National Research Council ILAR Committee on Animal Models and Genetic Stocks. He is currently on the Editorial Board of *Transplantation, Transplantation Science,* and *Comparative Medicine.* Among his numerous awards received are the Pfizer Award, Merck Annual Scholarship Award, Research Career Development Award from NIH, and the Alexander von Humboldt Foundation Research Fellowship from Federal Republic of Germany. Dr. Cramer is the author of more than 150 manuscripts, chapters, and books. Dr. Cramer's current research activities are focused on the identification of the pathogenetic mechanisms responsible for the rejection of vascularized allografts and xenografts.

Luis G. Podesta, M.D., is Associate Director of Transplantation Services at Cedars-Sinai Medical Center. Dr. Podesta trained under Dr. Thomas E. Starzl at the University of Pittsburgh from 1986 through 1989. During the first year of that period Dr. Podesta completed a research fellowship in transplantation, with major emphasis on hepatic preservation, hyperacute rejection, and multivisceral transplantation. After the first year he became a visiting Professor on the medical staff performing numerous donor liver harvesting procedures and a large number of liver transplantations, including several children. At the same time he was co-investigator in research related to new preservation solutions and the initial trials on large animals using FK506. Dr. Podesta was also involved in the abdominal organ cluster procedure and multivisceral transplantation in the clinical phase. Born in Argentina, Dr. Podesta completed all of his medical and surgical training in that country and was staff member and Assistant Professor of the University of Buenos Aires School of Medicine Surgical Department. During his time at Cedars-Sinai Medical Center Dr. Podesta has headed multiple research programs and been co-investigator on a number of trials including a pig-to-human transplantation which was performed in 1992 and the most recent being the first successful use of the bioartificial liver machine on liver patients.

Leonard Makowka, M.D., Ph.D., is Chairman, Department of Surgery, Director of Transplantation Services, Cedars-Sinai Medical Center, and Professor of Surgery, UCLA School of Medicine, Los Angeles, California. He graduated in 1969 from University of Toronto, Ontario, Canada, with a M.S. degree in Pathology and obtained his M.D. degree in 1977 and his Ph.D. degree in 1982 also from University of Toronto, Ontario, Canada.

Dr. Makowka is a member of Academy of Medicine Ontario, Alpha Omega Alpha Honor Medical Society, American Association of Immunologists, American Association for the Advancement of Science, American Association for the Study of Liver Diseases, American College of Surgeons, American College of Surgeons — Southern CA Chapter, American College of Physician Executives, American Council on Transplantation, American Federation for Clinical Research, American Liver Foundation, American Management Association, American Medical Association, American Physicians Fellowship, Inc. for Medicine in Israel, American Society of Liver and Pancreas Surgery, American Society of Transplant Physicians/Surgeons, American Surgical Association, Association for Academic Surgery, Association of Program Directors in Surgery, Canadian Medical Association, Cell Transplant Society, Central Surgical Association, College of Physicians and Surgeons of Ontario, European Association for the Study of the Liver, European Society for Organ Transplantation, Federation of American Societies for Experimental Biology, and numerous other professional organizations.

Some honors and awards received by Dr. Makowka are the Alpha Omega Alpha Honour Medical Society, Charles E. Frosst Scholarship (Bronze Medal), Second Annual Assembly of General Surgeons (First Place), Gallie Bateman Surgical Essay Award (First Place), Medalist in Surgery, Royal College of Physicians and Surgeons of Canada, Davis and Geck Surgical Essay Award, Schering Scholarship Award, American College of Surgeons, The Canadian Foundation for Ileitis and Colitis Research Award, Graham Campbell Fellowship in Pathology, Faculty of Medicine, University of Toronto, and Medalist in Surgery, Royal College of Physicans and Surgeons of Canada.

Some research grants awarded Dr. Makowka include Canadian Liver Foundation (Co-investigator), Clinical Center for NIH Liver Transplantation Database, NIDDK Contract (Principal Investigator), DuPont Merck Pharmaceutical Company (Principal Investigator), Prevention of Reperfusion Injury in Cadaveric Renal Transplantation with Recombinant Human Cu/Zn Superoxide Dismutase (r-hSOD), Phamacia-Chiron Partnership (Principal Investigator and Study Chairman), "Study of an Artificial Liver: Charcoal Hemoperfusion in a Dog Model of Acute Hepatic Failure", Ash Medical Co., Renewal, Clinical Center for NIH Liver Transplantation Database (Principal Investigator) and "Effectiveness of PEG-SOD in Preserved Rat Kidney Function", Sterling Drug Company.

Dr. Makowka has presented over 60 invited lectures at international and national meetings. He has published more than 300 papers and authored and co-authored numerous books.

CONTRIBUTORS

Robert Black, M.D.
Department of Surgery
University Hospital
University of Western Ontario
London, Ontario, Canada

Christoph E. Broelsch, M.D., Ph.D.
Professor and Chief of Section
Department of Surgery
The University of Chicago
Chicago, Illinois

Frances A. Chapman, M.D.
Transplantation Biology Research Laboratory
Department of Surgery
Cedars-Sinai Research Institute
Cedars-Sinai Medical Center
Los Angeles, California

David K. C. Cooper, M.D., Ph.D., F.R.C.S.
Baptist Medical Center
Oklahoma Transplantation Institute
Oklahoma City, Oklahoma

Carlos A. Cosenza, M.D.
Department of Surgery
Cedars-Sinai Medical Center
Los Angeles, California

H. Joachim Deeg, M.D.
Transplantation Biology Program
Clinical Research Division
Fred Hutchinson Cancer Research Center
Seattle, Washington

Donald V. Cramer, D.V.M., Ph.D.
Director, Transplantation Biology Research Laboratory
Department of Surgery
Cedars-Sinai Medical Center
Los Angeles, California

Achilles A. Demetriou, M.D., Ph.D.
Director, Department of Surgical Research
Cedars-Sinai Research Institute
Cedars-Sinai Medical Center
Los Angeles, California

John J. Fung, M.D., Ph.D.
Department of Surgery
Transplantation Institute
University of Pittsburgh
Pittsburgh, Pennsylvania

Philip C. Gazzetta, M.D.
Division of Pediatric Surgery
University of Texas Southwestern Medical Center
Dallas, Texas

David R. Grant, M.D., F.R.C.S.C.
Department of Surgery
University Hospital
University of Western Ontario
London, Ontario, Canada

Rosemary Hickman, M.D., Ch.M.
Department of Surgery
University of Cape Town Medical School
Cape Town, South Africa

Allen Hoffman, M.D.
Department of Surgery and Hepatobiliary
Cedars-Sinai Medical Center
Los Angeles, California

Todd Howard, M.D.
Assistant Professor of Surgery
Department of Surgery
Washington University School of Medicine
St. Louis, Missouri

Oscar Imventarza, M.D.
Transplant Institute
University of Pittsburgh
Pittsburgh, Pennsylvania

Delawir Kahn, Ch.M., F.C.S.
Department of Surgery
University of Cape Town Medical School
Cape Town, South Africa

Michael Knoop, M.D.
Department of Surgery
University Hospital Rudolf Virchow
Berlin, Germany

Robert Korngold, Ph.D.
Department of Microbiology
and Immunology
Jefferson Medical College
Philadelphia, Pennsylvania

William Lopatin, M.D.
Department of Surgery and Liver
Transplantation
Cedars-Sinai Medical Center
Los Angeles, California

Leonard Makowka, M.D., Ph.D., F.R.C.S., F.A.C.S.C.
Chairman, Department of Surgery
Director, Department of
Transplantation Services
Cedars-Sinai Medical Center
Los Angeles, California

Edgar L. Milford, M.D.
Department of Medicine
Brigham and Women's Hospital
Boston, Massachusetts

Arnold Mixon, M.D.
National Cancer Institute
National Institutes of Health
Bethesda, Maryland

Albert D. Moscioni, Ph.D.
Department of Surgical Research
Cedars-Sinai Research Institute
Cedars-Sinai Medical Center
Los Angeles, California

Noriko Murase, M.D.
Pittsburgh Transplantation Institute
University of Pittsburgh
Pittsburgh, Pennsylvania

Kazuaki Nakajima, M.D.
Department of Surgery
Chiba University School of Medicine
Chiba, Japan

Marek Niekrasz, D.V.M.
Division of Animal Resources
Oklahoma University Health Sciences
Center
Oklahoma City, Oklahoma

Hiroji Noguchi, M.D.
Pediatric Division
Loma Linda University
Medical Center
Loma Linda, California

Marcos Nores, M.D.
Department of Surgery
Cedars-Sinai Medical Center
Los Angeles, California

Alejandra Oks, M.D.
Transplant Institute
University of Pittsburgh
Pittsburgh, Pennsylvania

Mark S. Orloff
Department of Surgery
Section of Transplantation
Strong Memorial Hospital
University of Rochester
Rochester, New York

Marshall J. Orloff, M.D.
Department of Surgery
University of California, San Diego
UCSD Medical Center
San Diego, California

Hennie Pienaar, F.R.C.S.
Department of Surgery
University of Cape Town
Medical School
Cape Town, South Africa

James B. Piper, M.D.
Department of Surgery
University of Chicago
Chicago, Illinois

Luis G. Podesta, M.D.
Associate Director, Transplantation
Services
Department of Surgery
Cedars-Sinai Medical Center
Los Angeles, California

Shiguang Qian, M.D.
Department of Surgery
Transplantation Institute
University of Pittsburgh
Pittsburgh, Pennsylvania

Camillo Ricordi, M.D.
Professor of Surgery
Diabetes Research Institute
University of Miami School of Medicine
Miami, Florida

Horacio L. Rodriguez Rilo, M.D.
Transplant Institute
University of Pittsburgh
Pittsburgh, Pennsylvania

Jacek Rozga, M.D., Ph.D.
Department of Surgical Research
Cedars-Sinai Research Institute
Cedars-Sinai Medical Center
Los Angeles, California

David H. Sachs, M.D.
Director, Transplantation Biology Research Center
Massachusetts General Hospital
Boston, Massachusetts

Ryo Saito, M.D.
Department of Thoracic Surgery
Institute of Development, Aging and Cancer
Sendai, Japan

Mohamed H. Sayegh, M.D.
Department of Medicine
Brigham and Women's Hospital
Boston, Massachusetts

Craig V. Smith, M.D.
Department of Surgery
Harbor-UCLA Medical Center
Torrance, California

Thomas E. Starzl, M.D., Ph.D.
Professor of Surgery
Director of Transplantation Institute
Presbyterian Hospital
Pittsburgh, Pennsylvania

Rudolf Steffen, M.D.
Department of Surgery
University Hospital Rudolf Virchow
Berlin, Germany

Larry H. Stevens, M.D.
Methodist Hospital
Methodist Professional Center
Indianapolis, Indiana

Rainer Storb, M.D.
Professor of Medicine
Department of Medicine
University of Washington
and
Member and Head of Transplantation Biology Program
Fred Hutchinson Cancer Research Center
Seattle, Washington

John Terblanche, Ch.M., F.C.S., F.R.C.S., F.R.C.P.S.
Department of Surgery
University of Cape Town Medical School
Cape Town, South Africa

Paul F. Waters, M.D., F.R.C.S.
Professor of Surgery
Director, Lung Transplantation
Director, General Thoracic Surgery
UCLA School of Medicine
Los Angeles, California

Yong Ye, M.D.
Baptist Medical Center
Oklahoma Transplantation Institute
Oklahoma City, Oklahoma

Robert Zhong, M.D.
Department of Surgery
University Hospital
University of Western Ontario
London, Ontario, Canada

TABLE OF CONTENTS

Section I Kidney Transplantation

Section II Liver Transplantation

Section III Pancreas Transplantation

Section IV Heart Transplantation

Section I

Kidney Transplantation

Chapter 1

Renal Transplantation in the Rat

Mohamed H. Sayegh and Edgar L. Milford

CONTENTS

I. INTRODUCTION

The rat is considered to be a major animal model for the study of vascularized organ allografts.[1] The size of the animal, their availability in large numbers, low cost, and the existence of inbred strains with defined histocompatibility antigens make them a very useful tool in transplantation research.[2,3] The first description of the technique of microvascular surgery and renal transplantation in the rat was presented in 1961 at the Annual Congress of the American College of Surgeons by Lee.[1] This was followed by several important publications describing in detail the surgical techniques as well as the immunological aspects of renal transplantation in the rat. This chapter represents a summary of the important aspects of renal transplantation in the rat including the gross anatomy of the rat kidneys, surgical technique of renal transplantation, and experimental applications of renal transplantation in the rat.

II. TECHNIQUES

A. ANATOMY

Figure 1 shows the normal gross anatomy of the rat kidney. There are a few important points to remember when performing donor nephrectomy for renal transplantation. The relative position of the right and left kidneys is the reverse of that found in humans. In the rat, the right kidney lies more cephalad, and the renal vessels show a corresponding difference. In the rat, the renal arteries arise from the descending aorta very close to the superior mesenteric artery, the right usually higher than the left and frequently above the superior mesenteric artery. The renal artery crosses behind the inferior vena cava and runs somewhat anterior and dorsal to the renal vein. Both renal arteries give rise to the inferior suprarenal branches before dividing into the two main renal branches and entering the hilus of each kidney. The renal veins open into the inferior vena cava with the longer left

0-8493-3629-5/94/$0.00+$.50

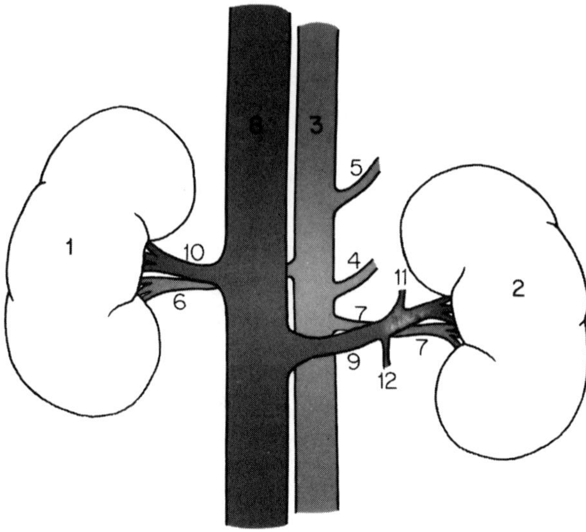

Figure 1 Gross anatomy of the rat kidneys and vasculature: (1) right kidney, (2) left kidney, (3) aorta, (4) superior mesenteric artery, (5) celiac artery, (6) right renal artery, (7) left renal artery, (8) vena cava, (9) left renal vein, (10) right renal vein, (11) phrenic vein, and (12) spermatic vein.

one crossing ventral to the aorta and opening well below the right one. The left renal vein receives the combined inferior suprarenal and phrenic veins anteriorly, as well as the testicular or ovarian veins posteriorly. The right renal vein receives only the inferior phrenic vein.[4]

B. DONOR NEPHRECTOMY

After anesthetizing the rat using either pentobarbital or chloral hydrate and shaving the skin over the abdomen, the animal is placed on a towel on an operating board with the ventral side up and the tail toward the surgeon. A long midline abdominal incision is made, using a scalpel to cut the skin and fascia and using scissors and an Adson forceps to open the abdominal cavity. The intestines are then wrapped in a saline-moistured gauze pack and retracted to the left, exposing the right kidney. After opening the retroperitoneum, the great abdominal vessels are exposed by gentle dissection using cotton-tipped swabs. Vessels draining into the renal vein are ligated using 5-0 silk ties and cut. The kidney is freed and the ureter is mobilized down to the bladder. Extreme care should be taken not to injure the periureteral adventitia, which contains the blood vessels, in order to avoid subsequent ureteral ischemia and necrosis. It is best to apply two holding sutures through the fat of the upper and lower poles of the kidney fixed with a straight clamp, thus making possible handling the graft without injuring it with fingers or instruments. The aorta is mobilized approximately 1 cm above and 1 cm below the origin of the renal artery. The superior mesenteric, the left renal, and often lumbar arteries originating from the aorta should be ligated and cut. A 3-0 silk ligature is then applied loosely around the aorta 2 to 3 mm below the renal artery. The aorta is also clamped 6 to 8 mm above the renal artery. Thereafter, the infrarenal aorta is punctured with a 0.7-mm Butterfly cannula, which is connected to a 2-ml syringe containing ice-cold Ringer's solution. The loose ligature is then tied around the cannula and the kidney is perfused. The cannula is then removed, the ligature is tied, and the aorta is cut distal to the aortic clamp and the ligature. Hence, the renal artery originates from a 5 to 6 mm aortic cuff which can be easily used to anastomose to the recipient's aorta. The renal vein is excised out of the inferior vena cava with a large patch of the vena cava attached. The ureter is transacted as close as possible to, or with a small piece of, the bladder wall and the graft is then placed in an ice-water bath while the recipient is being prepared (Figure 2).

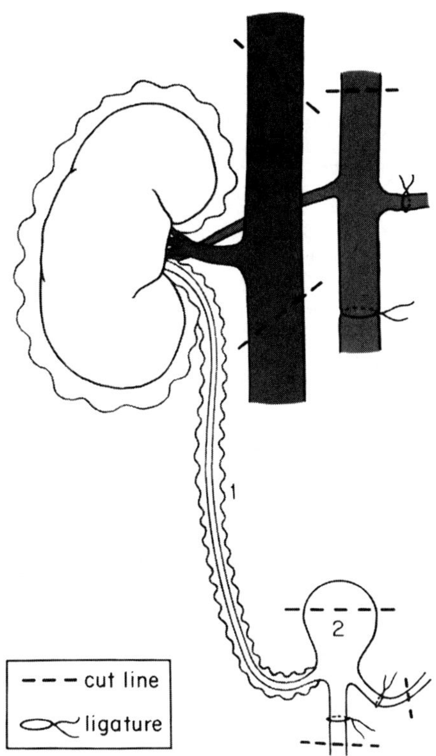

Figure 2 Schematic illustration of donor nephrectomy: (1) ureter, (2) bladder.

C. RECIPIENT TRANSPLANTATION

The recipient animal is anesthetized and prepared in a manner similar to the donor. After opening the abdomen, the intestines are retracted to the left, thus exposing the right retroperitoneal area. The great abdominal vessels are then exposed after opening the retroperitoneum. Both the aorta and the inferior vena cava are mobilized approximately 2 cm below the renal vessels and clamped in order to cut the blood supply over a length of approximately 1 cm. The right side of the vena cava is incised for a length of 3 mm, and an elliptical opening equal in size to the donor's aortic cuff is excised out of the ventral surface of the descending aorta. The end-to-side anastomosis of the donor's vena caval patch is performed using 8-0 or 9-0 monofilament sutures and a 6-mm curved atraumatic needle. The aortic anastomosis is done in exactly the same manner. Two stay sutures are usually placed in the opposite ends of the anastomosis which can then be completed using a continuous suture technique. After completing the anastomosis, the clamps are removed. Bleeding at the anastomosis site usually stops with tamponade using a gauze pad. Occasionally, a suture has to be placed at the bleeding point.

The ureteral anastomosis is slightly more complicated and several techniques can be used successfully. If ureterovesical anastomosis is going to be performed, then the fundus of the bladder is clamped with atraumatic forceps and displaced forward to expose the posterior surface. The posterior bladder wall is then opened 2 to 3 mm and the tip of the donor ureter is loosely fixed in the bladder with an 8-0 suture. The entrance of the ureter into the bladder is narrowed with a z-suture so that the ureter remains mobile in order to avoid ureteral strictures. Alternatively, a small piece of the donor's bladder which was removed with the ureter can be sutured to the recipient's bladder using 7-0 atraumatic sutures. This technique may be associated with a higher incidence of urinary reflux.

Finally, in certain experiments if applicable, the recipient's native kidneys are removed as described in Section II.B.

III. APPLICATIONS

The major applications of experimental renal transplantation in the rat model have been to investigate the following:

1. The role of the major histocompatibility complex (MHC) class I and class II molecules in mediated allograft rejection.
2. The immunological mechanisms of allograft rejection.
3. The role of new immunosuppressive agents in transplantation.
4. The mechanism of tolerance induction.

MAJOR HISTOCOMPATIBILITY COMPLEX OF THE RAT

Figure 3 Schematic illustration of loci within the RT1 (MHC) region. Relative positions of individual genes of the RT1 complex on chromosome 14 are shown as boxes. Open boxes represent class I MHC genes and black boxes represent class II MHC genes. Names of individual loci are noted in text below the boxes.

A. THE RAT MHC

The MHC of the rat is homologous to that of the mouse and humans and is called the RT1 complex.[2] This set of genes is found together in linkage group IX of chromosome 14, and encode the glycoproteins which constitute the major barriers to transplantation. There are at least six class I loci and two class II loci distributed over a 3000 to 3200 kb expanse of DNA, which is similar to the size of the human MHC region (Figure 3).

Class I genes are called RT1.A, Pa, F, E, G, and C. The RT1.A and RT1.E loci encode gene products that are homologous to the mouse H-2K and H2-D, respectively.

Rat class II loci are RT1.B, RT1.D, and the recently described RT1.H. RT1.B and D are homologous to human HLA-DQ and DR, respectively.

The class I molecules are prominently expressed on virtually all nucleated cells. In the rat, some class I molecules are also expressed on mature erythrocytes, in contrast to humans who lack class I on their red cells. Class I antigens stimulate both the cellular and humoral immune response and are the primary target for CD8+ cytotoxic T lymphocytes upon *in vivo* immunization through engraftment of skin or solid organs.

Rat class II molecules are well expressed on B cells, monocytes, some vascular endothelial cells, and proximal tubular epithelial cells. They are also expressed on dendritic cells. RT1.B and D antigens primarily stimulate the CD4+ T cell proliferative response to alloantigen as seen in the standard *in vitro* mixed lymphocyte response (MLR) assay, an *in vitro* model of the alloimmune response.[5,6] It is noteworthy, however, that CD4+ cytotoxic T lymphocytes directed at class II antigen can also be generated, and that selective production of antibody against class II antigen can occur *in vivo* after engraftment with organs or tissues which are incompatible for RT1.B or D.[7]

The discovery of the structure of the MHC molecule, and the availability of sequence data for the variable domains of MHC molecules has made it possible to synthesize peptides representing various portions of the native cell surface molecules and to use these peptides for the study of the alloimmune response. Although the mechanisms of processing and recognition of "conventional" antigens (viral proteins, for example) by the immune system has been fairly well established, the mechanisms of recognition of alloantigens are poorly understood. Understanding these mechanisms is very important for the development of specific immunotherapies to induce a state of long-term immune hyporesponsiveness or tolerance. The possible mechanisms of alloreactivity include T cell recognition of intact allo-MHC allopeptides presented on autologous antigen-presenting cells. Preliminary observations indicate that lymphocytes from rats immunized with polymorphic class II MHC allopeptides recognize and proliferate to specific polymorphic amino acid sequences on allogeneic cell surface MHC molecules *in vitro* and *in vivo*. The use of synthetic oligopeptides for understanding the mechanisms of allorecognition of specific MHC epitopes may provide novel approaches for specific immune interventions in organ transplantation.[6]

B. MECHANISMS OF RENAL ALLOGRAFT REJECTION

The rat is the major model for vascularized allograft rejection because of the availability of well-characterized inbred and congenic strains, and the ease with which vascular anastomosis can be accomplished (see above). In completely class I and class II incompatible (allogeneic) strains the unmodified renal allograft survival ranges from 7 to 14 days. In semiallogeneic, 1-haplotype matched combinations (F1 hybrid donor into parental recipient), the graft survival can be up to 18 days. This variability in graft survival is related to recipient immune responsiveness, donor antigen immunogenicity, antigen dose (i.e., homozygous vs. heterozygous), and presence vs. absence of minor histocompatibility antigen differences not encoded by the MHC (see below). Congenic strains which share the same non-MHC background and which differ only in their MHC genotype are widely available and represent a useful tool for understanding the contribution of the MHC to graft rejection. In addition, a variety of inbred strains which are recombinants between particular RT1 loci have been used to determine the relative roles of the individual RT1 loci in acute allograft rejection.[3]

Below is a listing of the most commonly employed inbred rat strains used in renal transplantation experiments and their RT1 (MHC) allotype designations. It is common for investigators to utilize F1 hybrids of these parental strains as "semiallogeneic" donors when a lesser degree of incompatibility and exclusion of donor vs. host effects are desired:

Rat strain	RT1 allotype
DA, ACI	a
BUF	b
PVG, AUG	c
KGH	g
SHR, WKY	k
LEW, F344	l
BN, MAXX	n
WF	u

Of the non-MHC antigens involved in kidney graft rejection, a so-called "endothelial/ monocyte" antigenic system apprise to play an important role in specific strain combinations (such as MAXX into Brown Norway). In addition, the F344 (Fisher) and LEW (Lewis) strains, which differ only slightly in their class I MHC antigens, do not suffer from acute allograft rejection and provide a useful model for the study of chronic renal allograft rejection, the most common cause of chronic graft loss.[8]

In unmodified animals, acute allograft rejection is accompanied by a progressive fall in renal blood flow over a period of 5 to 7 days, and in nephrectomized rats, this is accompanied by decreasing urine output and elevation of serum creatinine and blood urea nitrogen. The animals ultimately die from progressive uremia. Cytotoxic antibodies against peripheral blood lymphocytes are detectable as soon as Day 7 post-engraftment. By 48 hours after renal transplantation, there are nests of mononuclear leukocytes infiltrating the donor organ, with evidence of *in situ* proliferation of these cells. While these infiltrates are initially in the perivascular areas of the cortex, they affect the cortical interstitium more diffusely by Day 7. At this time, there may be massive interstitial edema, tubulitis, and frank tubular necrosis. Vascular endothelial cells become swollen and sometimes undermined by mononuclear cells which may plug the lumina of vessels. Immunohistological evaluation of rejecting rat renal allograft sections stained by immunoperoxidase using monoclonal antibodies against specific cell markers indicate that mononuclear cell infiltrate consists of CD4+ T cells, CD8+ T cells, and monocytes/ macrophages. In the grafts, there are also a significant number of activated T cells that express the IL-2 receptor (CD25). The infiltrating cells can be recovered to study their phenotypic and functional characteristics, including CD4+ T helper cell function and CD8+ T cytotoxic cell function. In addition, small sections from the grafts, as well as infiltrating lymphocytes, can be recovered and used for mRNA analysis of expression of certain antigens (Class II MHC molecules, for example), activation (IL-1, IL-2, IL-6, etc.) or regulatory (IL-4, IL-10) cytokines, and growth factors, either by Northern blot analysis or polymerase chain reaction. Such applications are very valuable when studying the molecular mechanisms of allograft rejection, the effects and mechanisms of allograft rejection, and the effects and mechanisms of action of specific immunotherapeutic interventions.

C. THE RAT RENAL ALLOGRAFT AS A MODEL FOR IMMUNOSUPPRESSION

The rat renal allograft model has been extensively used to study the safety, efficacy, and mechanisms of action of new immunosuppressive agents.[9-11] It is also a model for induction of specific immunologic unresponsiveness or tolerance.[12-19] In particular, pharmacologic agents such as azathioprine, steroids, cyclosporine, and more recently rapamycin, FK 506, deoxyspergualin, Brequinar sodium, and RS-61443, and others have all been extensively studied in the rat renal allograft model prior to use in primates or humans. The model has been used to study the interactions of ischemic injury at the time of engraftment with cyclosporine nephrotoxic injury, in addition to testing several pharmacological agents for prevention of cyclosporine toxicity (for example, fish oil and calcium channel blockers).[9,20] This model is also useful in assessing biological agents such as polyclonal anti-lymphocyte globulins, monoclonal antibodies, potentially immunosuppressive lymphokines, and prostaglandins.

D. THE RAT RENAL ALLOGRAFT AS A MODEL FOR TOLERANCE INDUCTION[12-19]

The rat renal allograft model has also been used for the detailed study of effective means for tolerance induction. Results from such studies have been extremely important in understanding the mechanisms of alloimmunity including the induction of specific tolerance. Below is a list of the different ways whereby specific immunologic unresponsiveness can be induced in the rat renal allograft model:

1. Active enhancement by administration of:
    - Blood transfusion
    - Purified lymphocytes (intravenous or intraportal)
    - Intrathymic alloantigen (glomeruli or mononuclear cells)
2. Passive enhancement
    - Polyclonal anti-donor antibody
    - Monoclonal anti-donor antibody
3. Total lymphoid irradiation
4. Induction with monoclonal antibodies
    - Anti-CD3
    - Anti-CD4
    - Anti-CD25 (IL-2 receptor)
5. Bone marrow transplantation

E. OTHER APPLICATIONS

The model of rat renal transplantation has been very valuable in studying the physiologic changes that occur in the renal tubules post-transplantation. In addition, the model has been very helpful in studying the mechanisms of post-transplant hypertension with or without cyclosporine.[20]

REFERENCES

1. **Jakuboswski, H.D.,** Renal transplantation in the rat, *Microsurgical Models in Rats for Transplantation Research,* Thiede, A., Deltz, E., Engemann, R., and Hamelmann, H., Eds., Springer-Verlag, Berlin, 1985, 47-50.
2. **Hedrich, H.J.,** List of mutant genes and polymorphic loci in the rat *(Rattus norvegicus), Rat Newslett.,* 25, 4-19, 1991.
3. **Altman, P.L. and Katz, D.D., Eds.,** *Inbred and Genetically Defined Strains of Laboratory Animals Part 1: Mouse and Rat,* Federation of American Societies for Experimental Biology, Bethesda, MD, 1975, 313-334.
4. **Olds, R.J. and Olds, R.J., Jr., Eds.,** *A Colour Atlas of the Rat: Dissection Guide,* Wolfe Medical Publications Ltd., London, 1988, 31-67.
5. **Frankel, A.H., Sayegh, M.H., Rothstein, D.M., Milford, E.L., and Carpenter, C.B.,** Requirements for the induction of allospecific CD8+ suppressor T cells in the rat primary mixed lymphocyte response. CD4+, CD45R+ T cells, or supernatant factor, *Transplant,* 48, 639, 1989.
6. **Sayegh, M.H., Khoury, S.J., Hancock, W.W., Weiner, H.L., and Carpenter, C.B.,** Induction of immunity and oral tolerance with class II MHC allopeptides in the rat, *Proc. Natl. Acad. Sci. U.S.A.,* 89, 7762, 1992.
7. **Hall, B.M.,** Cells mediating allograft rejection, *Transplantation,* 51, 1141, 1991.
8. **Tilney, N.L., Whitley, W.D., Diamond, J.R., Kupiec-Weglinski, J.W., and Adams, D.H.,** Chronic rejection — an undefined conundrum, *Transplantation,* 52, 389, 1991.
9. **Preico, N., Daddan, J., and Remuzzi, G.,** Endothelin mediates the renal vasoconstriction induced by cyclosporine in the rat, *J. Am. Soc. Nephrol.,* 1, 76, 1990.
10. New pharmacologic immunosuppressants, *ASHI Quart.,* 15, 1991.
11. **Wood, R.P., Katz, S.M., and Kahan, B.D.,** New immunosuppressive agents, *Transplant. Sci.,* 1, 34, 1991.
12. **Wood, K.J., Evins, J., and Morris, P.J.,** Suppression of renal allograft rejection in the rat by class I antigens on purified erythrocytes, *Transplantation,* 39, 56-62, 1985.
13. **Heffron, T.G. and Thislethwaite, J.R.,** New monoclonal antibodies, *Transplant. Sci.,* 1, 63, 1991.
14. **Sayegh, M.H., Kut, J.P., and Milford, E.L.,** Anti-CD4 monoclonal antibody effects cellular hypo-responsiveness and prolongs renal allograft survival in the rat, *Human Immunol.,* 26, 131-136, 1989.
15. **Wasowska, B., Baldwin, W.M., Howell, D.M., and Sanfilippo, F.,** The effects of donor-specific blood transfusion enhancement of rat renal allografts on cytotoxic activity and phenotypes of peripheral blood lymphocytes, splenocytes, and graft-infiltrating cells, *Transplantation,* 51, 451, 1991.
16. **Hamashima, T., Yoshimura, N., Matsui, S., Lee, C.J., Ohsaka, Y., and Oka, T.,** The effects of perioperative portal venous inoculation with donor lymphocytes on renal allograft survival in the rat, *Transplantation,* 49, 171, 1990.

17. **Hutchinson, I.V. and Morris, P.J.,** The role of major and minor histocompatibility antigens in active enhancement of rat kidney allograft survival by blood transfusion, *Transplantation,* 41, 166, 1986.
18. **Sayegh, M.H., Perico, N., Imberti, O., Hancock, W.W., Carpenter, C.B., and Remuzzi, G.,** Thymic recognition of class II major histocompatibility complex allopeptides induces donor-specific unresponsiveness to renal allografts, *Transplantation*, 56, 461, 1993.
19. **Yoshimura, N. and Kahan, B.,** Suppressor cell activity of cells infiltrating rat renal allografts prolonged by preoperative administration of extracted histocompatibility antigen and cyclosporine, *Transplantation,* 40, 708, 1985.
20. **Curtis, J.,** Hypertension after renal transplantation: cyclosporine increases the diagnostic and therapeutic considerations, *Am. J. Kidney Dis.,* 13, 28(6 Suppl. 1), 1989.

Chapter 2

Renal Transplantation in the Rabbit

William Lopatin, Allen Hoffman, Hiroji Noguchi, and Leonard Makowka

CONTENTS

I. INTRODUCTION

The first consideration in the use of different vascularized allograft animal models is the relevance of the species to the problem under investigation, and whether the data obtained from the study can be applied to, or extrapolated to, the human clinical situation.[1] Although the rabbit represents a species of relative infrequent use in experimental transplantation, the similarity of its renal structure and function to the man makes this animal an appropriate model for kidney transplantation.[2]

The advantage of a small animal transplant model is related to the preoperative, operative and postoperative care, and the associated cost implications. The rabbit offers a physical size that is applicable for most surgical interventions, yet remains within the economical and animal care constraints of a small-animal model.

Experiments in immunologic regulation for solid organ transplantation has often overlooked the rabbit model.[3] Yet, in other immunologic investigations, such as allergenic response, the rabbit has played a very important role in the development and understanding of the interactions of the human immune system. The rabbit has been shown to reproducibly exhibit hyperacute allograft rejection following sensitization to donor antigens.[4] The rabbit kidney allograft model has been used as a small animal vascularized allograft that displays a hyperacute organ rejection similar to man.[5] The rabbit kidney isotransplantation, due to its functional similarity with the man, has been frequently employed for the study of cold ischemia-reperfusion damage.[8] The following chapter describes the technique for renal orthotopic transplantation and autotransplantation in the rabbit.

0-8493-3629-5/94/$0.00+$.50

II. TECHNIQUES

A. ANIMAL SELECTION AND ENVIRONMENT

The best results have been obtained with young adults of a large breed. New Zealand White rabbits, weighing 2.5 to 3.0 kg, have large vessels and little adipose tissue, and they have been the most frequently used. The French lop-eared rabbits, weighing 3.0 to 3.5 kg, are also an acceptable breed. The animals are maintained in an air-conditioned environment at a temperature of 23°C with a 12-hour dark-light cycle. Rabbits can be fasted up to 12 hours preoperatively but should be allowed access to water prior to anesthesia.[1]

B. ANATOMY

The kidney consists of the renal cortex, medulla, and pelvis and is located in the retroperitoneal space, surrounded by Gerota's fossa. Unlike most mammals, the rabbit has unipapillate kidneys. The kidneys are commonly imbedded in retroperitoneal fat on the dorsal wall of the abdomen. The ureter can be identified behind the peritoneum, connecting the renal pelvis to the urinary bladder. The adrenal glands are located on the medial aspect of the superior pole of each kidney and must be carefully handled during renal dissection.

C. ANESTHESIA

The operation is performed under clean but not sterile conditions. The operating board should contain a heating element of pads to maintain the body temperature between 37.5 and 38.5°C. All procedures are performed under general anesthesia. Handling before anesthesia induction must be carefully done as two adverse events may occur at the time of anesthesia. First, fear may easily produce an intense release of cathecolamines that predispose the animal to cardiac arrhythmias and arrest during anesthesia. Second, the animal may struggle sufficiently to induce a fracture of vertebrae and an irreversible posterior paralysis. Approximately 30 min before the operation, the rabbits are premedicated with intramuscular atropine, 0.2 mg/kg body weight, to decrease pulmonary secretions. The rabbits are anesthetized with ketamine, 50 mg/kg body weight, intramuscularly, which provides between 60 to 90 min of operating time. Occasionally, it may be necessary to administer small amounts of ketamine during the procedure to maintain adequate anesthesia or concurrent administration of diazepam (1 mg/kg) or xylazine (1 mg/kg) to provide more sedation and surgical relaxation. The repeated doses are given until respiration is mildly depressed and the pulse, eye movements, and response to painful stimulus are diminished. Loss of corneal reflex indicates dangerously deep anesthesia. Since respiratory depression is a problem that is frequently encountered,[1] during the procedure the animals are supplied with oxygen at 2 l/min via an open face mask.

D. ORTHOTOPIC RENAL TRANSPLANTATION

Kidney transplantation in the rabbit requires the use of selected microsurgical techniques.[7] In general, the vascular anastomoses may require a Zeiss operating microscope (magnification × 12.5). Other microsurgical instruments include dissecting scissors, jewelers forceps, arterial and venous vascular occlusion clamps, and a pair of fine needle holders. All instruments are cleaned thoroughly in antiseptics and rinsed in sterile distilled water.

1. Donor Procedure

Either kidney may be used as the donor organ. The left kidney, however, is usually preferred because of the longer renal vein. Anesthesia is administered as described earlier. After induction, the rabbit is placed on the operating board with the limbs secured and well retracted. The abdomen is shaved and prepared with alcohol or an iodine-containing skin cleanser. A full-length midline incision is made from the xiphoid process to the pubic symphysis. The intestinal contents are placed outside the abdominal cavity and covered by a wet saline sponge/gauze to prevent injury to the bowel. The posterior peritoneum overlying the aorta is incised. Dissection is carried anteriorly to expose the renal vein and subsequently the renal artery. To gain the full length of the renal vessels and to prevent vasospasm, mobilization is performed in a careful manner so that the small venous branches are not avulsed. In addition, 1% lidocaine (without epinephrine) or papaverine may be applied topically to the renal vessels to lessen the vasospasm. The gonadal vein is identified, ligated, and transected. It is not necessary to enter the hilum of the kidney. Vessel length is attained by dissecting them medially to their point of origin, to the vena cava for the renal vein and to the aorta for the renal artery.

The final step in the organ procurement is cold perfusion of the kidney *in situ*. The renal artery is occluded with a small vascular clamp and 15 to 20 ml of cold (4°C) heparinized saline or Eurocollins solution is infused using a 27-gauge needle. The kidney should uniformly blanch, indicating good perfusion of the renal parenchyma. The renal vein is clamped in close approximation to the inferior vena cava and divided, but not ligated, after instillation of approximately 10 ml of additional solution. The remainder of the perfusate is infused until a clear effluent from the renal vein is seen. The kidney is then fully mobilized from its bed with approximately 3 cm of the proximal ureter. Extreme care is taken during the ureteral dissection such that the ureter with its surrounding tissue remains intact in an effort to avoid the damage of the ureteral blood supply. The kidney is placed into a sterile beaker with cold storage (4°C) solution. The storage conditions and the container should be of adequate size and shape to permit further dissection if required.

The renal bed and vessel ligatures are inspected for adequate hemostasis. Clamps are removed and the intestinal contents are replaced in the abdominal cavity. The wound is closed in two layers, taking care to incorporate the peritoneum together with the linea alba. The main disadvantage of the midline incision in the rabbits may be stretching of the scar, and the subsequent development of a ventral hernia due to the friable nature of their abdominal wall.

2. Recipient Operation

The animal is anesthetized as described earlier, and placed on the operating board such that the limbs are retracted and secured. A preoperative hydration with 100 to 120 ml of saline solution intravenously is recommended to avoid intraoperative dehydration and hypotension. Orthotopic renal transplantation is performed on the left side because of the longer left renal vein and the proximity of the aorta. The abdomen is shaved and prepared with alcohol or an iodine-containing skin cleanser. The midline incision is made from the xiphoid process to the symphysis. The intestinal contents are placed outside the abdomen covered by a wet saline gauze. The posterior peritoneum overlying the abdominal aorta is incised and dissection is carried anteriorly. The left renal artery and vein are exposed and fully mobilized. These vessels are then occluded by microvascular clamps at their point of origin. The renal artery and vein are transected at the renal hilum thereby preserving the maximum length of each vessel. The recipient left kidney is removed from its bed, and the proximal ureter is carefully dissected from the lower pole of the kidney avoiding to compromise the blood supply of the ureter. The ureter is then transected in close proximity to the lower pole of the kidney.

The cold-stored donor kidney is then placed in the left renal bed and covered by a cold, wet saline gauze. The donor and recipient vessels are placed over a colored piece of plastic to facilitate suture placement. Appropriate vessel length is measured by loosely stretching both the donor and recipient vessels in an overlapping manner, and the excess in vessel length removed to prevent kinking. The vascular anastomoses are performed in an end-to-end manner, with low level magnification. The ends of the renal arteries and veins are cleared from their adventitial attachments, providing clean vessel edges for a more precise suture placement. The arterial side is approached first, using two 9-0 prolene sutures to align and secure the anastomosis, one at the anterior edge and the other at the posterior edge of the arteries. The shorter end of the suture is used to provide tension while the longer end is used to begin the anterior anastomotic line. A continuous line of full thickness, close and precise suture bites are made. After completion of the anterior suture line, the suture is tied, and the artery is turned 180° by means of rotation of the anterior and posterior ends of the anastomosis. The posterior anastomotic line is subsequently completed. Just prior to the venous anastomosis, 5 mg of furosemide is administered intravenously to the recipient animal to improve the arterial renal flow before reperfusion. The end-to-end venous anastomosis is then completed in the same manner as described for the renal artery. The vascular clamps are then removed and bleeding from the suture line is controlled by subocclusive pressure. The donor kidney should rapidly regain its normal color and become firm in its consistency. If the renal artery has been carefully handle, arterial vasospasm should not be observed. However, if this phenomenon is seen, the topical administration of papaverine may reverse in a few minutes. The vascular phase of the recipient procedure should take no longer than 45 min.

The ureteral anastomosis may be approached in two fashions, either an end-to-end ureteroureterostomy or an end-to-side ureterovesicostomy. The end-to-end anastomosis requires an internal stent to avoid postoperative ureteral stenosis. Many methods have been described, the most common is an internal polyethylene stent of 1.0 mm diameter and 1.5 cm in length. The ureteroureterostomy is completed with 7-0 or 8-0 polypropylene sutures. Each stitch should be full thickness and, in most cases, interrupted sutures are required. The ureterovesicostomy is performed by mobilization of the ventrolateral aspect of the bladder. The seromuscular layer of the bladder is dissected to exposed the musculomucosa and a small cystostomy is made. Using 7-0 polyglyconate suture, an interrupted end-to-side ureterovesicostomy anastomosis is completed using the full thickness of the ureter and the musculomucosa layer of the bladder. The seromuscular layer of the bladder is loosely approximated over the ureterovesicostomy creating a 1.5-cm submucosal tunnel that will prevent postoperative vesicoureteric reflux.

E. AUTOLOGOUS RENAL TRANSPLANTATION

1. Nephrectomy

The left kidney is preferred because of the longer renal vein. A well hydrated animal is anesthetized and secured on the operating board. The nephrectomy is performed in a manner similar to that described in the text for the donor procedure in the orthotopic model. A full midline incision is made and the posterior peritoneum overlying the aorta is incised. Dissection is carried anteriorly, identifying the left renal vein as it crosses the aorta. Posterior and usually anterior to the renal vein is the left renal artery. Careful mobilization is performed such that the small venous branches are not avulsed. Solutions of 1% Lidocaine (without epinephrine) or papaverine may be applied topically to the vessels to minimize vasospasm. It is important during the dissection not to enter the renal hilum. Vessel length is attained by dissecting these vessels medially to their point of origin.

The next step requires the flushing of the kidney with cold (4°C) preservation solution to cool the graft. The renal artery is occluded with a small vascular clamp and approximately 25 to 30 ml of the cold solution infused using a 27-gauge needle. The renal vein is clamped at its point of origin, and its distal portion divided to vent the flushing solution. After adequate perfusion of the kidney, the kidney is removed with 3 cm of proximal ureter from the renal bed using a combination of blunt and sharp dissection. Extreme care is taken during the ureteral dissection not to disturb its blood supply. With this technique, the warm ischemia time should be below 10 min. The storage conditions are defined by the design of the experiment. If the kidney cold preservation period is longer than 2 hours, the abdominal wound is carefully closed in two layers and the animal is allowed to recover as described in the postoperative section.

2. Transplant Procedure

The rabbit is anesthetized and appropriately prepared as previously described. Following the established cold storage period, the midline incision is reopened. The intestinal contents are wrapped in wet gauze and placed outside the abdomen. The vena cava is followed superiorly to the right renal vein. Should there be difficulty in identifying the right renal vein by this approach, then the bowels are retracted to the left side of the animal and the right kidney is approach laterally. Dissection is carried through the Gerota's fascia (perirenal fascia), the kidney is identified and lightly retracted laterally. The renal vessels are subsequently identified medial to the renal hilum, and the artery and vein are ligated and transected. The kidney is removed from its bed, and the ureter is divided at the level of the iliac artery. Attention is directed now to the abdominal aorta and inferior vena cava. With an appropriate dissection to remove the overlying adipose and fibrous tissue, the lumbar arteries and veins are ligated and transected. The animal must be hemodynamically stable and well hydrated at this point of the procedure to avoid subsequent revascularization problems.

The preserved kidney is removed from the cold storage and placed heterotopically in the left side of the retroperitoneum. The vena cava is partially crossclamped with a pediatric Satinsky vascular clamp, and an appropriate anterior venotomy is created. After alignment of the renal vein, an end-to-side anastomosis is begun. Two 7-0 polypropylene sutures are used to align and secure the anastomosis, one at the anterior edge and the other at the posterior edge of the venous anastomosis. A continuous line of full thickness, close and precise suture bites are made. After completion of the medial suture line, the kidney is placed in the right retroperitoneum to complete the lateral anastomosis line. At this point, 5 mg of furosemide is intravenously administered to the animal. The arterial aorta-renal end-to-side anastomosis is performed in similar fashion but approximately 5 mm below the venous anastomosis. The arterial anastomosis is made posterior to the venous in order to provide adequate exposure to the medial aspect of the arterial anastomosis line. The vascular clamps are then released and bleeding from the suture line is controlled by subocclusive pressure. The preserved kidney should regain its normal color and become firm in its consistency. The vascular phase of this procedure should not take longer than 45 min.

A ureterovesicostomy is performed by identifying the ventrolateral aspect of the bladder. The seromuscular layer of the bladder is incised exposing the muscularis mucosa, and an appropriate sized cystotomy is made. An interrupted end-to-side ureterovesicostomy is completed by using 7-0 polyglyconate suture. The seromuscular layer over the anastomosis is loosely approximated with 7-0 suture creating a submucosal tunnel. It is essential to replace fluid loss during the operation to avoid hypotension. Fluid is lost in considerable amounts in small animals due to evaporative loss from the intestine. This is restored by intravenous injection of isotonic saline and dextrose solutions.

F. POSTOPERATIVE CARE

An animal is at high risk immediately after surgery and while recovering from anesthesia. It should be protected from hypothermia, respiratory obstruction, and cardiovascular failure due to hypotension. Immediately after surgery, the rabbit is kept warm in its cage with a room temperature at approximately 32 to 33°C. The use of ketamine hydrochloride in rabbits provides certain advantages, since it tends to stimulate both the cardiovascular and respiratory system. Its effects usually last for 40 to 60 min and animals are fully awake within 2 to 3 hours following the operation. Postoperative analgesia may be provided by injecting subcutaneously buprenorphine (0.05 mg/kg) every 8 to 12 hours for up to 48 hours. The rabbits are allowed to eat a standard diet, and are weighed before the operation and on each postoperative day. If a weight loss of more than 50 gm/day is observed, routinely replacement with intravenous isotonic saline and dextrose solution through the marginal vein of the ear can be supplied. It is important to prevent dehydration in the early postoperative period to assure a good renal flow. Blood samples are easily drawn from the marginal vein of the ear for assessment of electrolyte abnormalities and renal function every 24 hours.

The nature and timing of the contralateral nephrectomy is dependent upon the experiment. It is usually performed, however, 24 to 72 hours following orthotopic renal transplantation. The abdominal wound is reopened and the right renal artery and vein are identified by following the inferior vena cava superiorly. Ligature ties are placed and the vessels are transected. The right kidney is subsequently mobilized and the ureter is identified, ligated, and transected. Hemostasis is carried out and the wound is closed in two layers.

III. APPLICATIONS

Following experimental kidney transplantation in the rat, the rabbit represents the second most frequent small animal model used for renal transplantation. There are two major areas in organ transplantation where this model has been applied. The first is related to the preservation of the organ outside its normal milieu. That is, the time the organ is removed from its blood supply until it is reperfused and is established in its new environment. Preservation is a fragile period with many coinciding variables having associated and separate consequences upon cellular function and viability. Experimental kidney perfusion and preservation studies require that the assessment of renal function is changed by events reflecting loss of normal blood supply, organ storage, and reperfusion. Therefore, in order to evaluate and investigate physiologic interactions, the animal must be of a size that is suitable for autotransplantation and allows repeated access of blood and urine. Autotransplantation in the rabbit has been a reliable method that permitted reproducible success in technique as well as ease of blood and urine analysis in a docile small animal.[9-11]

The second concern in solid organ transplantation is the immunologic reaction. In the allograft renal transplant model the body recognizes nonself and initiates a cascade of immunologic events that results in organ rejection and subsequent failure. The current means of investigation require that the animal model has immunologic regulation for solid organ transplantation that is similar to that in man. The rabbit orthotopic allograft renal transplant model has been used to test immunosuppressive drugs, mainly cyclosporin A,[12] and as a model of pharmacologically induced tolerance.[13] In addition, the rabbit kidney allograft model reproducibly exhibits hyperacute rejection following sensitization.[14] The characteristics of this model reliably demonstrate sensitization to donor antigens and reliably reproduce the humoral rejection observed in sensitized kidney transplant patients.[15,16]

IV. CONCLUSION

Renal transplantation in the rabbit offers a surgical transplantation model that is intermediate in cost and complexity when compared to the same procedure performed in rodents or larger species, such as the dog and pig. The rabbit does not require special surgical or housing facilities and the animals care personnel are usually familiar with the procedures necessary for sample collection and postoperative care. The surgical procedure requires experience with microvascular techniques albeit less extensive than the same procedure in rodents. Due to its similarity with human kidney physiology, this model is appropriate for the study of preservation-reperfusion damage conditions. One special consideration for the use of the rabbit may be the facility and consistency with which the hyperacute rejection of the kidney can be induced by prior sensitization of the recipient.

REFERENCES

1. **Lumley, J.S.P., Green, C.J., Lear, P., and Angell-James, J.E.,** *Essentials of Experimental Surgery,* 1st ed., Butterworths, London, 1990.
2. **Kraus, A.L., Weisbroth, S.H., Flatt, R.E., and Brewer, N.,** Biology and diseases of rabbits, in *Laboratory Animal Medicine,* Fox, J.G., Cohen, B.J., and Loew, F.M., Eds., Academic Press, New York, 1984, chap. 8.
3. **Francis, D.M.A., Millar, R.J., Dumble, L.J., and Clunie, G.J.A.,** Surgical research, model of orthotopic renal transplantation in the rabbit, *Aust. N.Z.J. Surg.,* 60, 45-49, 1990.
4. **Holter, A.R., McKearn, T.J., Neu, M.R., Fitch, F.W., and Stuart, F.P.,** Renal transplantation in the rabbit, development of a model for study of hyperacute rejection and immunological enhancement, *Transplantation,* 3, 13, 1972.
5. **Bayuk, J.M. and Schmidlapp, C.J.,** Kidney transplantation in the rabbit, a method for the study of transplantation immunology, *Invest. Urol.,* 4(4), 378-388, 1967.
6. **Heron, I.,** Kidney transplantation in the rabbit, a new method, *Acta Pathol. Microbiol. Scand.,* 78, 90-95, 1970.
7. **Dunn, D.C.,** Orthotopic renal transplantation in the rabbit, *Transplantation,* 22(5), 427-433, 1976.
8. **Jacobsen, I.A.,** Renal transplantation in the rabbit, a model for preservation studies, *Lab. Anim.,* 12, 63-70, 1978.
9. **Fuller, B.J., Lunec, J., Healing, G., Simpkin, S., and Green, C.J.,** Reduction of susceptibility to lipid peroxidation by desferrioxamine in rabbit kidneys subjected to 24-hour cold ischemia and reperfusion, *Transplantation,* 43, 604-606, 1986.
10. **Green, C.J.,** Rabbit renal autografts as an organ preservation model, *Lab. Anim.,* 7, 1-6, 1973.
11. **Green, C.J., Healing, G., Lunec, J., Fuller, B.J., and Simpkin, S.,** Evidence of free-radical-induced damage in rabbit kidneys after simple hypothermic preservation and autotransplantation, *Transplantation,* 41, 161-165, 1986.
12. **Green, C.J. and Allison, A.C.,** Extensive prolongation of rabbit kidney allograft survival after short-term cyclosporin A treatment, *Lancet,* 1, 1182-1183, 1978.
13. **Green, C.J., Allison, A.C., and Precious,S.,** Induction of specific tolerance in rabbits by kidney allografting and short periods of cyclosporine A treatment, *Lancet,* 2, 123-125, 1979.
14. **Ueno, A., Nicastri, A.D., and Friedman, E.A.,** Hyperacute renal allograft rejection in rabbit, passive transfer, specificity, and prevention, *Transplantation,* 14, 574-576, 1972.
15. **Terada, Y. and Ueno, A.,** A microangiographic study of renal allograft rejection and the effects of immunosuppression in the rabbit, *Transplantation,* 37, 443-446, 1984.
16. **Terada, Y. and Ueno, A.,** Hyperacute renal allograft rejection in the rabbit, *Transplantation,* 35, 205-208, 1983.

Chapter 3

Kidney Transplantation in Yucatan Miniature Swine

Todd Howard, Carlos A. Cosenza, Donald V. Cramer, and Leonard Makowka

CONTENTS

I. INTRODUCTION

The following is a summary of the technique developed for transplantation of the kidney in the miniature (Yucatan) swine. This animal model is appropriate for the study of kidney transplantation because of the ease in handling this animal and its small size.[1] The adult animal does not usually exceed 40 kg and the animals remain easy to handle even with prolonged postsurgical survival. The species has been well studied and sufficient information about the normal animal is available to allow study of most physiologic functions.[2] The details of the procedure are intended as a guide to the investigator so that the model can be easily and reliably reproduced.

The Yucatan swine is capable of mounting a strong cellular rejection.[3] The typical histologic picture is that of tubular injury and glomerular preservation. Because of the inbred nature of small colonies of these animals, however, we have found a high frequency of major histocompatibility complex (MHC) identity among donor/recipient pairs.[5] Recipient animals that share MHC haplotypes with the donor may not mount a strong cellular reaction to the kidney graft and may exhibit prolonged allograft survival even as untreated controls. Accordingly, we have relied on mixed lymphocyte responses when selecting donor and recipient pairs to ensure mismatching of MHC histocompatibility genes and a more consistent vigorous allograft rejection response. Alternatively, a second breed of small pigs, such as the Hanford pig, can be used as the donor. In our experience, these two breeds reliably exhibit strong histocompatibility differences.

0-8493-3629-5/94/$0.00+$.50

II. TECHNIQUES

A. ANATOMY

Several differences between human and porcine anatomy should be mentioned as important landmarks of the surgical procedure. The entry of the renal veins into the vena cava is found just below the point at which the vena cava enters the liver. The pancreas and duodenum both lie to the left of the cava and are not retroperitoneal. The infrahepatic vena cava, the entire right renal vein and kidney, and the origin of the left renal vein are therefore easily visualized. The left renal vein passes behind the mesentery of the small bowel and emerges on the left side. The distal left renal vein and the left kidney also lie just below the surface of the peritoneum and are easily visualized. The proximal colon is twisted upon itself and lies in a spiral within the abdomen. The descending colon arises from the base of the spiral and lies to the left of the midline as it dips into the pelvis.

B. ANESTHESIA

To minimize the content of the gastrointestinal tract, the animals are fasted for 12 hours before the operation. The animals are injected with ketamine, 10 mg/kg i.m., and atropine, 0.05 mg/kg i.m., after anesthesia induction.

1. The Donor

Anesthesia is induced with isoflurane administered by mask. The animals are then ventilated with 100% oxygen and 2% isoflurane to maintain anesthesia during preparation and surgery. The larynx is intubated and the animal is placed in a supine position and restrained. The normal heart rate under anesthesia is 100 to 120 beats per minute. Spontaneous ventilation may be maintained with lighter levels of anesthesia or mechanical ventilation at a rate of 6 to 10 breaths per minute and a tidal volume of 400 to 600 ml may be used to maintain normal oxygenation and acid-base status.

The abdomen, chest, and neck are shaved in preparation for surgery and vascular access is obtained through cut down (see below). Crystalloid solution is administered to maintain stable blood pressure and perfusion of the organs in both donor and recipient. In a well-anesthetized animal, a mean arterial blood pressure of 40 to 60 torr is satisfactory for maintenance of normal organ function and homeostasis. Arterial pressure monitoring is optional, particularly in the donor. If arterial pressure monitoring is not available, the clinical appearance of the viscera may be a guide to the adequacy of perfusion. In addition, fluid may be administered as necessary to maintain strong pulsations in the abdominal aorta as assessed by the operating surgeon.

We have found that approximately 30% of the animals will appear to be hypoxemic based on the dark, desaturated appearance of the arterial blood. Analysis of blood gases reveals that most of these animals have normal arterial oxygenation and acid-base status, and have no adverse consequences associated with this abnormal clinical finding. We have elected to perform arterial blood gas analysis in animals that appear to be hypoxic to differentiate between normally oxygenated dark blood and hypoxemia resulting from inadequate ventilation or other technical problems.

Stable anesthesia is maintained during the donor procedure. Once the surgeon is ready to cool and perfuse the kidneys, 200 U/kg of heparin is given to anticoagulate the animal. In addition, 1000 U of heparin is added to each 2-l bag of Eurocollin's solution. The objective of perfusion is to rapidly cool and remove the blood from the organ. The preservation solution should be kept cold (approximately 4°C) until the last possible moment and hung just before the donor is exsanguinated. When the vena cava has been transected, the aortic flush is opened maximally. Two liters should be infused over 5 to 10 min. Once the animal is exsanguinated and the organs removed, anesthesia may be terminated.

2. The Recipient

The animals are prepared and induced as described for the donor procedure. Endotracheal intubation, electrocardiogram (ECG) monitoring, venous access, and arterial pressure monitoring are provided. Anesthesia is maintained with Isoflurane and 100% oxygen.

Excessive fluid administration can lead to marked visceral edema, difficulty closing the abdomen, and high intraabdominal pressure. An excessively "tight" abdomen may also have adverse consequences for ventilation and renal function postoperatively. For this reason, it is well to keep intravenous fluid administration to a minimum and, under normal conditions, a total of 2 l of crystalloid will be sufficient for most recipient operative procedures. Should unusual blood loss occur, fluid administration may be increased to maintain blood pressure. Since the right atrial pressure is low, even in the euvolemic state, central venous pressure monitoring is of limited value as a guide for fluid administration.

Before clamping the vena cava for the venous anastomosis, 100 U/kg of heparin is administered to reduce the risk of thrombosis during the performance of the vascular anastomoses. With application of the aortic clamp for the arterial anastomosis, the arterial pressure may rise, and once the clamp is removed, may drop to very low levels. Dopamine may be required to maintain arterial pressure after the clamp is removed. It may be helpful for the surgeon to manually occlude the aorta distal to the renal artery anastomosis to maintain perfusion of the graft while the animal is stabilizing. Ordinarily, a mean arterial pressure of 40 to 60 torr is adequate for perfusion of the allograft.

As the operative procedure is nearing completion, the anesthetic may be reduced and finally interrupted to provide a smooth emergence from anesthesia. The animals will begin to awaken within several minutes and may have the tracheal tube removed once airway protective reflexes and spontaneous ventilation have returned. The animal is wrapped in a warm blanket and removed to the recovery area.

Hypothermia can represent a substantial problem and therefore, attention to heat conservation may result in greater hemodynamic stability. At the conclusion of the operation, warm irrigation of the abdominal cavity will assist in the resumption of normothermia and hemodynamics. Hypothermia is also an intense stimulus to renal vasoconstriction and may reduce renal function during the emergence and recovery phases.

C. SURGICAL TECHNIQUE

1. The Donor Procedure

A vertical incision is made in the neck parallel to the trachea and medial to the edge of the sternocleidomastoid muscle. The internal carotid artery and internal jugular vein are found deep to the platysma muscle and medial to the sternocleidomastoid muscle in the tracheal-esophageal groove. The carotid sheath is quite deep in the neck and will require careful exploration to locate the artery and vein. These vessels may be cannulated for venous access and arterial pressure monitoring. The cut end of a standard, sterile i.v. tubing is fixed in the jugular vein, and a similar cut end of sterile pressure monitoring tubing is introduced into the carotid artery. A smaller artery is usually found just superficial to the carotid sheath, but this vessel is ordinarily not suitable for cannulation with the large pressure tubing. A small plastic i.v. cannula may be inserted in this vessel, but is difficult to reliably fix in place. An alternative site for venous access is the external jugular vein. This vein is larger than the internal jugular and is found superficial and lateral to the sternocleidomastoid muscle.

A midline incision is made from the xiphoid process to the pubis. The abdomen should be routinely explored. The descending colon is mobilized from the spiral portion of the colon by dividing the peritoneum overlying the junction. This maneuver exposes the full course of the left renal vein. The mesentery remains undivided. Once the colon is mobilized, the small intestine and colon are most conveniently handled by wrapping them together in a moist towel. The following may then be completed in any convenient order:

1. The vena cava is isolated below the renal veins by dividing the overlying peritoneum. This is most easily accomplished at approximately the level of the inferior mesenteric artery. The cava should be surrounded with a heavy silk tie so that the distal cava may be ligated later. When dissecting the posterior aspect of the cava, care should be taken to avoid injury to the lumbar veins. Inadvertent injury may cause troublesome bleeding that is difficult to control. Direct sustained pressure is the most effective measure to control bleeding from lumbar veins.
2. The aorta is isolated at the same level by encircling with heavy suture. Care should be taken to avoid entering the large lymphatic vessel overlying the aorta. A large volume of lymph will be continuously encountered if this vessel is torn, making subsequent dissection difficult. This lymphatic vessel is most easily preserved by dissecting the areolar tissue anterior to the aorta from the right. The porcine aorta, unlike the human, is not surrounded by a well-defined plane of dissection. The dissection is therefore more difficult and should be accomplished with some care. The lumbar arteries are quite small and fragile. Accordingly, traction on the aorta should be minimized to avoid avulsing these vessels.
3. The left and right renal veins are mobilized by dividing the peritoneum overlying the vena cava between the renal veins and the liver. If the full length of the left renal vein is to be mobilized before the aortic flush, care should be taken to avoid injury to the left adrenal vein that enters the midportion of the left renal vein. The left adrenal vein is very short and thin-walled. The adventitia of the renal vein, in contrast, is quite tough, increasing the risk of accidental injury and brisk blood loss. Attempts to repair this injury *in vivo* will often result in a narrowing of the renal vein, potentially impairing the cold perfusion of the organ as well as deforming the vessel for implantation. As an alternative, we recommend that the final mobilization of the left renal vein be delayed until the aortic flush has begun. Injury to the adrenal vein at this point in the procedure is less serious, and careful repair on the back-table can be quickly and easily accomplished.
4. The superior mesenteric artery is mobilized and ligated. The best approach to this vessel is from the right; lateral and posterior to the portal vein, at the point where the vessel emerges from behind the crus of the diaphragm.
5. The celiac trunk is mobilized and ligated. Again, this vessel is best approached as it emerges from behind the anterior edge of the right crus of the diaphragm either anterior or posterior to the portal vein.
6. The kidneys may be mobilized before exsanguinating and perfusing the organs. Alternatively, this may be reserved until the aortic flush has begun. Dissecting in the retroperitoneum after flushing the aorta minimizes blood loss and hypothermia-induced hemodynamic instability before perfusion and shortens the operative procedure. The renal capsule is loosely attached to the kidney and careless handling may initiate stripping of the capsule. This may result in troublesome bleeding once the organ is reperfused. It should be noted that the renal arteries should not be dissected or disturbed during the donor operation. These vessels are prone to intense vasospasm and handling should be minimized.

Once these steps have been completed, heparin is administered. The distal aorta is ligated and the proximal end of the infrarenal aorta is cannulated. A flared tip on the cannula is useful to prevent dislocation of the cannula during removal of the organs. The cannula is fixed in place with heavy ligatures. Once the aorta is cannulated, the surgeon should notify the rest of the team that perfusion is about to begin. The assistants should ensure that the cold preservation solution is prepared and that all flush tubing is ready. Air should be evacuated from the perfusion lines to prevent air embolism and subsequent preservation injury. In rapid succession, the diaphragm is opened just anterior to the suprahepatic vena cava and, and through this opening, the vena cava is divided just below the right atrium. The aorta is clamped in the thorax to prevent warm blood from entering the abdominal viscera and to prevent loss of perfusate into the thorax, upper extremities, and head. While the abdominal aorta is flushed with 2000 cc of Eurocollin's solution at 4°C, the abdomen may be packed with ice or cold irrigation solution to provide for abdominal surface cooling. The infrarenal vena cava is ligated distally, to probably prevent entrance of warm blood from the lower extremities into abdominal cava and the cava is then divided. Once the kidneys are cold and exsanguinated, and while perfusion of the kidneys continues, the superior mesenteric artery is divided and the intestines lifted off the retroperitoneum.

The ureters are isolated at the brim of the pelvis as they pass over the iliac arteries. A generous margin of fatty tissue is removed with the ureters to avoid devascularization. The aorta, vena cava, and both ureters are then controlled with clamps and lifted by an assistant. If not completed before perfusion, the kidneys should be mobilized and brought to the midline. The most difficult portion of this dissection is anterior, where the adrenal gland is found. The posterior attachments of the great vessels, kidneys, and ureters are divided lifting the kidneys off the spine. The ureters may be injured during this phase of the operation if their location is not constantly monitored. The suprarenal aorta and infrahepatic vena cava are then divided and the kidneys are removed *en bloc*. Minimal dissection of the anterior attachments should be required. The kidneys are then placed in a basin of iced crystalloid solution.

2. The Back-Table Operation

A sterile back table should be prepared before perfusion of the donor organs. A large basin should be filled with ice-cold physiologic crystalloid and crushed frozen crystalloid. As much as possible the kidneys should be kept submerged in this ice slurry during the back-table operation. The kidneys are inspected and placed with their ventral surface facing down. This will position the dorsal surface of the aorta facing the surgeon and all other structures will be deep to the aorta. Using fine scissors, the dorsal aorta is then divided longitudinally between the lumbar branches, thus avoiding injury to the renal arterial orifices. Although we have not encountered this anomaly in pigs, the left renal vein may occasionally pass dorsal to the aorta, placing it at risk of transection during division of the aorta. Care should therefore be exercised when opening the dorsal portion of the aorta. With the aorta open, the renal arterial orifices can be identified and the ventral surface of the aorta is carefully divided. The left renal vein should be identified and kept in view to avoid accidental injury as the ventral surface of the aorta is divided.

The kidneys are then turned so that their ventral surface faces the surgeon. The ventral wall of the vena cava is then divided longitudinally. Again, with the renal vein orifices in clear view, the dorsal wall of the vena cava can be divided. Once the vessels are safely divided, the adventitial tissue between the great vessels can be safely divided and the kidneys separated. Throughout this procedure, the ureters should be kept in sight to avoid accidental laceration or division.

With the kidneys separated, the fat can be trimmed and the proximal portions of the renal vessels cleaned. The renal arteries should be cleaned of excess connective tissue, but dissection directly on the wall of the vessels is to be avoided. Since a patch of aorta or vena cava is used for the arterial and venous anastomoses, these patches can be carefully cleaned. However, it is not necessary to dissect close to the renal vessels or past the first bifurcation of either the renal artery or renal vein. Furthermore, the renal arteries are prone to intense vasospasm and they should be handled as little and as gently as possible.

This is best done by retaining the complete length of aorta with the renal artery until the back-table is completed. Sutures placed in the ends of the aortic tissue, well away from the renal arterial orifices will facilitate manipulation of the renal arteries. Care should be taken to avoid twisting the renal artery around its long axis. Such rotations are difficult to identify when the vessel is not distended with blood and, if discovered after reperfusion, may result in early thrombosis and require a period of warm ischemia while the rotation is corrected.

Several small, non-renal branches arise from each renal vein and these should be identified and ligated. On the right, the adrenal vein is closely adherent to the junction of the renal vein and vena cava, and removing this tissue may be difficult. Once the vessels are satisfactorily cleaned, the excess vena cava and aorta may be trimmed leaving a 2- to 3-mm patch around the orifice of each vessel. It may be helpful to mark the superior pole of the renal artery with a fine suture. Once in spasm, with the Carrel patch cut, a rotation of the aortic patch twisting the renal artery can be difficult to recognize.

The capsule of the kidney is loosely attached to the organ and can be easily stripped off accidentally. Although the graft will function well without the capsule, troublesome bleeding may result if the capsule is stripped. This argues for careful handling of the kidney throughout the donor procedure and back-table preparation. The ureters can also be cleaned of excess fat but this is not necessary and care should be taken to leave a generous amount of tissue with the ureters to avoid devascularization and subsequent necrosis or stricture of the ureters. The organs can be kept in the iced solution until the recipient is ready for implantation. We have found no difficulty keeping kidneys in the cold slurry for the time necessary for recipient preparation.

3. The Recipient Procedure

As for the donor procedure, the jugular and carotid cut-down is performed to provide for venous access and arterial monitoring. Arterial monitoring is more important in the recipient as perfusion of the transplanted organ requires adequate blood pressure. Furthermore, the administration of pressors required to maintain the blood pressure should be guided by arterial pressure monitoring.

The abdomen is opened in the midline to give good exposure to the kidneys in the mid-abdomen. A xiphoid to pubis incision is not necessary and may increase morbidity and loss of animals. Special care should be taken to avoid injury of the long, and narrow friable spleen that may lie in an anterior position. The spleen may be positioned in the left upper quadrant with packs for the duration of the procedure. The intestines may be wrapped in a moist towel to minimize fluid loss from the surface of the gut and facilitate retraction.

The aorta and vena cava are mobilized at the level of the inferior mesenteric artery. Again, care should be taken to avoid injury to the lymphatic vessels adjacent to the aorta. Brisk oozing of lymph may make aortic anastomosis very difficult. The aorta and vena cava are mobilized enough to allow placement of Satinsky clamps on the vessels. Lumbar branches of the aorta should be divided to avoid avulsion. Unlike in the donor, bleeding lumbar veins or arteries may be impossible to find and ligate or to control with direct pressure, particularly after the administration of heparin. We have found that the animals do very poorly once they become anemic. Accordingly, careful hemostasis is critical and even minor bleeding, because of heparinization, may persist and result in severe anemia and stress for the recipients.

Both kidneys are then mobilized by dividing the overlying peritoneum. The hilum is identified and doubly clamped with curved clamps. The hilum is divided distal to the second clamp and the native kidney removed. The hilum is ligated proximal to both clamps and then suture ligated between the clamps. Careful hemostasis in the bed of each kidney will minimize bleeding later when the animal is heparinized.

Whether the left or right kidney is selected for implantation is usually of little consequence. The left renal artery is shorter than the right and the right renal vein is shorter than the left. When performing two renal transplants, the particular anatomy of each recipient may suggest that one organ may be more suitable for use in a particular recipient. For example, the aorta is occasionally positioned almost behind the vena cava. With its longer renal artery, the right kidney may be used to advantage in this situation.

The recipient is heparinized and the donor kidney is wrapped in ice and gauze to minimize rewarming. Sutures of 6-0 polypropylene are placed in the upper and lower poles of the donor vein patch. The clamp is placed on the vena cava and a longitudinal venotomy is performed. The donor kidney is brought to the field, and simple upper and lower pole sutures are completed and tied. An end-to-side running anastomosis is completed. The back wall of the anastomosis may be completed from inside the vein with the kidney retracted to the right or from the outside with the kidney retracted to the left. A generous amount of tissue can be safely included in the anastomosis to facilitate its construction, and to aid in hemostasis since a patch of vena cava is used. The renal vein is then clamped close to the anastomosis and the vena cava flow is restored. Bleeding points should be controlled at this point as they may be difficult to visualize once the renal artery is implanted. Care should be taken to avoid dislodging the clamp on the renal vein and accidentally reperfusing and rewarming the kidney.

The aorta is now clamped and an aortotomy performed. The site of the anastomosis should be selected so that the renal artery lies without kinking or impinging on the renal vein. The anastomosis should not be placed at the same level as the renal vein, so that the renal artery does not obscure the view of the renal vein anastomosis, and *vice versa.* Care should be taken to correctly orient the renal artery to avoid twisting. The upper and lower poles of the aortic patch are affixed with simple sutures of 6-0 polypropylene. A second set of tacking sutures can be placed half-way along the anastomosis and used to hold the aorta open and provide better exposure. The right side of the anastomosis is most easily completed from within the vessel, since the renal vein obscures the view from the right side. The arterial anastomosis should be carefully performed to avoid leakage, as the site is often difficult to identify once the clamp is removed. The clamp is removed from the aorta and then quickly from the renal vein and the kidney is reperfused. If the blood pressure drops when the aortic clamp is removed, digital occlusion of the distal aorta may improve perfusion of the graft.

Once reperfused, the kidney should not be manipulated so that the organ can be rewarmed and reperfused and twisting of the vessels is avoided. Reversal of heparin should not be necessary, and a modest incidence of arterial thrombosis can be expected even without reversal of heparin. Careful anastomotic technique will minimize bleeding which may require manipulation of the kidney to visualize and control. The careful handling of the renal artery is an important component of the entire procedure. Routinely, topical papaverine and i.v. manitol (50 ml) are used to reduce the spasm frequently seen in the renal artery and to ensure adequate diuresis after the first postoperative hours.

The renal vascular anastomoses may now be packed and the neoureterocystostomy can be performed. The ureter is trimmed to a length that will allow a comfortable path between the renal pelvis and the dome of the bladder with plenty of slack left to allow for movement of the kidney. However, it is important to minimize the length of the donor ureter to reduce the risk of twisting and inducing ischemic necrosis. The section of the ureter should be done over a well-vascularized area to avoid subsequent ureteral leaks or ureterocyctostomy stenosis. The end of the ureter is then spatulated to provide an anastomotic length of approximately 1.5 cm. The muscular coat of the bladder is incised with electrocautery, a Satinsky clamp is placed, and the mucosa of the bladder is opened sharply. A careful mucosal closure with running 6-0 monofilament absorbable suture is then performed. Inadvertent closure of the ureteral orifice may be avoided by placing a

20-gauge plastic i.v. catheter in the ureter while the heel of the anastomosis is performed. The muscular wall of the bladder is then loosely closed over the anastomosis with two or, at most, three simple sutures of 6-0 absorbable suture. An additional ureteral tunnel of approximately 1.5 to 2 cm can be performed with single sutures to avoid vesicoureteral reflux.

The abdomen should then be carefully inspected for hemostasis and gauze and packs removed. In particular, the sites of the superior renal pole of each native kidney should be inspected as should the hilar stumps of the native kidneys and hilum of the transplanted kidney. Electrocautery in the hilum of the transplanted kidney may cause additional arterial spasm and should be used with care.

The abdomen may be closed using braided 2-0 absorbable suture. The skin is closed with interrupted absorbable sutures as is the jugular and carotid cut-down site after the removal of the cannulae.

D. POSTOPERATIVE CARE AND GRAFT MONITORING

After surgery, recovery can best takes place in the animal's cage. If surgery has not been prolonged (1 to 2 hours), recovery will be uneventful. In cases where surgery has been prolonged (more than 2 hours), and especially if the animal is fat, recovery may be correspondingly longer. During prolonged anesthesia, the body temperature may fall several degrees. As mentioned before, hypothermia is a frequent complication in this model and intra- and postoperative measures should be taken to avoid excessive loss of body heat. We routinely use a warming blanket during surgery, and as soon as the wound is closed, the pig is covered by a blanket and additional heat is provided by a heating lamp. The lamp has to be placed approximately 18 in. above the animal's skin to avoid burns. The endotracheal tube should be left in place until swallowing or body movements are evident. The first postoperative hour is crucial, and body temperature, as well as breathing movements, have to be monitored frequently. Animals generally fully recover from anesthesia after 1 or 2 hours postsurgery.

If good sterile techniques are used during the surgery, postoperative infections will be rare. In this model we administer 1 gm of a long-acting cephalosporin immediately before closure, and 12 hours postoperatively.

The pigs generally remain quiet after surgery, but in order to increase the comfort of the animal, intramuscular Buprenex can be administered every 8 hours during the first 24 postoperative hours. Usually drinking and eating will resume in a normal fashion within 24 hours.

Besides the body temperature and the general condition of the animal, the most important parameter to be recorded in this model is urine production. If micturition has not been visualized within 24 hours of surgery, a blood creatinine level is recommended as a test for renal artery thrombosis, one of the most common complications of this surgery. Additionally, the wound should be examined for the presence of urine, a sign of failure of the ureteric anastomosis. If detected early, this complication can be successfully repaired. We routinely record urine production, and measure serum urea and creatinine levels beginning on postoperative Day 3. At this time, creatinine levels in the normal range (1.2 to 1.6 mg/ml) generally reflect a normally functioning graft. Elevated creatinine levels at postoperative Day 3 should be considered to be suspicious for a technical complication such as renal artery thrombosis.

III. APPLICATIONS

A. IMMUNOSUPPRESSION

Kidney transplantation in swine is an effective alternative to the use of other large species, particularly the dog, for studying the influence of immunosuppressive drugs on the allograft reaction. With an appropriate choice of donor/recipient pairs, the exchange of kidneys results in a brisk and predictable rejection reaction that results in kidney failure at approximately 10 to 12 days posttransplantation.[6] The miniature pig is of sufficient size as to not require the use of microsurgical techniques, yet small enough to permit easy handling postoperatively should long-term graft survival result from the procedure. Finally, the animals are readily available and are easily managed with the use of appropriate restraints.

The use of the miniature pig for kidney transplantation requires special attention in order to avoid the problems of histocompatibility matching and postoperative renal artery thrombosis. As described above, the development of commercial herds of miniature swine has increased the incidence of shared haplotypes within an individual herd.[5] This MHC matching can result in prolonged graft survival between untreated individuals and may therefore, reduce the sensitivity of the model to immunosuppressive drugs. This issue can be avoided by using a second small breed of pigs, such as the Hanford pig, as the donor of the kidney grafts. Alternatively, the donor/recipient pair can be purposely mismatched using the mixed lymphocyte response. This survival and the mixed lymphocyte response (MLR) testing requires additional handling and experimental costs.

The second area of special consideration is the potential for a high incidence of renal artery thrombosis should the surgical procedure be performed without sufficient care. The renal artery of the pig is susceptible to intense vasospasm when handled. This spasm, combined with postoperative hypotension, may result in arterial thrombosis. To reduce this complication to an incidence of less than 10% requires careful handling of the vessel during the operation and following of the procedures described above. The occurrence of renal artery thrombosis should be considered likely if significantly elevated serum creatinine levels are observed 3 days postoperatively and/or the recipient fails to micturate. The diagnosis of renal infarction (vs. acute rejection) can be readily accomplished at autopsy with histopathological examination of the graft.

B. ORGAN PRESERVATION

The miniature pig provides an excellent model for *in vivo* and *in vitro* studies of kidney function, organ preservation, and organ perfusion. The kidney of the pig is considered to be anatomically and functionally similar to the human kidney.[6] The kidneys can be harvested easily and tolerate cold ischemia well. In our experience, acute tubular necrosis has not been a problem provided that the ischemia time does not exceed 2 hours.

IV. CONCLUSIONS

The Yucatan miniature swine is an easily managed large animal that is suitable for kidney transplantation. The animal is large enough for surgical procedures to be easily performed and yet remains small enough to be easily cared for following the transplant. The miniature pigs tolerates the surgical procedure well, and represents a good model of kidney transplantation and response to therapy for evaluating new drugs and procedures for application towards clinical transplantation. Immunologic and preservation experiments can be performed with this animal model with a reasonable degree of reproducibility and a minimum of specialized training.

REFERENCES

1. **Panepinto, L.M. and Phillips, R.W.,** The Yucatan miniature pig: characterization in utilization in biomedical research, *Lab. Anim. Sci.,* 36, 344, 1986.
2. **Radin, M.J., Weisner, M.G., and Fettman, M.J.,** Hematologic and Serum biochemical values for Yucatan miniature swine, *Lab. Anim. Sci.,* 36, 425-427, 1986.

3. **Calne, R.Y.,** Allografting in the pig in *Immunological Aspects of Transplantation Surgery,* Calne, R.Y., Ed., Medical and Technical Publishing, London, 1973.
4. **Lunney, J.K., Pescovitzz, M.D., and Sachs, D.H.,** The swine major histocompatibility complex: its structure and function, *Swine in Biomedical Research,* Tumblesen, M.E., Ed., Plenum Press, New York, 1986, 1821.
5. **Hreha, G., Hough, K., Cramer, D.V., Hill, D., Cosenza, C., Ulker, N., Chapman, F., and Makowka, L.,** Evidence for inbreeding, MHC haplotype matching, and prolonged kidney graft survival in miniature swine, *Transplantation,* 55, 224-227, 1993.
6. **Kirkman, R.L., Colvin, R.B., Flye, M.W., Leight, G.S., Rosenberg, S., Williams, G.M., and Sachs, D.H.,** Transplantation in miniature swine. VI. Factors influencing survival of renal allografts, *Transplantation,* 28, 18, 1979.
7. **Hatfield, P.J., Cameron, J.S., and Cadenhead, A.,** Renal biopsy in the pig, *Res. Vet. Sci.,* 19, 88, 1975.

Section II

Liver Transplantation

Chapter 4

Liver Transplantation in Mice

Shiguang Qian and John J. Fung

CONTENTS

I. INTRODUCTION

Orthotopic liver transplantation (OLT) has been performed and widely used in experimental studies in the rat. This operation has not been possible in inbred mice because the procedure has been technically difficult.[1,2] Although the size of the mouse is only one-tenth of that of the rat, there are many advantages in using the mouse OLT model for transplantation immunogenetic studies. The mouse genome has been more thoroughly characterized than the rat and any other species of mammals. Large numbers of genetically defined inbred strains of mouse and commercial monoclonal antibodies are available. The striking homology between the mouse H-2 system and the human HLA system further propelled the mouse as the ideal animal model to study human transplantation immunogenetics. The experience gained with the mouse system has been invaluable for the progress of the human studies.

A key development in rat orthotopic liver transplantation procedure developed by Lee, Zimmerman, and Kamada[3-5] was the use of a cuff technique instead of sutures for the portal-vein anastomoses. This shortened the clamping time of the portal vein and increased the surgical survival rate. This technique was successfully applied to the model of mouse liver transplantation developed in our laboratory. The surgical success rate has approached 83% which has allowed for the performance of a variety of investigational studies.

II. TECHNIQUES

A. ANATOMY OF THE MOUSE LIVER

The mouse liver is divided into four lobes. The median lobe, which bears a fissure for the falciform ligament anteriorly and a gallbladder posteriorly, is the largest lobe. The left lobe sits between the median lobe and the small caudate lobe, which is partially divided into an anterior and posterior lobule fitting around the junction of the esophagus and stomach. The right lobe, which is partially divided into superior and inferior

0-8493-3629-5/94/$0.00+$.50

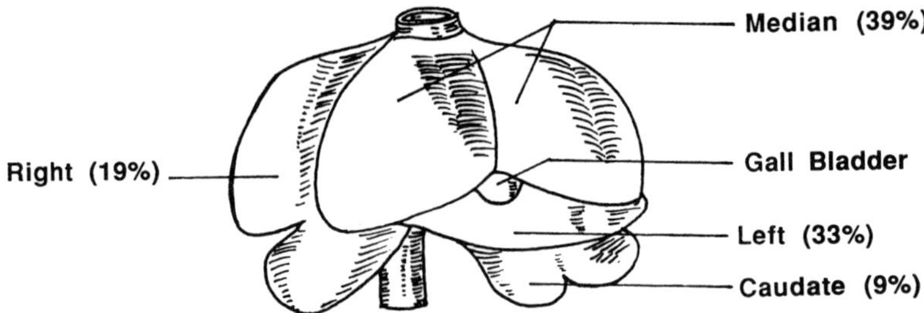

Figure 1 The mouse liver comprises four main lobes. The number shows the relative size of each lobe. The gall bladder is located at the base of the deep bifurcation of the median lobe.

lobes, encircles the inferior vena cava. The relative weight of each lobe is illustrated in Figure 1.

Unlike the rat, the mouse has a gallbladder which lies on the visceral side of the median lobe of the liver. The cystic duct connects the gallbladder to the hepatic duct. The common bile duct, usually made up of three hepatic ducts from the four lobes of the liver (Figure 2), is joined by the main pancreatic duct and together they open into the duodenum.[6]

The course of the hepatic artery in mice is quite variable and seems to be strain dependent.[7] The mouse liver is usually supplied from the hepatic artery branch of the coeliac artery, but in some cases the hepatic artery comes from the left gastric branch, arising as this vessel reaches the lesser curvature of the stomach and runs along the

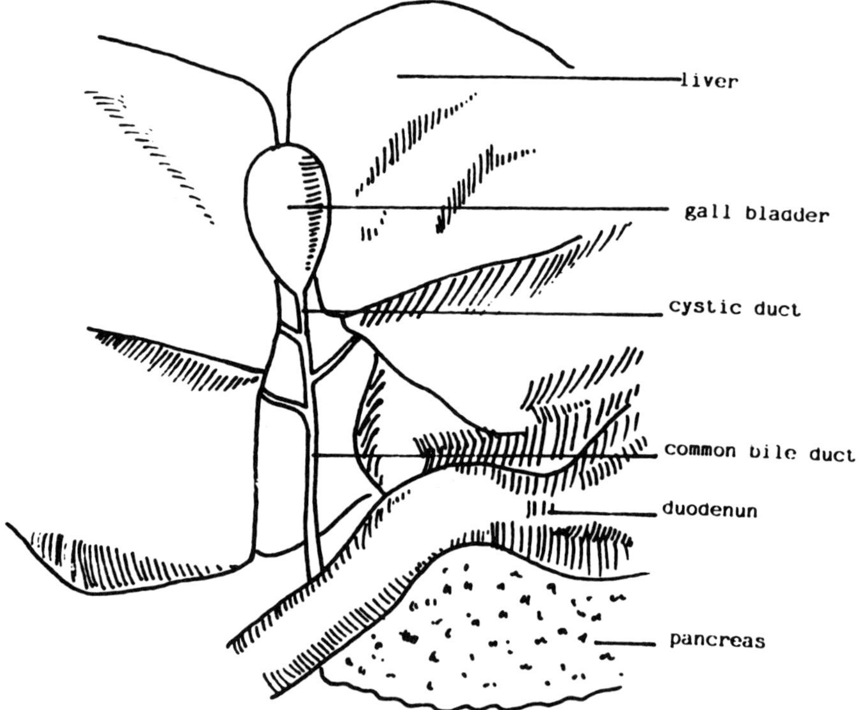

Figure 2 The hepatic ducts from the liver and the cystic duct from the gall bladder unite to form the common bile duct which, joined by the pancreatic duct, opens into the duodenum.

esophagus to the liver. In some mice, both types of the hepatic artery are present, one from the common hepatic branch to the right lobes of the liver and one from the left gastric to the left lobes. Occasionally, the common hepatic artery originates from the superior mesenteric artery.

B. SURGICAL TECHNIQUE

Clean operative technique is used for both donor and recipient operations and all procedures are performed under the operating microscope (with 4 to 6.4 × magnification). Both donor and recipient mice are anesthetized with methoxyflurane (2,2-dichloro-1,1-difluoroethy methyl ether). Male inbred mice (10- to 12-weeks old; 25 to 32 gm) are used as donors and recipients.

1. Donor Operation

The skin of the abdomen is shaved and disinfected with Betadine or alcohol and the abdominal cavity is entered through a longitudinal midline or a transverse incision. Exposure is maintained with small retractors and one clamp with which the xiphoid is grasped and retracted. During the entire procedure the liver and surrounding tissues are kept cool with frequent rinses of cold Ringer's lactate solution. First, the bile duct is transected distally and the pyloric and splenic vein, hepatic artery, right adrenal vein, and right renal vein are ligated. The donor is then heparinized by slowly injecting 100 U of heparin (diluted to 0.3 ml with Ringer's) through the penile vein. The liver is perfused with about 1 ml cold Ringer's lactate, given through the portal vein after making an outlet in the vena cava to allow the perfusate to escape. The portal vein, right renal vein, and right adrenal vein are then transected. The suprahepatic vena cava is cut as near as possible to the diaphragm.

The infrahepatic vena cava is divided below the level of the left renal vein, leaving a sufficient length for cuff preparation. The freed liver is then placed into a container of cold Ringer's lactate.

2. Back-Table Preparation

The back-table preparation of the graft is performed in a container of cold lactated Ringer's solution. The cystic duct is ligated and the gallbladder removed with scissors. The infrahepatic vena cava and the portal vein are dissected free of the attached connective tissue. The cuff for both the infrahepatic vena cava and portal vein are made from Teflon tubes (outer diameter of 1.7 mm for vena cava; of 1.2 mm for portal vein). The length of the cuff is adjusted to 1.5 mm and an equal length extension is used for holding. The end of the vein is passed through the cuff tube, the cuff extension and the vein is then held in place by a Satinsky or miniature Bulldog clamp. The end of the vein is then folded over the cuff tube and secured with a silk suture (Figure 3). The infrahepatic vena cava is cross-clamped by a fine vessel clamp to prevent blood leakage after the transplanted liver is revascularized.

3. Recipient Operation

A long midline or transverse incision is made and the exposure procedures are the same as those outlined above for the donor. The bile duct is transected proximally and the hepatic artery is ligated and divided. The right adrenal vein is ligated and divided. After the ligaments around the liver are separated, a suture is passed beneath the suprahepatic vena cava, which is used to pull the liver and diaphragm down enough to provide for clamping. The infrahepatic vena cava and the portal vein are cross-clamped in order. After gently pulling down the preplaced traction suture beneath the suprahepatic vena cava, the suprahepatic vena cava, including a portion of diaphragm, is cross-clamped with

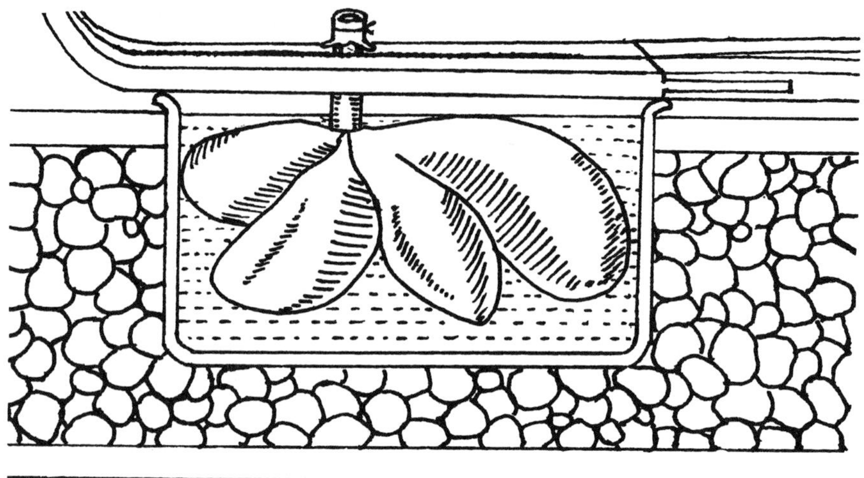

Figure 3 Cuff preparation of the portal vein and the infrahepatic vena cava of donor mouse liver graft which is sinked in ice-bathed saline. Cuff extension and the vein is held by a Satinsky clamp. The end of the vein is folded over the cuff and secured with a suture.

a Satinsky clamp. The suprahepatic vena cava and portal vein are then divided near the liver margins. The infrahepatic vena cava is divided. The recipient liver is then removed.

The donor liver is placed orthotopically. The suprahepatic vena cava anastomosis is performed first using a one guide suture technique to provide better exposure of the posterior vessel walls with running 10-0 nylon suture (Dermalon, Davis & Geck Inc., Manati, PR) (Figure 4). The end of the portal vein is retracted using a mosquito clamp and a venotomy is made in the front wall. The lumen of the vein is irrigated with saline through the venotomy site with a blunt angled needle connected to a syringe. The cuffed donor portal vein is then inserted into the recipient portal vein through the venotomy and secured with a silk suture (Figure 5). The suprahepatic vena cava and then the portal vein clamps are removed. The blanched liver quickly turns pink after revascularization. The total clamping time of the portal vein is kept to a maximum of 20 min. The cuffed donor infrahepatic vena cava is inserted into the recipient vena cava, which is secured by a silk suture. The bile duct continuity is achieved with a 4-mm long polyethylene tube (outer diameter of 0.61 mm) (Beckton Dickinson & Co., NJ) which is inserted into the lumen of bile ducts, and secured with silk sutures (Figure 6). Blood loss during the operation is replaced with Ringer's lactate solution.

The abdomen is enclosed in two layers with 4-0 sutures. All mice are given a single injection of Cefamandole Nafate (5 mg) and Bicillin (60,000 U) i.m. immediately after the surgery and kept under a warming lamp until normally active. Food and water are allowed *ad libitum* postoperatively.

III. APPLICATIONS

A. SYNGENEIC LIVER TRANSPLANTATION IN MICE

In our pilot studies, over 40 syngeneic mouse liver transplants had been performed before six long-term survivors were achieved. As our techniques improved, we felt confident that other surgeons experienced in microvascular techniques, especially in rat liver transplantation, could be successfully instructed to perform mouse liver transplantation. In our experience, the surgical success rate in syngeneic mice liver transplantation has reached 83%. The behavior of mouse syngeneic liver grafts were studied biochemi-

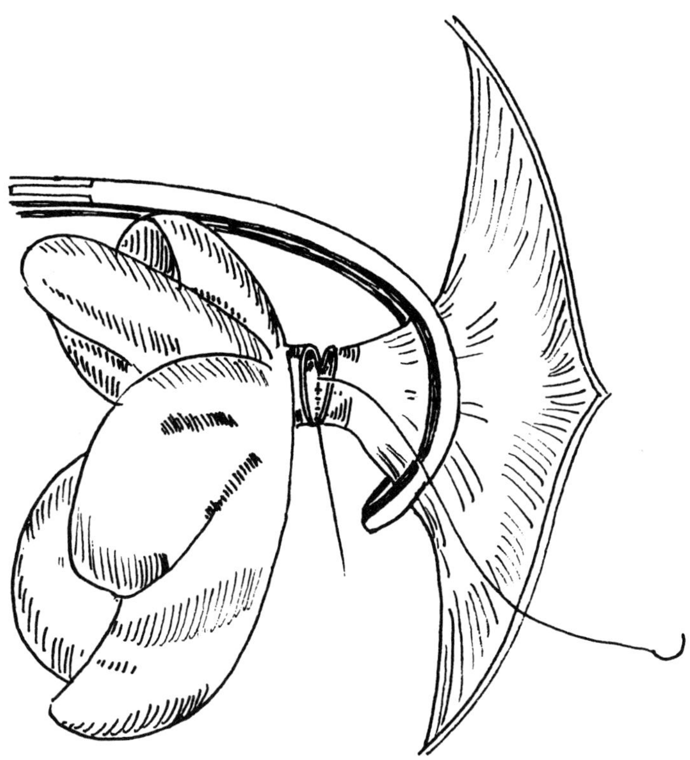

Figure 4 Anastomosis of suprahepatic vena cava, showing the one guide suture anastomosis technique which provides better exposure of the posterior vessel walls.

cally and histologically. Except for the persistent elevation of alkaline phosphatase (AKP), the remaining liver functions, serum bilirubin, and glutamic oxaloacetic transaminase (SGOT) usually returned to normal in 2 to 3 weeks after surgery. Total protein and albumin levels were normal in all the animals except in a mouse that died from bile duct complications, where the level of protein and albumin were lower than normal. Histologically there were no major changes in liver structure in long-term survivors. However, there were usually focal abnormalities, as in the rat, such as mild fibrosis, mild ductular proliferation, and a light lymphoplasmocytic infiltrate of the triads. The ducts, in some cases, showed degenerative changes in the form of apoptosis and vasculization. The duct degenerative changes, elevation of AKP, and other subtle histopathologic changes in the portal triads have been considered to be the consequence of the failure of providing an arterial blood supply. However, as in the rat, these changes it did not apparently interfere with long-term survival.[1]

B. ALLOGENEIC LIVER TRANSPLANTATION IN MICE

The model of orthotopic liver transplantation in mice has also been used for research of allogeneic liver graft transplantation at the University of Pittsburgh. Orthotopic liver transplantation was performed in 13 genetically defined histoincompatible strain combinations of mice with no immunosuppression. Liver grafts transplanted in two strain combinations with MHC class I, II and minor disparity had 20 and 33% survival of >100 days but the other 11 combinations, including 4 that were fully allogeneic, yielded 45 to 100% >100 days survival. The C3H strain accepts C57BL/10 liver grafts although they are fully allogeneic for both major histocompatibility

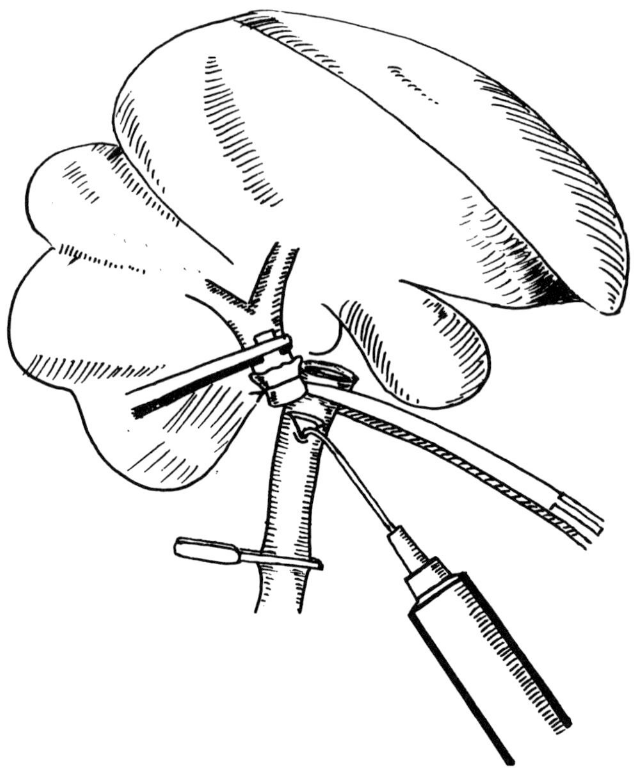

Figure 5 Portal vein anastomosis. The cuffed donor portal vein which was held by forceps is inserted through a venotomy into the recipient portal vein and secured with a circumferential suture.

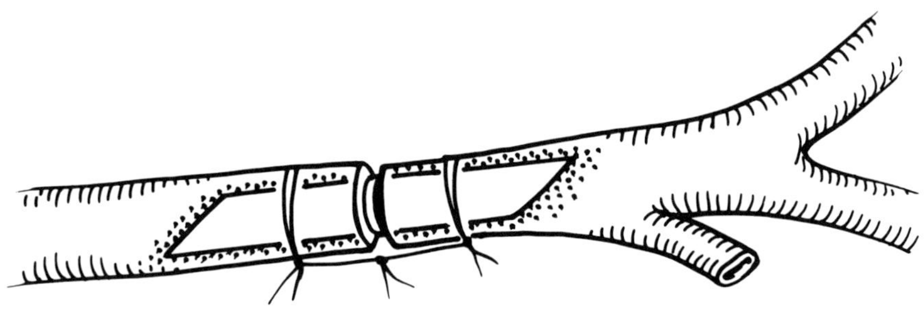

Figure 6 Bile duct anastomosis. A polyethylene tube is inserted into both bile ducts and secured with sutures.

complex and minor histocompatibility complex (MHC) antigens. The C57BL/10 strain accepted B10.BR liver grafts (mismatched for both class I and class II MHC loci). Long-living recipients tolerated subsequent donor skin or heart graft, in spite of detectable anti-donor *in vitro* activity with MLR and CML testing.[2,8]

C. OTHER PLANNING PROJECTS RELATED TO THIS MODEL

Several interesting projects related to the mouse liver transplantation model are now being conducted in this center. It has become possible to analyze the immunogenetics of the rejection of liver graft and the tolerance induced by liver grafting using the hundreds of inbred and congeneic strains of mice commercially available. We are interested in more precisely defining the role of antigen presenting cells, such as interstitial dendritic cells, in liver allograft rejection and tolerance induction developing treatment strategies to reduce or eliminate the antigenicity of the liver allografts by eliminating the allogeneic stimulatory activity. It is also a good model to investigate the mechanisms of the specific tolerance induced by liver allografts in the mouse.[8]

REFERENCES

1. **Qian, S., Fung, J.J., Demetris, A.J., Ildstad, S.T., and Starzl, T.E.,** Orthotopic liver transplantation in the mouse, *Transplantation,* 52, 562, 1991.
2. **Qian, S., Fung, J.J., Sun, H., Demetris, A.J., and Starzl, T.E.,** Transplantation unresponsiveness induced by liver allografts in mouse strains with various histocompatibility disparities, *Transplant. Proc.,* 24, 1605, 1992.
3. **Lee, S., Charter, C., Chandler, J.G., and Grloff, M.J.,** A technique for orthotopic liver transplantation in the rat, *Transplantation,* 16, 664, 1973.
4. **Zimmerman, F.A., Butcher, G.W., Davies, H.S., Brons, G., Kamada, N., and Turello, O.,** Techniques for orthotopic liver transplantation in the rat and some studies of the immunologic response to fully allogeneic liver grafts, *Transplant. Proc.,* 11, 571, 1979.
5. **Kamada, N. and Calne, R.Y.,** Orthotopic liver transplantation in the rat: technique using cuff for portal vein anastomosis and biliary drainage, *Transplantation,* 28, 47, 1979.
6. **Cook, M.J.,** *The Anatomy of the Laboratory Mouse,* Academic Press, London, 1965, 71.
7. **Froud, M.D.,** Studies on the arterial system of three inbred strains of mice, *J. Morphol.,* 104, 441, 1959.
8. **Qian, S., Demetris, A.J., Murase, N., Rao, A.S., Fung, J.J., and Starzl, T.E.,** Murine liver allograft transplantation. Tolerance and donor cell chimerism, *Hepatology,* in press, 1993.

Chapter 5

Orthotopic Rat Liver Transplantation

Michael Knoop, Rudolf Steffen, Donald V. Cramer, and Leonard Makowka

CONTENTS

I. INTRODUCTION

Orthotopic liver transplantation has become a widespread, accepted form of treatment for terminal hepatic diseases. Since the first experimental heterotopic liver transplantation in the dog by Welch,[1] numerous animal models of hetero- and orthotopic liver transplantation have been developed and thoroughly investigated.[2,3] While techniques for kidney and heart transplantation in the rat were developed in the early 1960s, it was not until 1966 that Lee and Edgington[4] described a model of partial heterotopic liver transplantation that was designed to study liver regeneration after grafting. The technique consisted of a 70% partial hepatectomy in the donor with the removal of the median and lateral lobes. The graft in the recipient, was revascularized by several different ways using portal venous or arterial blood supply reconstruction. The bile duct was inserted into the duodenum. In 1972 Hess et al.[5] described a model of heterotopic grafting in which a portal (70%) donor liver was placed into the recipient who underwent a 70% hepatectomy and right nephrectomy. Revascularization was achieved by an end-to-end anastomosis of the recipient to the donor portal vein and an end-to-side anastomosis of the donor inferior vena cava (IVC) to the infrahepatic recipient IVC. No attempt was made to reestablish arterial blood flow. The bile duct was implanted into the duodenum. To create a nonauxiliary graft, the operation was completed by removal of the host liver. Kort et al.[6] later reported a modified model. Mueller[7] described a model of heterotopic auxiliary liver grafting in which the grafts portal vein was supplied with arterial blood via the left renal artery after left nephrectomy. Venous outflow was to the left renal vein and the bile duct was anastomosed to the left ureter. In contrast to the previous models, Marni and Ferrero[8] used a sutureless cuff technique for heterotopic liver transplantation. A further discussion of these heterotopic models, including their physiological peculiarities and experimental use is beyond the scope of this chapter (for overview see Hess,[9] Zelder,[10] and Lee and Scott[11]).

The later development of an orthotopic replacement of the recipient liver by the graft required three venous (superior vena cava [SVC], portal vein [PV], and IVC), a bile duct, and an optional arterial anastomosis. The existence of hepatic arterial reconstruction

0-8493-3629-5/94/$0.00+$.50

characterizes the graft as an arterialized graft and the absence of the arterial anastomosis is considered as a venous (nonarterialized) graft.[12] Most of the difficulties in the development of the orthotopic procedure were encountered in the microsurgical suture techniques used to complete the suprahepatic vena cava and the fragile end-to-end portal vein anastomosis within an ischemia time of less than 25 min. If the anhepatic time exceeds this limit, irreversible liver damage precipitates shock and death of the recipient. To overcome this problem Lee et al.[12] described a technique for orthotopic liver transplantation with extracorporal portojugular bypass to decompress the splanchnic bed during the anhepatic phase. Arterial blood flow was reestablished by connecting the donor aortic segment with the celiac axis and the hepatic artery to the recipient's aorta. This technique was difficult and complex, however, because of a requirement for sutured anastomoses of all liver vessels. A refinement of the technique to avoid clotting of the bypass resulted in a simplified model without an extracorporal bypass and rearterialization.[13] The donor bile duct in both techniques was directly implanted with a pull-through maneuver into the recipient's duodenum.

Several modifications of the basic orthotopic procedure involving new microvascular techniques and bile duct reconstruction were introduced in the following years. Zimmermann et. al.[14] developed a bile duct reconstruction with an end-to-end choledocho-choledochostomy. A small plastic tube was inserted into the donor bile duct and fixed with a ligature. The recipients bile duct was pulled over the free end of the tube and fixed with one or two stitches. Finally, omentum was wrapped around the anastomosis. The high incidence of duodenal bleeding, failure of the choledocho-duodenal anastomoses, and ascending cholangitis with the pull-through technique of Lee could be reduced using this splint technique of Zimmermann, thus increasing the survival rate. In the same year, Kamada and Calne[15] introduced the cuff technique for the portal vein and infrahepatic vena cava in their venous model with an anhepatic time that could be held to under 15 min. The bile duct anastomosis was accomplished by telescoping a donor bile duct tube into the larger recipient bile duct tube. An extension of this technique used an additional cuff for the suprahepatic vena cava anastomosis.[16-18] This technique, however, does not represent a simplification at this site since additional dissection steps are necessary. A properly handsewn SVC anastomosis is fully sufficient and easier to perform.[19] In both the Zimmermann and Kamada technique, the graft hepatic artery was ligated. The first practical physiological model with preserved arterial perfusion was presented by Engemann et al.[20] The graft received its arterial supply through a long donor aortic segment, reaching from the diaphragm to the aortic bifurcation, with anastomosis to the recipients infrarenal aorta. The bile duct anastomosis was an end-to-end choledocho-choledochostomy as described by Zimmermann.[21]

Following the description of these techniques, additional modifications of rearterialization have been reported. Table 1 lists the most important models and their features. Hasuike et al.[22] suggested a cuff anastomosis of the donor aorta to the recipient's right renal artery after right nephrectomy. Howden et al.[23] choose the same site for a microsurgical suture. Both techniques require the removal of the right kidney, i.e., they are not physiological in a strict sense. Steffen et al.[24] described a simple technique that included reconstitution of the arterial blood flow by placing a cuff on the recipient's common hepatic artery, over which the donor celiac artery was slipped and secured (Figure 3). Patent arterial anastomoses were found in 32 out of 36 grafted rats at the time of autopsy by cutting the arterial vessel distally to the cuff anastomosis. Chaland et al.[25] used two telescoping tubes inserted into the donor and recipient hepatic artery resembling Kamada's technique of bile duct reconstruction. In general, cuff or tubing techniques for the arterial reconstruction are associated with a lower incidence of thrombosis than handsewn anastomoses and have become the most popular means of providing arterial reconstruction for the liver graft.

Table 1 **Models of Arterialized Orthotopic Liver Transplantation in the Rat**

Surgical anastomosis		Arterialization	Bile duct	Ref.
D: Aortic segment R: E/S to infrarenal aorta microsuture	Microsuture SVC, PV, IVC	+	Pull-through technique	12
D: Long aortic segment R: E/S to infra renal aort microsuture	Microsuture SVC, PV, IVC	+	Splint technique	21
D: Short hepatic-aort segment R: E/S to aorta, microsuture	Microsuture SVC, PV, IVC	+	Pull-through technique	29
D: Aortic segment R: E/E to right renal artery, cuff technique	Microsuture: SVC Cuff, PV, IVC	+ (right nephrectomy)	Splint technique	22
D: Celiac artery R: E/E to right renal artery, microsuture	Microsuture: SVC Cuff, PV, IVC	+ (right nephrectomy)	Splint technique	23
D: Celiac artery R: E/E to common hepatic artery, cuff	Microsuture: SVC Cuff, PV, IVC	+	Splint technique	24
D: Celiac artery R: E/E to hepatic artery, telescoping technique	Microsuture: SVC Cuff, PV, IVC	+	Splint technique	25

Note: D = donor; R = recipient; E/S = end-to-side anastomosis; E/E = end-to-end anastomosis; SVC = suprahepatic vena cava; PV = portal vein; IVC = infrahepatic vena cava.

II. TECHNIQUES

A. ANATOMY

The anatomy is presented with particular emphasis on how it concerns the surgical procedure. The rat liver consists of five distinct lobes. The median or cystic lobe, subdivided into the right and left central lobes by a deep fissure in the midportion, is located between the left lateral lobe and the diaphragm (Figure 1). The removal of these two lobes, which represent 70% of the liver volume, results in a partial hepatectomy. The caudate or Spigelian lobe fits around the esophagus and is interposed between the latter and the stomach. The right lateral lobe and the triangulated lobe arise from a common base. The triangulated lobe almost completely encircles the IVC and caps with its lateral part the upper end of the right kidney. The triangulated lobe joins the caudate lobe behind the portal vein. The falciform ligament, the left lateral ligaments, the retroperitoneal attachments, and the vascular pedicles keep the liver fixed in the right hypochondrium.

The liver has a dual blood supply, receiving oxygenated blood from the hepatic artery and venous blood from the splanchnic region via the portal vein. The hepatic artery, the terminal branch of the common hepatic artery after giving off the gastroduodenal artery, runs parallel to the portal vein on its left side in the hepatoduodenal ligament before arborizing into the lobar arteries. A small arterial branch communicates between the hepatic and esophageal arterial vessels (Figure 2). The portal vein collects blood from the splenic and pyloric tributaries before it enters the liver. In the hilar region, it divides into the right and left main branches. The right branch supplies the right median, the right lateral, the triangulated, and the caudate lobes, while the left branch enters the left median and left lateral lobes. The hepatic veins collect the blood from the central veins of the liver lobules and converge into the suprahepatic vena cava just below the diaphragm forming a short, common venous channel. The left inferior phrenic vein drains into it at the level of the diaphragm. The infrahepatic vena cava receives the right suprarenal vessels and runs embedded in hepatic tissue at the posterior aspect of the liver to collect, now as the

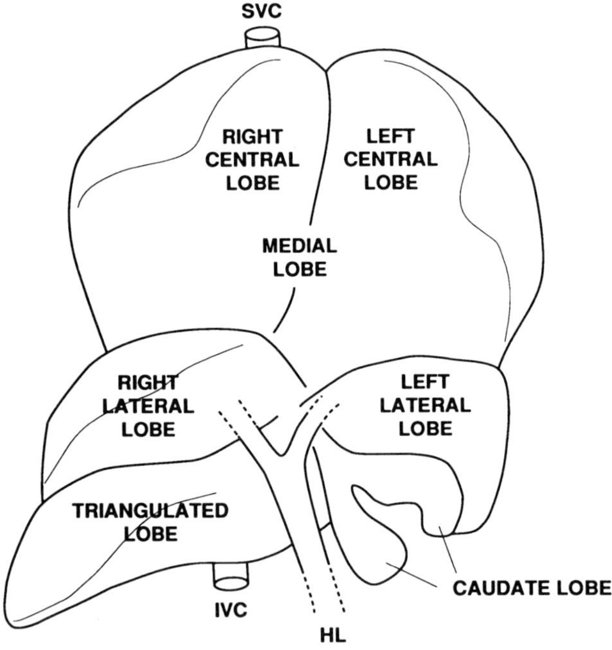

Figure 1 Schematic illustration of the rat liver. HL, hepatoduodenal ligament.

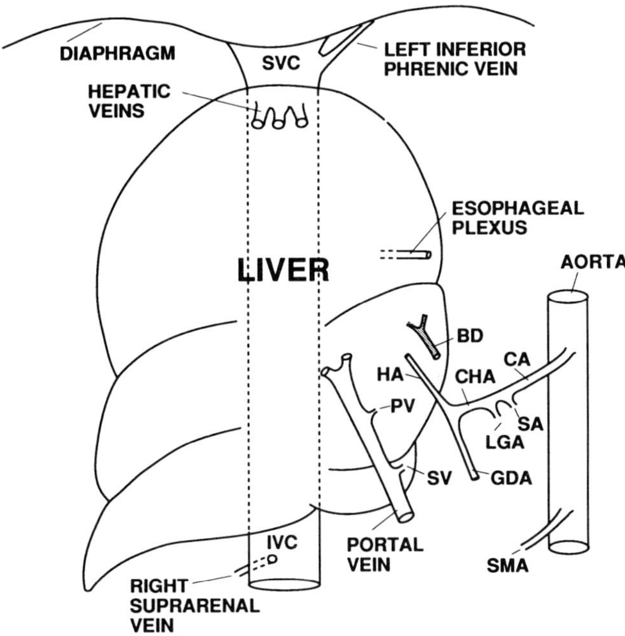

Figure 2 Schematic diagram of the venous and arterial blood supply to the rat liver. HA, hepatic artery; PV, pyloric vein; SV, splenic vein; BD, bile duct; SMA, superior mesenteric artery; CA, celiac artery; SA, splenic artery; LGA, left gastric artery; CHA, common hepatic artery; GDA, gastroduodenal artery.

suprahepatic vena cava, the hepatic veins. Bile ducts from each lobe come together to form the common bile duct, which passes in the hilar region over the portal vein and overlies the hepatic artery in the hepatoduodenal ligament to the left side. The common bile duct merges about 1 cm from the liver into the pancreas and joins the pancreatic duct. A gallbladder is absent in the rat and the bile ducts also appear to lack the ability to concentrate bile. The low tonus of the sphincter of Oddi precludes ductal bile storage, resulting in continuous bile flow into the descending duodenum at about 2.5 cm from the pylorus.

B. SURGICAL TECHNIQUE

This description contains the method for orthotopic arterialized liver transplantation as currently applied by the authors. Rearterialization is done according to the method described by Steffen et al.,[24] and the bile duct anastomosis as described by Zimmermann.[14]

Male rats of 250 to 300 gm bodyweight are anesthetized either by ether or methoxyflurane inhalation in a glass box, shaved, and fixed to a surgical board with rubber bands on the paws and the rats tail facing the surgeon. Anesthesia is maintained with a mixture of room air and an inhalation agent which evaporates from a plastic cone containing a gauze soaked with the anesthetic. Intraperitoneally injected agents that are metabolized by the liver are not suitable for liver transplantation. Conventional microsurgical instruments are used for a clean, nonsterile technique. An operating microscope with variable magnification (6 to 25×) is desirable, although most steps can be safely performed with magnifying glasses. For all ligatures 6-0 silk is used. All visceral organs, and especially the liver should be handled with moist gauze or cotton buds to avoid permanent membrane damage.

C. DONOR OPERATION

The donor abdomen is opened by a longitudinal midline laparotomy from the xiphoid process to the bladder. The incision is extended bilaterally to the costal margins, taking care not to cut the epigastric vessels. The musculocutaneous flaps created by this exposure are fixed with two needles to the board. A curved clamp pulls the xiphoid cranially. A saline-soaked gauze is placed on the liver and gently retracted caudally to expose the falciform ligament which is cut towards the suprahepatic vena cava. Now the left lateral ligaments of the liver are severed. The left inferior phrenic vein is separated from the SVC, double ligated with 6-0 silk, and cut. Depending on the rat strain used, the vein is sometimes inseparable from the SVC and cut at the time of liver removal. The upper part of the caudate lobe is freed from its peritoneal layer and rotated to the rat's right side to expose the hepatoesophageal plexus, which is ligated and cut. The lower caudate lobe is partially freed from the front and the portion located behind the stomach is mobilized after division of the gastrolineal ligament. The small intestine is wrapped in moist gauze and placed, without traction, on the rats left side. Next, the infrahepatic vena cava is freed from the surrounding fatty tissue. The right suprarenal vessels are located behind the triangulated lobe with a twisting movement using a splinter forceps. A single ligature is placed. The liver is carefully pulled to the left side to clear the retrohepatic space of connective tissue and then returned to its normal position. The preparation of the portal area is commenced by double ligation and division of the pyloric vein. Between the ligature knot on the portal vein and the splenic vein, a fine clamp is passed through the fatty tissue under the portal vein. At the time of flushing, the portal vein is clamped at this site. To obtain the celiac segment, the hepatic artery is separated from the gastroduodenal artery which is double ligated and divided. After ligation and division of the left gastric and splenic arteries, the hepatic artery is dissected to the celiac trunk, mainly by blunt separation. The bile duct is set under slight tension by a distal ligature as it enters the pancreas, freed from adjacent pancreatic tissue, and partially cut open. Skeletonizing the bile duct toward the liver hilum should be avoided, in order to preserve arterial perfusion for the bile duct. A stent, approximately 8-mm long and prepared from a 22- to 24-gauge i.v. catheter, is inserted to one-half of its length into the bile duct and secured with a circular ligature. The bile duct is now transected. Before starting hypothermic perfusion with 4°C saline, the infrahepatic (IVC) is clamped from the right side with two curved hemostats and cut close to the lower clamp. The portal vein is clamped with a fine tubed hemostat from the left side. The anterior wall of the portal vein is lifted with fine forceps and opened. A 16-gauge i.v. catheter, connected to an infusion solution with 25 cm of water pressure, is inserted into the portal vein while dripping slowly. To provide sufficient run-off for the saline perfusate, the diaphragm and the intrathoracic IVC are cut. The perfusion flow volume is now adjusted to its maximum flow. The celiac artery is transected where it emerges from the aorta. The suprahepatic IVC is cut just below the diaphragm. Two 6-0 nylon stay sutures are placed at the opposite corners of the vessel. The perfusion is stopped, the portal vein is cut above the clamp. The right suprarenal vessels are transsected in such a way that the single ligature remains with the graft. The liver is now removed with the hemostat which is still attached to the infrahepatic IVC and serves as a handle. The liver is stored in a 4°C saline bath.

D. RECIPIENT OPERATION

The recipient is entered via a midline incision without lateral extension. Two hooks with blunt edges pull the abdominal walls craniolaterally. The preparation of the left infraphrenic vein, the ligaments, and the right suprarenal vessels is the same as the donor operation. The small intestine is placed to the left side of the rat. The pyloric vein is double ligated and divided. The hepatic artery is single ligated and cut towards the liver. The gastroduodenal artery is double ligated and divided leaving the common hepatic artery ending in a T-shape with the stumps of the hepatic and gastroduodenal artery. The common hepatic artery is freed of a lymph node that is consistently present, down to the splenic artery. The bile duct is occluded with a single ligature and cut below its bifurcation. The recipient's situs is now prepared for the clamping and subsequent removal of the liver. The infrahepatic IVC is clamped above the right renal vein and the portal vein below the knot of the pyloric vein. The ends of both hemostats show to the right side. The liver is squeezed to direct blood toward the thoracic cavity, and the SVC is clamped with a Satinsky clamp from the left side, grasping a small margin of diaphragm. The SVC is cut just above the liver, the portal vein at the level of its bifurcation, and the IVC above the clamp in a way that the single ligature of the suprarenal vessels stays in the recipient. The liver is removed and the rat is rotated 180°C, the head now facing the surgeon. During the anhepatic time, the rat is very sensitive to the anesthetic due to reduced cardiac output. Anesthesia can be reduced during for this time without danger that the animal will awaken. A saline-soaked gauze is placed into the situs. The graft is positioned and the anastomosis of the SVC is performed. The two prepositioned 6-0 nylon sutures of the graft are sewn to the corresponding corners of the recipient's SVC. The short ends of both sutures are held under slight tension to straighten out the vessel walls. First, the back wall is sewn from inside using the right corner suture with a standard running microsurgical technique. After the back wall is completed, the suture is tied twice to the short end of the left corner suture. The front wall is completed with the same suture technique. Before the final knot is tied, the anastomosis is gently rinsed with saline to remove air bubbles and prevent air embolism. The animal is turned back into the normal position and the portal vein of the graft is identified. The portal vein stumps are approximated with two 8-0 angle sutures which hold the anastomosis in place. The back wall is completed from the inside with 3 to 4 stitches of running suture. After the suture has been tied at the corner, the front wall of the portal vein is completed with 4 to 5 stitches. Now the venous blood supply of the liver is restored. If the anhepatic time does not exceed 20 min, the liver soon returns to its normal color, and anesthesia should be readministered. While access to the suprahepatic anastomotic site is difficult once the Satinsky clamp is removed, anastomotic bleeding of the portal vein can usually be repaired. Reclamping should be avoided due to the risk of thrombosis. Gentle pressure with cotton swabs will stop smaller hemorrhages at this site. Next, the infrahepatic IVC is reconnected with 8-0 nylon suture as described for the portal vein. Once blood flow is reestablished, the cardiac output increases and this results in a significant improvement in intestinal perfusion. It may be necessary to increase the level of anesthesia at this time. The previously dissected common hepatic artery is clamped at its base with a microclip and stripped of connective tissue. The vessel is then cut at its bifurcation, creating a funnel-shaped opening. The stump is rinsed with saline and a single 10-0 nylon stay suture is fixed at the margin of the arterial opening. The vascular cuff is prepared from a 24-gauge i.v. catheter (Figure 3). The 10-0 stay suture and the attached artery are pulled through the cuffs carefully without any tension on the artery. The position of the cuff is held firmly by a clamp that is attached to a piece of modeling clay used to ensure a constant position for the cuff. The arterial edges are everted, slipped over the cuff, and secured with a 8-0 ligature. The donor celiac artery is pulled over the cuff and fixed with a 6-0 ligature. Arterial blood flow is then reestablished. The bile duct anastomosis is accomplished by stitching the free margin of the donor bile duct distally to the ligature with a long 8-0 suture, which is then pulled through the edge of the recipient's bile duct. The 8-0 suture is not tied but both ends are pulled in a crossover manner to approximate both bile ducts (Figure 4). The splint of the donor bile duct is slipped into the recipient's bile duct, and the margins of the bile ducts are fixed to each other with one 8-0 stitch. The positioning suture is then tied with itself so that the donor and recipient bile duct are finally secured to each other with two 8-0 stitches. If good hemostasis has been

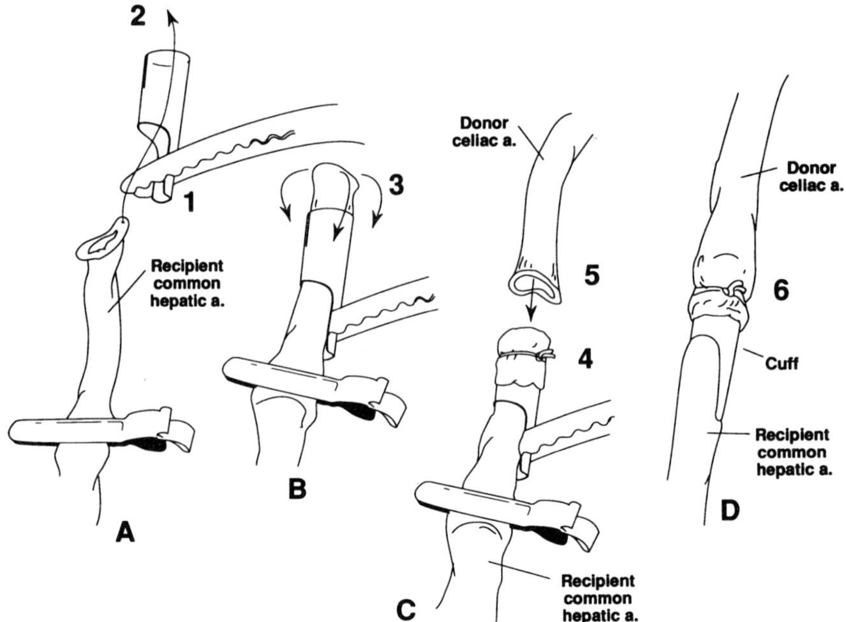

Figure 3 Donor/recipient cuff preparation for anastomosis. (From Steffen et al., *Transplantation*, 48, 166, 1989. With permission.)

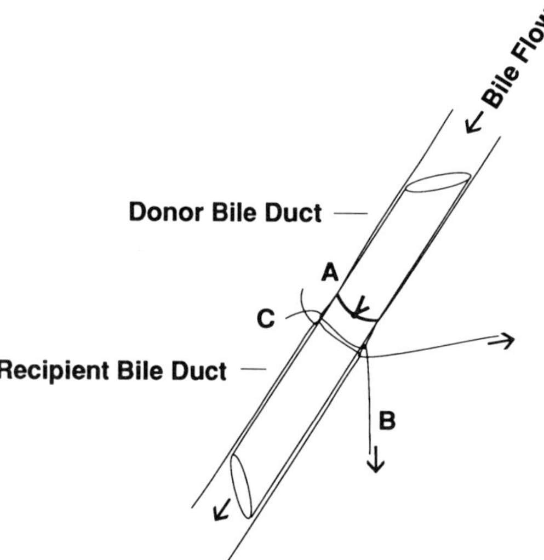

Figure 4 Diagrammatic illustration of the bile duct reconstruction.

provided, 1 or 2 ml of saline are injected into the penile vein. Before the abdomen is closed, the abdominal cavity is rinsed with warm saline to remove blood clots and bile present in the abdomen. A two layer suture of the abdominal wall and the skin with 3-0 absorbable suture finishes the recipient operation. Successfully grafted rats recover and start drinking water in 1 to 2 hours.

III. APPLICATIONS

The model of rat liver grafting has been intensively used for different investigational purposes. While early studies were performed without rearterialization due to the difficulty of the anastomosis, recent reports show an increasing number of groups that reestablish arterial blood flow. In the following section the important applications and findings of rat liver grafting are discussed.

A. GENETIC AND IMMUNOLOGICAL CONSIDERATIONS

The rejection of liver allografts in the rat demonstrates some unique immunological features when compared to other primarily vascularized allografts. These grafts are rejected less vigorously than other organs, they exhibit spontaneous, prolonged survival in selected, fully allogeneic strain combinations, and are capable of inducing a state of systemic tolerance to the transplantation of other tissues.[26] Observations demonstrating a similar biological advantage of liver allografts have been made earlier in pigs where prolonged, spontaneous graft survival has been observed.[27] The reduced susceptibility for rejection in untreated liver recipients is mirrored by the apparent ease with which immunosuppressive drugs, like Cyclosporin A or FK 506, can induce prolonged or permanent liver graft survival.[28] Systemic tolerance apparently develops with a transition from an early, nonspecific induction phase to a late period of specific tolerance to donor tissues.[29] The mechanism(s) underlying this tolerance in long-term surviving liver recipients is a matter of controversy. Soluble antigen and immune complexes, generation of suppressor cells,[30] nonspecific serum factors and enhancing antibodies,[29] clonal deletion of alloreactive recirculating T-cells,[26,31] and anti-class II antibodies[32] have all been suggested to be responsible for the observed hyporeactivity. It is still not clear to what extent the different proposed mechanisms create unresponsiveness and how these findings are related to the type of surgical procedure or donor-recipient combination used.

Most liver graft experiments have been conducted without rearterialization.[33] Recent experimental evidence has suggested that reconstruction of the hepatic artery may be an important factor in immune-mediated rejection of the graft. Engemann et al.[20] found that in the fully allogeneic BN-to-LEW strain combination 70% of the recipients of arterialized liver grafts survived without immunosuppression for 90 days or longer. None of the recipients of venous grafts survived longer than 22 days. Liver recipients not only accept their grafts permanently but also develop a donor-specific transplantation tolerance as demonstrated by the acceptance of donor type skin grafts and rejection of third party grafts. Previously only semiallogeneic, nonrearterialized grafts were shown to be accepted in this combination.[34]

The mechanism by which the reconstruction of the hepatic artery results in prolonged graft survival is not known. One potential mechanism for this phenomenon is the generation of donor-specific T-suppressor lymphocytes after grafting. Using adoptive transfer assays splenic T-suppressor cells have been demonstrated in long-term surviving recipients of fully allogeneic liver grafts.[35,36] A possible explanation for the observed long-term survival of LEW rats receiving arterialized BN allografts might be the physiological advantage of rearterialization that allows it to overcome the host's immune response and subsequent development of cellular suppressive mechanisms. Ischemic damage, which may be more pronounced in venous liver grafts, and hepatic regeneration following surgical trauma induce expression of MHC class II antigens.[37] This nonspecific activation of the immune system may be present in a nonarterialized liver graft and provide a better target for rejection.

B. THE EFFECT OF REARTERIALIZATION

Prior to the introduction of the cuff technique, rearterialization was a technically difficult procedure, substantially prolonging the recipient operation. Unacceptably long operation times and the satisfactory survival rates for venous graft recipients did not support the need for rearterilization.[38] Accordingly, most of the experimental studies of liver transplantation in this species have utilized the simpler venous surgical model. There are, however, a variety of biochemical and morphological alterations in the venous transplantation model, which may have the potential of markedly altering the results of studies using this technique for liver transplantation.

Arterialization of orthotopic liver grafts in the rat results in a higher survival rate when compared to venous liver grafts. Engemann et al.[20] demonstrated in the LEW-to-LEW strain combination an improvement in 90 day survival rates from 45% in the strictly venous group to 80% in the rearterialized group. Similar increases in the survival rate of syngeneic grafts after rearterialization have been reported in other studies.[17,23,39] Kamada and Calne[38] reported a higher rate of long-term survival (100 days for 95%) for venous syngeneic PVG grafts and even described a high rate of survival for liver allografts in the low responder DA/PVG strain combination.

The increased survival rate of arterialized liver grafts has been attributed to a lack of bile duct necrosis and subsequent biliary peritonitis.[21] These findings have been confirmed by other studies that demonstrated that the abrogation of biliary tract complications was the key to long-term graft survival.[22-23] A venous liver graft in the rat does not stay dearterialized permanently after ligation of the hepatic artery in the donor. The rate of collateral rearterialization was examined by Isai et al.[40] using angiography to visualize the arterial tree in the grafted liver via collaterals. In 75% of the grafts, newly established collaterals could be demonstrated in rats with long-surviving grafts. Despite reestablishment of collateral circulation, the bile ducts are subject to severe ischemic damage until collateralization is established. Even in primarily arterialized grafts, the blood supply to the donor bile duct is crucial and in our early experience bile duct necrosis occurred in 31% of the donor grafts despite a patent hepatic artery.[24] Avoiding disruption of nutrient arteries during the donor operation is essential to minimize the effect of ischemic damage to the bile duct. With care, bile duct necrosis has become a rare complication. Howden et al.[23] carried out liver function tests in recipients of venous and rearterialized grafts. All venous isografts displayed abnormal elevations of bilirubin, alkaline phosphatase, gamma-glutamyl transferase, and aspartate aminotransferase throughout the observation period. Arterialized grafts had a reduction of liver damage as measured by abnormal liver function tests. Postoperative transaminase levels exhibited a similar pattern in both venous and arterialized grafts. In venous grafts, higher alkaline phosphatase values were observed and these remained elevated throughout the experimental interval, suggesting an outflow resistance in the biliary tree. Most experimental studies have used either the venous or the arterialized procedure so that direct comparison of the two techniques is limited.[17,38] In general, measurement of liver function indicators did not prove to be of great predictive value for graft rejection, acceptance, or in differentiating venous from rearterilized grafts.

Arterialized liver isografts display a normal histological appearance from day 5 to day 200 post-transplantation. In the first 2 weeks only occasional mononuclear cells can be detected in the portal areas, the portal triads are well preserved, and no bile duct proliferation is visible.[41] Rearterialization produces a histologic appearance in long-term survivors which does not differ from normal liver tissue. In contrast, fibrosis of the liver and extensive periportal and intrahepatic mononuclear infiltrations have been seen in venous liver grafts.[20] The more pronounced inflammatory changes observed in venous isografts are associated with the emergence of MHC class II antigen expression on Kupffer cells. These cells are considered to be sessile macrophages and normally are negative for class II antigen expression. Class II antigen expression was absent in arterialized isografts.[42] Whether the appearance of MHC class II antigen expression renders the venous liver graft more immunogenic and thereby more susceptible to rejection than an arterialized one is not known.

Bile duct proliferation has been seen in venous liver grafts. It has been attributed to the technique of bile duct reconstruction, and acknowledged as a late complication of the transplantation procedure. There is, however, growing evidence that the lack of the hepatic arterial flow stimulates bile duct proliferation. We found bile duct proliferation to be a constant finding in grafts with thrombosed arterial anastomoses. Bile duct ischemia may lead to functional impairment or fibrosis of the biliary tree with an increased outflow resistance. This usually associated with elevated levels of canalicular enzymes (alkaline phosphatase, gamma-glutamyl transferase). Increased pressure in the biliary system can evoke bile duct proliferation with a histopathological pattern similar to that seen in venous liver grafts.[43]

C. ORGAN PRESERVATION

Although the difference between venous and arterialized rat liver isografts in preservation experiments may not be measurable by means of qualitative or quantitative liver function tests,[44] donor organ preservation time can be used as an indicator of improved graft function achieved with arterialization. Data from various groups indicate that rearterialized rat liver grafts can tolerate longer preservation periods,[23,24,45] similar to large animal or human liver transplantation. The use of University of Wisconsin solution as preservation fluid allows successful static ice storage for up to 24 hours, and subsequent recipient survival in rearterialized models. The cold storage time using hypertonic citrate is limited to 12 hours to provide safe limits for recipient survival. The influence of rearterialization and the use of the University of Wisconsin solution, however, have not been investigated as independent factors in most studies. While extended preservation times made possible by rearterialization per se is an acknowledge fact, the venous model is considered as fully sufficient since the quantitative differences seen when preservation solutions are compared are similar, irrespective of the use of the venous or the arterialized technique.[46] The physiologically compromised venous model may be even a stronger proof of the efficacy of a preservation solution.[46] Yet, rearterialization of rat liver grafts allows for preservation times comparable to large animal or human livers, a fact that would greatly enhance the acceptance of arterialized rat liver transplantation for preservation studies.

IV. CONCLUSIONS

The orthotopic transplantation of the liver can be performed with a high rate of success by an experienced microvascular surgeon. The most appropriate model includes a reconstruction of the hepatic arterial blood supply which allows for the establishment of a functional and relevant model for liver transplantation in larger species and humans. The rat orthotopic liver transplantation model has gained wide acceptance as an experimental tool for investigating the role of histocompatibility differences that mediate graft rejection, studies of the immunological characteristics of tolerance induction, testing for the efficacy of new immunosuppressive drugs, and the development of improved methods of graft preservation.

REFERENCES

1. **Welch, C.S.,** A note on transplantation of the whole liver in dogs, *Transplant. Bull.,* 1, 54, 1955.
2. **Lee, S.,** *Experimental Microsurgery,* Igaku-Shoin, New York, 1987, 213.
3. **Lee, S.H. and Fisher, B.,** Portocaval shunt in the rat, *Surgery,* 50, 668, 1961.
4. **Lee, S. and Edington, T.S.,** Liver transplantation in the rat, *Surg. Forum,* 17, 220, 1966.
5. **Hess, F., Juresalem, C., and v.d. Heyde, M.N.,** Advantages of auxiliary liver homotransplantation in rats, *Arch. Surg.,* 104, 76, 1972.
6. **Kort, W.J., Wolff, E.D., and Eastham, W.N.,** Heterotopic auxiliary liver transplantation in rats, *Transplantation,* 6, 415, 1971.
7. **Mueller, G.,** A simple technique for heterotopic auxiliary liver transplantation in the rat, *Transplantation,* 36, 221, 1983.
8. **Marni, A. and Ferrero, M.E.,** Heterotopic liver grafting in the rat — a simplified method using cuff techniques, *Transplantation,* 39, 329, 1985.

9. **Hess, F.,** Auxiliary heterotopic rat liver transplantation, in *Handbook of Microsurgery,* Vol. 2, Olszewski, W.L., Ed., CRC Press, Boca Raton, FL, 1984, 359.
10. **Zelder, O.,** Heterotopic auxiliary liver transplantation in rats, in *Handbook of Microsurgery,* Vol. 2, Olszewski, W.L., Ed., CRC Press, Boca Raton, FL, 1984, 377.
11. **Lee S. and Scott, M.H.,** Six models of heterotopic rat liver transplantation: introducing a reverse circulation model, *Microsurgery,* 7, 91, 1986.
12. **Lee, S., Charters, A.C., III, Chandler, J.G., and Orloff, M.J.,** A technique for orthotopic liver transplantation in the rat, *Transplantation,* 16, 664, 1973.
13. **Lee, S., Charters, A.C., III, and Orloff, M.J.,** Simplified technique for orthotopic liver transplantation in the rat, *Am. J. Surg.,* 130, 38, 1975.
14. **Zimmermann, F.A., Butcher, G.W., Davies, H.S., Brons, G., Kamada, N., and Tuerel, O.,** Techniques for orthotopic liver transplantation in the rat and some studies of the immunologic responses to fully allogeneic liver grafts, *Transplant. Proc.,* 11, 571, 1979.
15. **Kamada, N. and Calne, R.Y.,** Orthotopic liver transplantation in the rat — technique using cuff for portal vein anastomosis and biliary drainage, *Transplantation,* 28, 47, 1979.
16. **Settaf, A., Gugenheim, J., Houssin, D., and Bismuth, H.,** Cuff technique for orthotopic liver transplantation in the rat — a simplified method for the suprahepatic vena cava anastomosis, *Transplantation,* 42, 330, 1986.
17. **Miyata, M., Fischer, J.H., Fuhs, M., Isselhard, W., and Kasai, Y.,** A simple method for orthotopic liver transplantation in the rat — cuff technique for three vascular anastomosis, *Transplantation,* 30, 335, 1980.
18. **Tsuchimoto, S., Kusumoto, K., Nakajima, Y., Kakita, A., Uchino, J., Natori, T., and Aizawa, M.,** Orthotopic liver transplantation in the rat — a simplified technique using the cuff method for suprahepatic vena cava anastomosis, *Transplantation,* 45, 1153, 1988.
19. **Knoop, M. and Hutchinson, I.V.,** Suprahepatic vena cava cuffs in rat liver transplantation — a simplified method?, *Transplantation,* 47, 576, 1989.
20. **Engemann, R., Ulrichs, K., Thiede, A., Mueller-Ruchholtz, W., and Hamelmann, H.,** Value of a physiological liver transplant model in rats — induction of specific graft tolerance in a fully allogeneic strain combination, *Transplantation,* 33, 566, 1982.
21. **Engemann, R.,** Technique for orthotopic rat liver transplantation, in *Microsurgical Models in Rats for Transplantation Research,* Thiede, A., Deltz, E., Engemann, R., and Hamelmann, H., Eds., Springer-Verlag, Berlin, 1985, 69.
22. **Hasuike, Y., Monden, M., Valdivia, L.A., Kubota, N., Gotoh, M., and Nakano, Y.,** A simple method for orthotopic liver transplantation with arterial reconstruction in rats, *Transplantation,* 45, 830, 1988.
23. **Howden, B., Jablonski, P., Grossman, H., and Marshall, V.C.,** The importance of the hepatic artery in rat liver transplantation, *Transplantation,* 47, 428, 1989.
24. **Steffen, R., Ferguson, D.M., and Krom, R.A.F.,** A mew method for orthotopic rat liver transplantation with arterial cuff anastomosis to the recipient common hepatic artery, *Transplantation,* 45, 1153, 1988.
25. **Chaland, P., Braillon, A., Gaudin, C., Sekiyama, T., Bernuau, D., Adam, R., Bismuth, H., Benhamou, J.P., and Lebrec, D.,** Orthotopic liver transplantation with hepatic artery anastomoses — hemodynamics and response to hemorrhage in conscious rats, *Transplantation,* 49, 675, 1990.
26. **Kamada, N.,** The immunology of experimental liver transplantation in the rat, *Immunology,* 55, 369, 1985.
27. **Calne, R.Y., White, H.F.O., Yoffa, D.E., Binns, R.M., Maginn, R.R., Herbertson, R.M., Millard, P.R., Molima, V.P., and Davies, D.R.,** Prolonged survival of liver transplants in the pig, *Br. Med. J.,* 4, 645, 1967.

28. **Murase, N., Kim, D.G., Todo, S., Cramer, D.V., Fung, J.J., and Starzl, T.E.,** Suppression of allograft rejection with FK 506. I. Prolonged cardiac and liver survival in rats following short course therapy, *Transplantation,* 50, 186, 1990.
29. **Lie, T.S., Gulkowska, H., Jaeger, K., and Niehaus, K.J.,** Immunmechanismen nach Lebertransplantation, *Langenbeck's Arch. Chir.,* 360, 17, 1983.
30. **Houssin, D., Charpentier, B., Lang, P., Tamisier, D., Gugenheim, J., Gigou, M., and Bismuth, H.,** In vivo and in vitro correlates of the specific transplantation tolerance induced by spontaneously tolerated liver allografts in inbred strains of rats, *Transplant. Proc.,* 13, 619, 1981.
31. **Kamada, N., Davies, H.S., and Roser, B.,** Reversal of transplantation immunity by liver grafting, *Nature,* 292, 840, 1981.
32. **Kamada, N., Shinomiya, T., Tamaki, T., and Ishiguro, K.,** Immunosuppressive activity of serum from liver-grafted rats, *Transplantation,* 42, 581, 1986.
33. **Kamada, N.,** *Experimental Liver Transplantation,* CRC Press, Boca Raton, FL, 1988.
34. **Houssin, D., Gigou, M., Franco, D., Bismith, H., Charpentier, B., Lang, P., and Martin, E.,** Specific transplantation tolerance induced by spontaneously tolerated liver allograft in inbred strains of rat, *Transplantation,* 29, 418, 1980.
35. **Gassel, H.J., Hutchinson, I.V., Tellides, G., Knoop, M., Hackmann, J., Engemann, R., and Morris, P.J.,** Phenotypic characterization of T-suppressor lymphocytes induced by orthotopic rat liver transplantation, *Transplant. Proc.,* 21, 429, 1989.
36. **Knoop, M., Pratt, J.R., Pether, M.P., and Hutchinson, I.V.,** Immunosuppressive properties of sera and splenocytes from long-term surviving liver recipients, *Langenbeck's Arch. Chir. Suppl.,* 353, 1990.
37. **Jonjic, S., Radosevic-Stasic, B., Cuk, M., Jonjic, N., and Rukavina, D.,** Class II antigen induction in the regenerating liver of rats after partial hepatectomy, *Transplantation,* 44, 165. 1987.
38. **Kamada, N. and Calne, R.Y.,** A surgical experience with five hundred thirty liver transplants in the rat, *Surgery,* 93, 64, 1983.
39. **Lie T.S., Jaeger, K., and Niehaus, K.J.,** Microsurgical aspects of rat liver transplantation, in *Handbook of Microsurgery,* Vol. 2, Olszewski, W.L., Ed., CRC Press, Boca Raton, FL, 1984, 397.
40. **Isai, H., Miyata, A., Saito, M., Uchino, J., and Marshall, V.C.,** Angiographic evidence of collateral rearterialization of the grafted liver in the rat, *Transplant. Proc.,* 21, 2463, 1989.
41. **Gassel, H.J. and Engemman, R.,** Preservation of rat liver grafts, *Transplantation,* 44, 726, 1987.
42. **Gassel, H.J., Engemann, R., and Thiede, A.,** Major histocompatibility complex class II antigen expression on Kupffer cells after orthotopic rat liver transplantation: a consequence of nonspecific inflammation or allograft rejection?, *Transplant. Proc.,* 19, 3017, 1987.
43. **Slott, P.A., Lin, M.H., and Tavoloni, N.,** Origin, pattern, and mechanism of bile duct proliferation following biliary obstruction in the rat, *Gastroenterology,* 99, 466, 1990.
44. **Svensson, G. and Karlberg, I.,** Effect of rearterialization on short-term graft function in orthotopic rat liver transplantation, *Eur. Surg. Res.,* 21, 2, 1989.
45. **Gores, G.J., Ferguson, D.M., Ludwig, J.A., Steffen, R., and Krom, R.A.F.,** Effect of acidosis during cold ischemic storage on liver viability following transplantation in the rat, *Transplant. Proc.,* 22, 488, 1990.
46. **Yu, W., Coddington, D., and Bitter-Suermann, H.,** Rat liver preservation, *Transplantation,* 49, 1060, 1990.

Chapter 6

Liver Transplantation in the Dog

Larry H. Stevens, James B. Piper, and Christoph E. Broelsch

CONTENTS

I. INTRODUCTION

The use of the canine has been essential in the development of clinical liver transplantation. Although the canine model is technically demanding, the first report of successful liver transplantation was Welch's 1955 description of auxiliary liver transplantation into the pelvis of the dog.[1] This report of heterotopic liver transplantation, in which the new liver is placed in an ectopic position, was followed by Cannon's 1956 account of orthotopic liver transplantation,[2] in which the recipient's liver is excised and replaced with a new liver. From these early efforts with liver transplantation to the current development of techniques for reduced size liver transplantation and living related liver transplantation the use of canines has remained important.[3,4] Techniques cultivated in the animal laboratory have been applied in the clinics, and problems encountered in the clinics have been successfully addressed in the animal laboratories.[4,5]

In addition to the role played in the development of techniques for orthotopic, heterotopic, and segmental liver transplantation, the canine model has proven important in other areas of investigation. The process of rejection has been studied in canine liver transplants.[6] The role of various immunosuppressive agents in preventing rejection has also been studied in the dog.[7] The use of the dog has played a prominent role in the

0-8493-3629-5/94/$0.00+$.50

ongoing quest for better preservation of the donor liver.[8,9] In addition, liver transplantation in dogs has been important in improving our understanding of hepatic physiology and function. A great deal of the current knowledge of the importance of portal flow and hepatotrophic factors in sustaining a healthy liver was obtained from experiments done in the dog.[10,11]

This chapter reviews the techniques used for liver transplantation in dogs, including descriptions of orthotopic, heterotopic, and segmental transplants. The application of these techniques to the study of liver preservation, physiology, rejection, and immunosuppression is also discussed.

II. TECHNIQUES

A. ANATOMY

Unlike the human liver which consists of two lobes, the canine liver is comprised of six lobes. To the left of the umbilical fissure, the dog's liver is divided into a left lateral lobe and a left medial lobe. The left lateral lobe is the largest lobe of the liver. The free margin of the left lateral lobe is frequently notched. To the right of the umbilical fissure are the quadrate lobe, the right medial lobe, the right lateral lobe, and the caudate lobe. The caudate lobe is indistinctly separated from the rest of the liver, which lies dorsal and cephalad to the caudate lobe. The caudate lobe is further divided into two processes: the caudate process which lies to the right of the vena cava and caps the cranial portion of the right kidney, and the papillary process (or spigelian lobe) which lies to the left of the vena cava and is posterior to the lesser omentum. These two processes of the caudate lobe are separated by a constriction formed by the portal vein ventrally and the vena cava dorsally. Each lobe of the liver (except for the caudate) is separated from the others by prominent fissures. This is unlike the human liver in which the lobes and segments are not deeply cleaved. The gallbladder lies in a fossa between the quadrate lobe and the right medial lobe.[12-14] Figure 1 illustrates the morphology of the canine liver. The hepatic veins of the dog are somewhat unique. These veins, particularly the sublobular veins, are surrounded by a spiral-like arrangement of smooth muscle. Contraction of these muscular spirals produces a sphincteric effect, leading to stasis in the hepatic and mesenteric circulations.[15]

B. ANESTHESIA

Perfect anesthesia is one of the most important factors in achieving success with liver transplantation in the dog (or the human). Our current technique for canines uses thiamylal sodium (25 mg/kg) for induction. After endotracheal intubation, maintenance of anesthesia is provided by a 2:1 mixture of nitrous oxide and oxygen combined with 1% halothane. Pancuronium bromide provides paralysis. Ventilation is mechanically controlled. The electrocardiogram is continuously monitored. The right carotid artery and jugular vein are cannulated. This provides for continuous monitoring of arterial pressure and central venous pressure as well as a means to infuse fluids rapidly. Urethral catherization allows assessment of the urine output. Lactated Ringer's solution, dextran 70 in 5% dextrose, and packed red blood cells (obtained from other dogs) are used for volume replacement. The hematocrit is checked hourly and maintained over 30% with transfusion. Arterial blood gases are analyzed at regular intervals. The animal's temperature is monitored with an intraesophageal probe. A heating pad, intermittent peritoneal irrigation with warmed saline, and warmed intravenous fluids are used to minimize hypothermia.[4]

The Pittsburgh method is similar. Thiopental sodium (25 to 30 mg/kg) is used for induction. Rather than inhaled anesthetic agents, ketamine, 2 mg/kg every 20 to 30 min is used for maintenance, with 0.5 mg of pancuronium given to provide relaxation. No

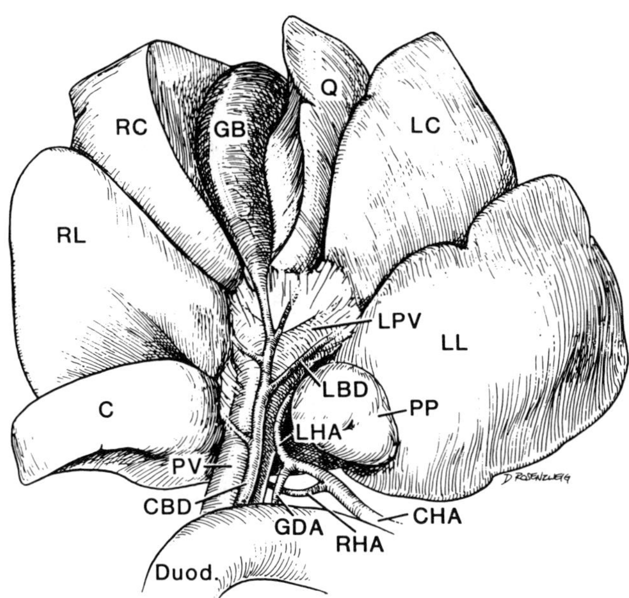

Figure 1 A view of the canine liver from below. Note the deep fissures between the lobes. C = caudate lobe, RL = right lateral lobe, RC = right central lobe, Q = quadrate lobe, LC = left central lobe, LL = left lateral lobe, PP = papillary process, GB = gallbladder, CBD = common bile duct, LBD= left bile duct, PV = portal vein, LPV = left portal vein, CHA = common hepatic artery, RHA = right hepatic artery, LHA = left hepatic artery, GDA = gastroduodenal artery, and Duod. = duodenum. (Reproduced with permission from Cherqui et al.[4])

ketamine is given after revascularization of the graft. Ventilation is controlled, with an FIO2 of 0.30 and 5 cm of positive end expiratory pressure being provided. About 2 to 3 l of crystalloid or colloid and 2 U of blood are given. One of these units of blood is routinely given during the anhepatic phase. Low dose dopamine is also initiated during the anhepatic phase. Calcium chloride and sodium bicarbonate are used after revascularization. Blood gases and electrolytes are monitored frequently. Avoidance of hypothermia is emphasized.[16]

C. ORTHOTOPIC

Cannon's initial report of orthotopic liver transplantation noted "several successful operations" but no survival of the recipients.[2] Moore and Starzl independently developed techniques for orthotopic transplantation of the canine liver, which resulted in maximal survival of 12 and 20.5 days, respectively.[17,18] These early efforts defined two factors essential for success: adequate preservation of the donor liver and decompression of the lower vena cava and portal vein during the anhepatic phase.

1. Donor Liver Preservation

In his first reports, Welch noted the liver's sensitivity to ischemia. He stated that "in no case did the recipient survive when the donor liver was anoxic for over thirty-three minutes".[19] Moore and Starzl subsequently recognized the importance of hypothermia in extending the liver's tolerance of ischemia. Moore cooled the animal by instilling cold saline into the peritoneal cavity.[17] Starzl immersed the donor in an ice bath and infused 1 l of cooled (5 to 10°C.) lactated Ringer's solution through the portal vein immediately before exsanguinating the animal, as illustrated in Figure 2.[18] Ischemic times of 45 min

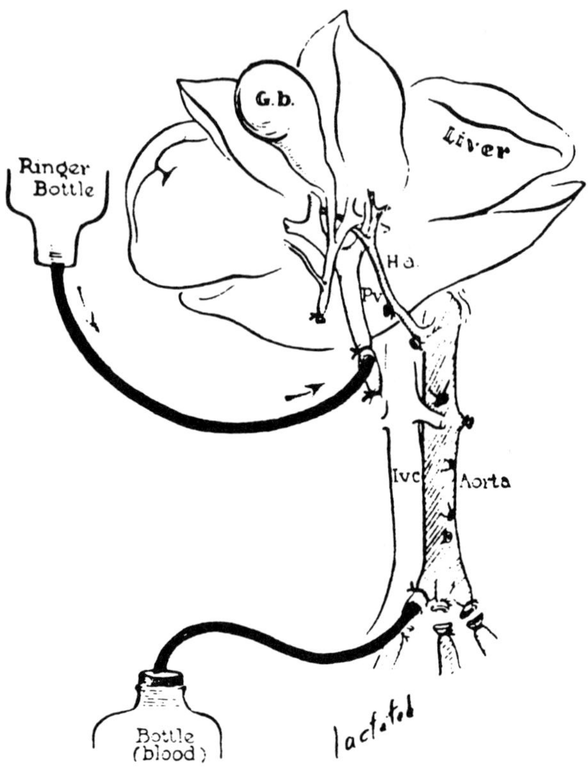

Figure 2 Preservation technique described by Starzl et al. Infusion of lactated Ringer's solution into the portal vein with collection of blood from the aorta. (Reproduced with permission from Starzl et al.[18])

to 2 hours were tolerated with this technique. Ischemic times greater than 2 hours resulted in livers which became tensely swollen and dark after revascularization. Histologic evidence of intense congestion was evident; death followed in a few hours from either hemorrhagic gastritis or hemorrhage from a distended, ruptured liver. Starzl proposed that the ischemic injury resulted in an "outflow block" in which the muscular, sublobular hepatic veins described by Arey[15] constrict; resulting in hepatic and visceral congestion.[18]

The technique of topical hypothermia combined with an intravascular flush of cold solution is still utilized for liver preservation.[20] Modifications of the original technique include: cannulation of the splenic vein rather than the portal vein (in order to minimize trauma to and loss of length for the portal vein) and the use of larger flush volumes (2 l of lactated Ringer's).[16] Others have added an arterial flush, through either the hepatic artery[21] or the aorta.[20] With this latter technique, as described by Jamieson, 1500 ml of chilled lactated Ringer's is perfused through the portal vein and 500 ml is perfused through the aorta after giving 3000 U of heparin to the donor.[20] Significant advances have been made in the composition of the solutions used for the flush. With utilization of the new University of Wisconsin solution, canine liver preservation times can be extended to 24 to 48 hours.[9] The role of the canine model in the testing of these solutions is described later. Although this solution greatly extends the viability of the donor liver, many investigators still rely on lactated Ringer's. For most experiments, the donor and recipient operations can be coordinated to minimize cold ischemia, and thus eliminate the need for expensive preservation solutions. This is not the case in clinical liver transplantation, where the donor and recipient are often separated by several hundred miles and a long cold ischemic time is unavoidable. Here, the preservation solutions are well worth their extra cost.

2. Venovenous Bypass

Successful completion of orthotopic liver transplantation in the dog requires venovenous bypass, if standard anastomotic techniques are used. The anhepatic phase of orthotopic transplantation requires occlusion of the inferior vena cava and the portal vein. Occlusion of these vessels results in pooling of blood in the abdominal viscera and lower extremities. Venous return to the heart is significantly decreased, leading to systemic hypotension. Products of anaerobic metabolism accumulate in the lower half of the body during this phase. If the period of vascular occlusion is longer than 20 to 30 min, reopening of the vascular clamps terminates in severe hyperkalemia, acidosis, cardiac arrhythmias, and death.[17] Moore used two plastic tubes as shunts, one from the infrahepatic cava to the right jugular vein and one from the portal vein to the left jugular vein.[17] Starzl used only one shunt, from the femoral vein (which decompresses the inferior vena cava) to the external jugular vein. The portal system was decompressed by constructing a preliminary side to side portocaval shunt, thus allowing the portal system to vent into the cava and be decompressed by the femoral catheter.[18] These shunts effectively decompress the inferior vena cava and portal vein, but problems with shunt clotting and pulmonary embolus are noted.[21] Systemic heparinization prevents clotting and embolization but leads to excessive hemorrhage. Venovenous bypass has become safer and more reliable with the advent of the centrifugal pump (Bio-Medicus, Inc.).[5,22] A "Y" shaped tubing system, with the centrifugal pump interposed in the long arm of the "Y" is used. The proximal femoral vein is cannulated to decompress the inferior vena cava. The distal femoral vein is ligated. The second short arm of the "Y" decompresses the portal system by direct cannulation of the portal vein or by cannulation of the central splenic vein after splenectomy. Venous return is achieved by inserting the long arm of the "Y" in the external jugular vein, as demonstrated in Figure 3. Standard chest tubes (12 french or larger) are used as cannulas. The cannulas are connected with 3/8-in. Tygon (Norton Industrial Plastics) tubing.[5,22] This system requires no systemic heparinization. Kam has demonstrated that utilization of this system, even with flows averaging only 575 ml/min, resulted in only 2 of 40 dogs sustaining a pulmonary embolus.[22] Profound hypotension and hemodynamic instability are avoided. The hemodynamic stability achieved with this system allows the operation to be conducted in a less hurried, more precise fashion. This system has therefore become standard for dog (and human) liver transplantation in many centers.

3. Donor Operation

The original technique for donor procurement is still used by the Pittsburgh group.[16] A long midline incision is used to enter the abdomen. Dissection is begun at the aortic bifurcation. All branches from the bifurcation of the aorta to the superior mesenteric artery (the lumbar arteries, the inferior mesenteric artery, and the renal arteries) are divided. The superior mesenteric artery is encircled but left intact. The diaphragmatic crura are divided, exposing the celiac axis. The splenic and left gastric arteries are divided. The hepatic artery is now skeletonized.[18]

The stomach and duodenum are retracted caudally to better expose the hilum of the liver. The gallbladder is incised at the proposed site of the cholecystoenterostomy. The bile is flushed from the gallbladder to prevent autolysis during the cold ischemia. The common duct and the gastroduodenal artery are divided just cephalad to the duodenum. The portal vein is freed of surrounding lymphatics and nerves. To obtain sufficient length for anastomosis, the portal vein is dissected further caudally: the pancreatic branch is divided and

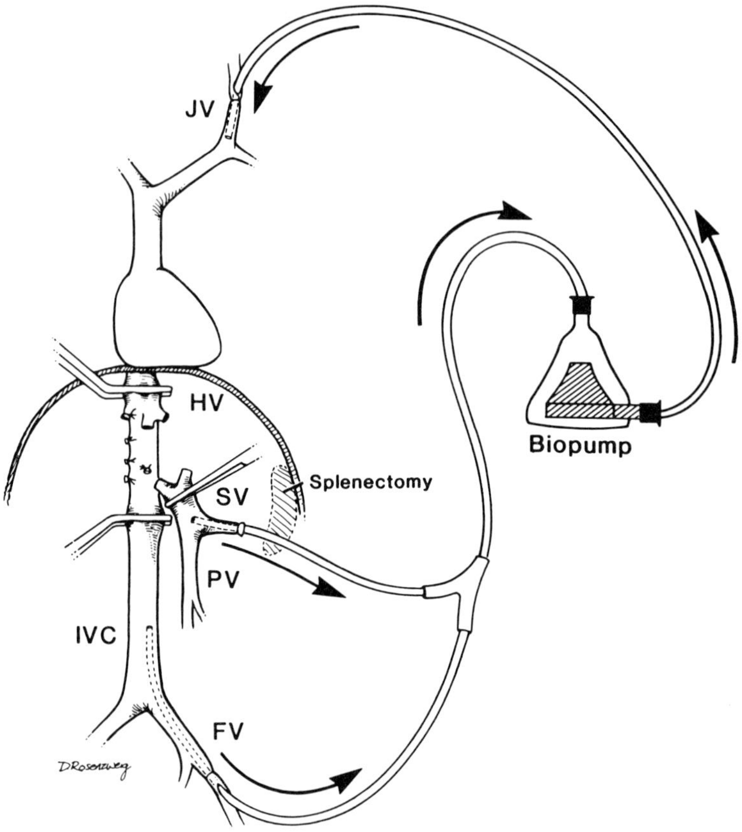

Figure 3 Illustration of venovenous bypass using a centrifugal pump. With the inferior vena cava (IVC) and portal vein (PV) clamped, blood passes from the femoral vein (FV) and splenic vein (SV) to the biopump. The pump then returns the blood to the jugular vein (JV). (Reproduced with permission from Cherqui et al.[4])

adjacent lymph nodes are removed. The vein is thus cleared of all surrounding tissue until the confluence of the splenic and superior mesenteric veins is reached. The splenic vein is cannulated for the perfusion.[16]

The falciform, gastrohepatic, left triangular, right triangular, and coronary ligaments are divided. The connective tissue posterior to the retrohepatic cava is dissected, taking care to be gentle with the retraction of the liver. The suprahepatic vena cava is encircled, as is the infrahepatic cava. The distal aorta is cannulated; allowing collection of blood for the recipient operation. The superior mesenteric artery and vein are ligated. While the animal is exsanguinated through the cannula in the aorta (see Figure 3), 2 l of cold lactated Ringer's solution are perfused through the splenic vein. The liver is topically cooled by instilling iced saline into the abdomen. The donor hepatectomy is completed by transecting the portal vein, the supraceliac aorta, and the inferior vena cava above and below the liver.[16]

Modifications of this technique include: dividing the hepatic artery at its origon instead of utilizing the entire abdominal aorta[23] and flushing the arterial system as well as the portal circulation.[20,23] The University of Wisconsin group emphasizes minimal dissection of the inferior vena cava prior to flushing. The renal veins are divided bilaterally, but no further caval dissection is done until the liver has been flushed. Once the flush has cooled the liver, rapid excision of the liver is performed. The liver is removed with a patch of diaphragm which surrounds the suprahepatic cava. Tissues posterior to the retrohepatic cava are now easily divided. The infrahepatic vena cava, caudal to the ligated renal veins, is now transected.[20]

4. Recipient Operation

The recipient operation is divided into three phases: the recipient hepatectomy, the anhepatic phase, and the post-revascularization phase. This standard recipient operation is modeled after the Pittsburgh technique.[16] The recipient operation is conducted through a long midline incision. The infrarenal aorta is exposed, taking care to avoid injury to the cisterna chyli. The recipient hepatectomy begins with division of the hepatic artery; this is done as far cephalad as possible. The bile duct is also divided as cephalad as possible. This allows the recipient hepatic artery and bile duct to be used in the reconstruction, if so desired. The portal vein is skeletonized, but left intact. The ligamentous attachments of the liver are divided, as described in the donor operation. Some surgeons divide the right adrenal vein, while others see this step as unnecessary.[23] Tissue posterior to the retrohepatic cava is divided. Once the liver remains attached by only the cava and portal vein, venovenous bypass is initiated, as described above. The portal vein, infrahepatic cava, and suprahepatic cava are clamped, in that order. These vessels are then transected and the recipient's liver is removed.[16]

The suprahepatic caval, infrahepatic caval, and portal venous anastomoses are constructed while on bypass, before revascularizing the liver. The suprahepatic caval anastomosis is completed first; this is done with a running, everting suture of 5-0 polypropylene. The portal vein of the donor is flushed with 500 ml of lactated Ringer's to wash out potassium and acid that have accumulated in the donor liver.[20] The infrahepatic caval and portal venous anastomoses are done with running 6-0 polypropylene. Particular care is taken to avoid narrowing the portal vein anastomosis. The clamps on the suprahepatic cava, infrahepatic cava, and portal vein are released in sequence. After release of each clamp a search is made for bleeding points and these bleeding points are controlled. The venovenous bypass is discontinued.

Once hemostasis is adequate, the arterial anastomosis is done. The donor aorta is brought through the right paravertebral gutter and anastomosed to the recipient's infrarenal aorta in an end-to-side fashion. This is done with running 6-0 polypropylene. While the supraceliac end of the donor aorta is still open, the donor aorta is flushed with blood to remove any air that may be present. The supraceliac end of the donor aorta is then closed. The final step of the post-revascularization phase is the biliary reconstruction. A bilioenteric anastomosis, often a two-layered cholecystoduodenostomy, is constructed. The completed operation is illustrated in Figure 4. After any residual bleeding is controlled, the abdomen is closed. Todo reports that the recipient operation requires about 4 hours. He also notes a 73% 24-hour survival and a 60% 5-day survival in the first 30 liver transplants done using this technique in dogs.[16]

There have been numerous modifications of the recipient operation. In one variant the portal anastomosis is done before the infrahepatic caval anastomosis. In this description, the portal vein is unclamped as soon as this anastomosis is completed, venting air and blood through the donor infrahepatic cava. The infrahepatic cava is then clamped. The suprahepatic caval clamp is now released, allowing portal flow through the liver.[23] This technique reduces the period of portal clamping, but results in very low flows through the venovenous bypass after the portal vein has been released.

For arterial revascularization, the donor common hepatic artery or celiac axis may be anastomosed to the recipient common hepatic artery end-to-end[21] or end-to-side.[24] This

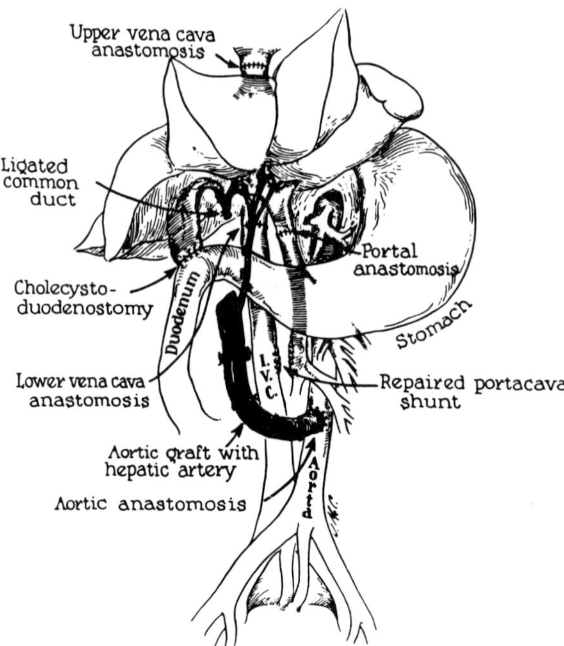

Figure 4 Completed orthotopic liver transplantation, using the technique described by Starzl et al. (Reproduced with permission from Starzl et al.[18])

eliminates the need to use the donor aorta as a conduit.

The cuff technique has been proposed as one method of simplifying the procedure. Cuffs made of polyvinyl chloride are used to connect the vessels, avoiding the need for time consuming, hand-sutured anastomoses. This technique has been used primarily for the venous anastomoses.[20,21] With this method, the donor vein is pulled through the cuff and then everted over the cuff. The vein is secured in place with a ligature. The recipient vessel is then pulled over the cuff and secured with additional ligatures.[21] The cuff technique has been used for the suprahepatic cava, the infrahepatic cava, and the portal vein. It was hoped that using the cuffed technique for these three anastomoses would eliminate the need for bypass, by minimizing the duration of the anhepatic phase. Removal of the left medial, left lateral, caudate, and right medial lobes is required to obtain sufficient length of suprahepatic cava for a cuffed anastomosis. With this method only 9.7 min of portal vein occlusion and 13.9 min of infrahepatic cava occlusion were required. Although 16 of 18 dogs reported survived for over 5 days, this method resulted in arterial pressures decreasing to a mean of 62.7 mmHg (46.5% of the initial value) during the anhepatic phase.[21]

Jamieson et al. has further modified the cuff technique.[20] Use of a cuff in the suprahepatic position requires partial hepatectomy (as described above) to provide an adequate length of vessel for the anastomosis. The cuffed suprahepatic anastomosis has been noted to kink and twist, resulting in decreased venous return to the heart and hypotension. For these reasons Jamieson has returned to a sutured suprahepatic caval anastomosis, reserving the cuffs for the infrahepatic cava and the portal vein. To avoid profound hypotension and reduce visceral congestion Jamieson uses a single lumen shunt to carry the portal flow to the jugular system. A mean operative time of 3 hours 10 min and a mean anhepatic phase of 25 min are reported with this technique. The anhepatic phase is noted to be well tolerated, with approximately a 30% drop in systemic pressure. Only one surgical failure was noted in over 30 cases.[20]

The method of biliary drainage is also subject to great variation. Cholecystoduodenostomy may be used, as described above. Other methods include: external drainage via a cholecystostomy tube,[19] Roux-en-y cholecystojejunostomy[21] or loop cholecystojejunostomy.[20]

5. Complications

Liver transplantation in dogs is a complex undertaking. The animals may experience postoperative problems despite a technically perfect operation. In early series, hemorrhagic gastroenteritis was a common complication. This was ascribed to splanchic venous stasis.18 Stuart noted gastrointestinal ulceration in 14% to 40% of dogs after liver transplantation, despite vagotomy and pyloroplasty or 70% subtotal gastric resection and gastrojejunostomy.[25] Other gastrointestinal complications include intussusception and pancreatic necrosis.[17] Frequent causes of death after liver transplantation in dogs include: myocardial infarction, intraperitoneal hemorrhage, splanchnic sequestration, hepatic artery thrombosis, gastrointestinal hemorrhage, hepatic insufficiency, and later, rejection.[24] With improvement in the preservation of donor livers, splanchnic sequestration, gastrointestinal hemorrhage, and hepatic insufficiency are seen less frequently. Operative survival rates of 70 to 75% or greater may be expected.[16,23]

D. HETEROTOPIC

The primary purported advantage of heterotopic liver transplantation is the elimination of the recipient hepatectomy. The native liver is left in place, undisturbed. This reduces the magnitude of the operation and obviates the venovenous bypass.[26] Unfortunately, heterotopic positioning of the liver results in other problems, which have limited its clinical utility. Welch's original technique for heterotopic transplantation of the liver in dogs places the donor liver in the lower abdomen of the recipient. The recipient's inferior vena cava is divided caudal to the renal veins. The distal aspect of the recipient's cava provides inflow to the donor's portal vein, while the proximal end of the inferior vena cava is used to establish hepatic outflow. A segment of the donor's abdominal aorta containing the celiac axis is interposed in the recipient's infrarenal aorta, restoring arterial flow to the liver. A modification of Welch's original technique is depicted in Figure 5. The transplanted livers produce bile for about 5 days. After the fifth day bile production ceases, the animals become toxemic, and necrosis of the liver is noted at exploration.[19]

Many variations of heterotopic liver transplantation have been studied in the dog. Mehrez placed the donor's liver in the left iliac fossa. The recipient's left iliac artery and vein are divided. End-to-end anastomoses of the iliac artery and vein to the hepatic artery

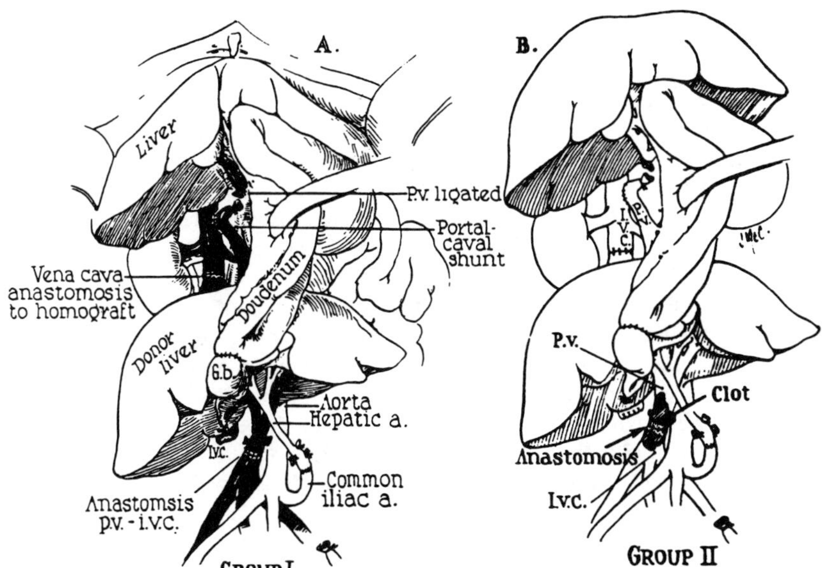

Figure 5 Heterotopic liver transplantation, using a modification of Welch's technique. Here the iliac artery is used to restore the arterial flow and a portocaval shunt has been added. In group II the donor livers received only arterial flow as the portal vein clotted. (Reproduced with permission from Halgrimson et al.[46])

and the suprahepatic cava, respectively, are done. The portal vein is connected to the recipient's splenic vein in an end-to-end anastomosis after performing a splenectomy. Pathologic examination of the transplanted livers reveals hepatocyte atrophy at 36 hours. After 9 days the hepatic architecture is completely destroyed, and at 22 and 45 days after transplantation there is almost complete disappearance of the graft.[27] Hagihara placed the transplant into left upper quadrant, in place of the spleen. The infrahepatic cava of the donor is anatomosed end-to-side to the recipient's infrarenal cava, and the donor portal vein is anastomosed end-to-side to the recipient's portal vein. The donor's hepatic artery is reconstructed with the recipient's abdominal aorta, splenic artery, or renal artery. Of 27 dogs, 12 survived at least 1 week after this operation.[28]

Marchioro performed what have become classic experiments with heterotopic liver transplantation. He evaluated the role of portal flow in sustaining the liver. The grafts are placed in the right paravertebral gutter. The donor vena cava is interposed in the recipient's inferior vena cava. The portal vein of the donor is anastomosed to the side of the recipient's superior mesenteric vein. The hepatic artery is sutured to the iliac artery of the recipient in an end-to-end fashion. Marchioro's three experimental groups differed in their handling of the recipient's portal vein. In group 1 the recipient portal vein is ligated above the most cephalad tributary, diverting all splanchnic flow to the graft. For group 2 the recipient's portal vein is ligated caudal to the splenic vein. Flow returning from the bowel is thus diverted to the donor liver, while flow from the spleen and pancreas is directed to the recipient's liver. In group 3 a preliminary portocaval transposition is performed before the transplant, and at the time of transplantation the recipient's portal vein is ligated cephalad to any tributaries. The allografts of group 1 retain their full size, while the host liver atrophies markedly in this group. In group 2 atrophy of the allograft is variable. Too few animals of group 3 survived to draw any conclusions. These observations refute the assumption that the atrophy of heterotopic livers is mediated purely by immunologic events. Marchioro correctly noted that the organ which has first access to the nutritive portal blood retains its function and integrity, while the other organ atrophies.[10]

Bengoechea-Gonzalez used a configuration similar to Marchioro's. The donor liver is again placed in the right paravertebral gutter. The infrahepatic cava is anastomosed to the inferior cava of the recipient. The celiac axis is sewn to the infrarenal aorta, and the portal vein is sewn to the recipient's superior mesenteric vein. All of these anastomoses are end-to-side. The recipient's superior mesenteric vein is constricted or ligated cephalad to the anastomosis. Again, it was noted that diverting the splanchnic blood to the transplanted liver reduces atrophy of the allograft.[26] Sigel has even transplanted a lobe of the liver into the neck. These grafts predictably atrophy, unless a portocaval shunt is done to divert some splanchnic flow into the systemic circulation.[29] This allows the hepatotrophic factors in the portal blood to reach the graft, albeit at low concentrations.

The various methods of heterotopic liver transplantation have been categorized into four types by Slapak. In type "A" the native liver receives its normal portal flow, while the transplanted liver receives portal inflow from the inferior vena cava. With type "B" the hepatic artery provides the only afferent flow to the transplant. Atrophy of the allograft is prominent in both of these types. The atrophy may be modified in types "A" and "B" by the addition of a portocaval shunt. Type "C" provides for sharing of the portal blood flow between the native and transplanted livers. This sharing can only occur if the vascular resistances of the two livers are equal. True sharing rarely exists, as the donor liver usually has an increased vascular resistance, secondary to preservation injury or rejection. Thus, this method does not eliminate atrophy of the graft. Type "D" diverts the portal flow away from the native liver to the transplant, providing the best maintenance of the transplant.[30]

Diverting portal flow from the native liver to the graft does not restore the heterotopic liver to normal. Jerusalem has shown that venous outflow is compromised by the heterotopic position. The further the hepatic venous outflow is from the right atrium, the higher the venous pressure. High outflow pressures are correlated with hyperacute graft destruction and a histologic picture of congestion and cirrhosis.[31] The optimal position for the liver appears to be adjacent to the diaphragm, where a very short distance from the hepatic outflow to the low pressure right atrium is obtained. The sum of these problems has limited the clinical utility of heterotopic liver transplantation, although valuable physiologic information has been gained.

E. SEGMENTAL

Segmental liver transplantation has become an important technique in the treatment of children with end stage liver disease.[32] Segmental liver transplantation utilizes a liver from a larger donor to transplant a child. The liver may be surgically reduced after procurement from a cadaver donor, or a portion of a living donor's liver may be excised. The use of canines has played an essential role in the development of these procedures.[3,4]

1. Cadaveric Donor

For the purpose of this discussion, if the entire liver is removed from the donor dog the procedure is considered equivalent to a cadaveric transplant. Welch was again a leader in the field, reporting survival of hepatectomized dogs after partial liver transplantation in 1966.[33] The technique described by this group provides for heterotopic transplantation of the left medial and left lateral lobes. The donor liver is removed in standard fashion. After excision of the donor liver the vascular supply to the right lobe is ligated. The parenchymal bridge between the right and left sides of the liver is divided. The right wall of the retrohepatic vena cava is excised, leaving a "button" of cava around the left hepatic veins. The recipient undergoes splenectomy and end-to-side portacaval shunt. The donor liver is placed in the recipient's pelvis. It is revascularized by an end-to-end anastomosis of the hepatic artery to the iliac artery and an end-to-side anastomosis of the "button" of donor vena cava to the recipient's inferior vena cava. The portal vein is ligated. Biliary drainage is achieved with a tube cholecystostomy or a cholecystojejunostomy. The native liver is excised to complete the procedure. Of 34 animals, 4 survived for more than 24 hours after this operation.[33]

Other attempts with heterotopic segmental liver transplantation include Aoki et al.'s 1967 report.[34] Left lateral lobe grafts were revascularized using either arterialization of the portal vein, splanchnic flow to the portal vein, or systemic venous flow to the portal vein. In some groups arterial reconstruction was added. Of 20 autotransplants, 2 dogs survived for at least 2 days. Of the 4 homotransplants, 2 survived.[34]

Lygidakis studied the use of segmental heterotopic transplants as an auxiliary liver.[35] Again, the donor liver is excised in standard fashion. During bench dissection the liver is divided preserving either the right lobes (60 to 70%) of the liver, or the left lobes (30 to 40%). The grafts are placed in the right paravertebral gutter of dogs weighing about 30 kg. The hepatic artery is sewn to the side of the infrarenal aorta, while the portal vein is sewn to the side of the recipient's superior mesenteric vein. A large "button" of donor cava is sewn to the side of the recipient's vena cava. A choledochojejunostomy completes the reconstruction. Of 13 recipients, 8 survived. The survivors were divided into two groups: group "A" received a liver segment weighing 195 ± 49 gm, while in group "B" the segments weighed 385 ± 85 gm. Group "A" animals experience acute portal hypertension when the portal flow is diverted to the graft. No significant changes in portal pressure are noted in group "B" after portal diversion.[35] This study demonstrated that approximately 10 gm of liver tissue per kilogram of body weight is sufficient to sustain normal portal hemodynamics.

Bax et al. evaluated the use of segmental transplants as orthotopic nonauxiliary grafts.[3] This model more closely approximates the current clinical usage of segmental transplants in children. Tissue-typed identical littermates are used. The left lateral and left medial lobes of the donor are removed during bench dissection, thus preserving the right half of the liver for transplantation. Care is taken in the hilar dissection to ligate the individual structures to the left lobes as peripherally as possible. The parenchyma is fractured with the back of a scalpel. The dissection is completed by ligating the left hepatic vein. The recipient operation is done in the standard manner for orthotopic transplantation. Long-term survival of 50% was achieved with this model.[3]

Taira et al. reported partial liver transplantation using a left lobe graft combined with preservation of the recipient's retrohepatic vena cava.[36] The recipient hepatectomy is done in the usual fashion, except the entire inferior vena cava is preserved. The donor liver parenchyma is divided between the left and right lobes, and the left hepatic vein is separated from the donor's cava. The reconstruction of the left lobe graft in the recipient utilizes an end-to-end anastomosis of the donor's left hepatic vein to the recipient's left hepatic vein. This is done with 5-0 polypropylene. The left portal vein is sewn to the recipient's portal vein end-to-end, and the common hepatic artery of the donor is sewn to the common hepatic artery of the recipient. A cholecystojejunostomy completes the procedure. Survival up to 25 days was reported with this method.[36] At our institution, this technique of preserving the retrohepatic cava of the recipient has become standard clinical practice for patients receiving a left lobe graft.[32]

2. Living Donor

The use of a portion of the liver from a living donor was proposed by Van Der Hyde et al. in 1966.[33] Smith further discussed segmental liver transplantation from a living donor in 1969.[37] The first clinical trial of this technique in the U.S. began 20 years later at the University of Chicago in November, 1989.[32] Before initiating the clinical trial the technique was evaluated in the animal laboratory.[4]

Dogs weighing 25 to 30 kg were used as donors, while the recipients weighed 10 to 15 kg. Both the donor and recipient operations are done through a bilateral subcostal incision with extension to the xyphoid in the midline. In the donor the falciform, left triangular, and gastrohepatic ligaments are divided. The left bile duct is divided. This duct usually has a long extrahepatic segment, and typically enters the common duct midway between the liver and the duodenum. The left portal vein is exposed and dissected proximally to the bifurcation of the main portal vein. Branches of the left portal vein entering the caudate lobe are divided. The left hepatic artery is isolated in continuity with the gastroduodenal artery, taking care to preserve the arterial supply to the remainder of the liver. The right gastric artery, which usually arises from the left hepatic artery, is divided. Branches of the portal pedicle to the caudate and quadrate lobes are divided after incising the capsule in these areas. The parenchyma is transected between the quadrate and left medial lobes. This is done by fracturing the tissue with a hemostat. Vessels are suture ligated as needed. The left hepatic vein is identified and dissected free for 1 cm. The graft is now attached solely by the vascular pedicles as illustrated in Figure 6. Note that the vessels are not occluded during the parenchymal transection, thus minimizing the ischemia that the liver sustains.[4]

Perfusion of the graft is achieved by inserting a 14F red rubber catheter into the left portal vein. The left portal vein is then clamped just beyond the bifurcation. The common hepatic artery is ligated just proximal to the left hepatic branch. The distal gastroduodenal artery is also ligated. The left hepatic vein is clamped as far cephalad as possible. Collin's

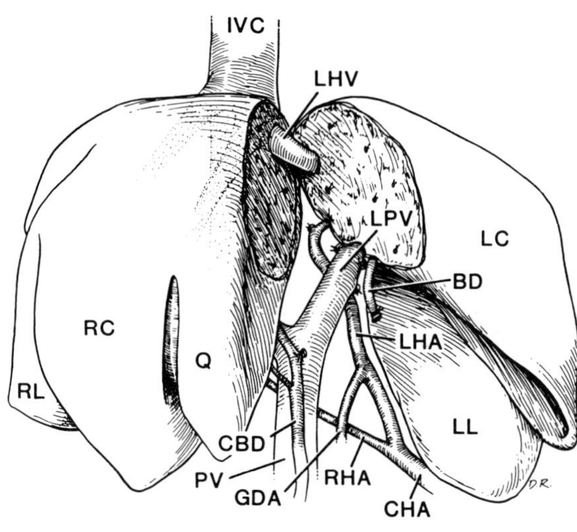

Figure 6 Living donor operation. The parenchyma between the left central (LC) and quadrate (Q) lobes has been divided, but the vasculature remains intact. (Reproduced with permission from Cherqui et al.[4])

solution (1 l) is infused through the left portal vein. The proximal gastroduodenal artery is cut to allow egress of the perfusion solution. The preservation technique is illustrated in Figure 7. The graft is now removed by dividing the left hepatic vein and the hepatic artery.[4]

For the recipient the left femoral vein and left external jugular vein are exposed for the venovenous bypass. The common bile duct and the hepatic artery branches are divided close to the liver. The portal vein is exposed from the level of the pancreatoduodenal branch (which is divided) to the bifurcation. The liver is now fully mobilized by dividing all the ligamentous attachments and the tissue posterior to the retrohepatic cava. The right adrenal vein is divided when necessary. Venovenous bypass from the femoral and splenic veins to the jugular vein is initiated. The portal vein, infrahepatic cava, and suprahepatic cava are clamped, achieving total vascular exclusion of the liver. The recipient's liver is excised, leaving the retrohepatic cava intact. The orifices of the middle and left hepatic veins are joined to form a common lumen for anastomosis. The small hepatic veins which enter the cava directly are sutured.[4]

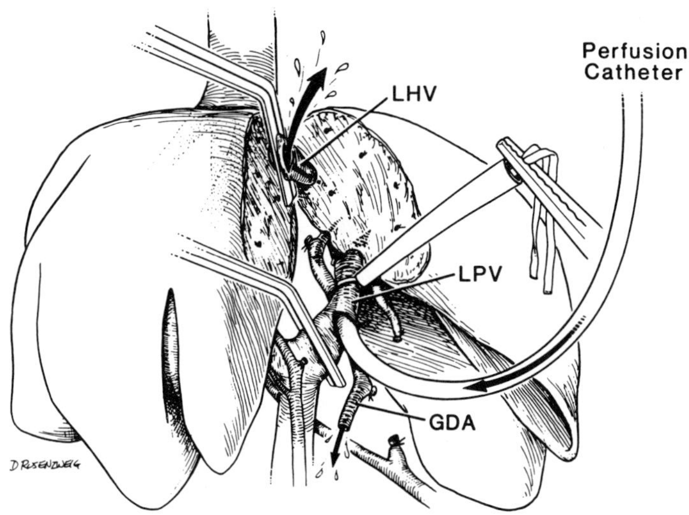

Figure 7 Preservation of the living donor graft. Collin's solution is infused through a catheter in the left portal vein (LPV). Egress of blood and Collin's solution occurs through the left hepatic vein (LHV) and the gastroduodenal artery (GDA). (Reproduced with permission from Cherqui et al.[4])

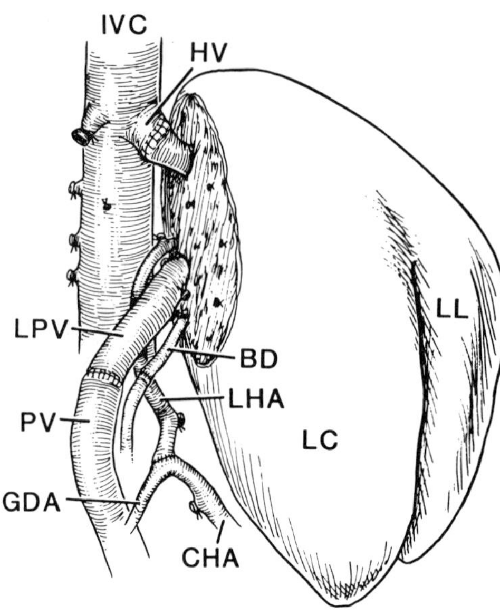

Figure 8 Completed orthotopic segmental liver transplantation from a living donor. Note the preservation of the recipient's retrohepatic vena cava. (Reproduced with permission from Cherqui et al.[4])

Implantation of the segmental graft is begun by anastomosing the hepatic vein of the donor to the orifice formed by the middle and left hepatic veins of the recipient. This is done with running 5-0 polypropylene. A small opening is left in the anterior wall to allow flushing of the preservation solution from the graft. The portal vein anastomosis is done end-to-end with running 6-0 suture. Before completing the anterior wall of the portal anastomosis, 500 ml of room-temperature saline is flushed through the graft. These anastomoses are completed, and then the graft is revascularized. The caval clamps are released first. The artery is reconstructed by anastomosing the donor gastroduodenal artery to the side of the the recipient's hepatic artery with 7-0 polypropylene. Biliary drainage is achieved by a duct-to-duct anastomosis using interrupted 6-0 sutures.[4] The final result is pictured in Figure 8.

The mean blood loss in the donors was 133 ± 98 ml. No donor required transfusion. The were no intraoperative deaths among the donors. The first donor died on the third postoperative day of an infected necrosis of the quadrate lobe. Subsequent donors underwent excision of the quadrate lobe and papillary process. These other nine donors survived without complications. Three recipients survived the immediate postoperative period.[4]

This technique, with some modification, has been applied in 22 patients at the University of Chicago as of February 1991.

III. APPLICATIONS

A. PRESERVATION

The liver is very sensitive to ischemia. Livers excised under normothermic conditions become nonviable in about 30 min.[19] The importance of hypothermia in protecting the abdominal organs from ischemic injury was recognized in the 1950s.[38,39] Lillehei et al. topically cooled intestine in their 1959 studies of small bowel transplantation.[40] Infusion of chilled lactated Ringer's was used to achieve core-cooling of the liver in the 1960 report by Starzl et al.[18] Combining topical hypothermia with an intravascular flush to achieve core-cooling has become the standard method of donor liver preservation.

The canine model has been important in the development and testing of improved preservation solutions. Collins developed a solution with potassium and magnesium present in concentrations which mimic the intracellular levels of these ions.[41] Benichou demonstrated 18-hour preservation of the dog liver using either Collins' solution or plasma for the flush. Lactated Ringer's achieved only 9 hours of preservation in these studies.[42] The University of Wisconsin has developed a complex preservation solution which contains: hydroxyethyl starch, lactobionate, raffinose, adenosine, glutathione, allopurinol, phosphate, potassium, and magnesium, as well as other items. Jamieson has shown that this solution allows preservation of canine livers for 24 to 48 hours.[9]

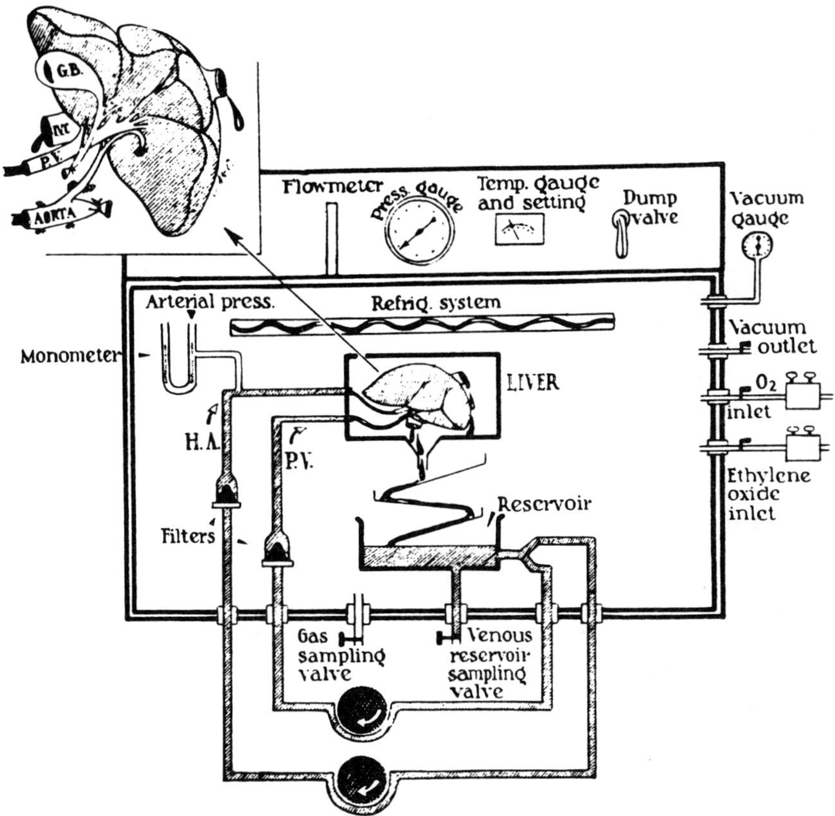

Figure 9 *Ex-vivo* perfusion system for liver preservation. (Reproduced with permission from Brettschneider et al.[45])

It has been proposed that continuous perfusion of the donor liver would be more physiologic and allow extended preservation times. Marchioro investigated the use of extracorporeal perfusion to achieve whole-body cooling of the donor dog. Kidney preservation up to 14 hours and liver preservation of 2 hours were obtained.[43] Slapak obtained 24-hour preservation of the dog liver by using continuous pulsatile perfusion and hyperbaric oxygen to sustain the donor organ after excision.[44] Brettschneider modified this technique (see Figure 9) of continuous perfusion and hyperbaric oxygenation by using blood for the perfusion. Satisfactory preservation of up to 24 hours was achieved in certain cases.[45] Although these techniques allow storage of the donor liver for up to 24 hours, they are not used clinically because of the complexity involved. Further efforts to understand the mechanisms of ischemic and reperfusion injury sustained by the liver should lead to continued improvement in the ability to store donor livers.

B. PHYSIOLOGY

We have already seen the important role played by experimental heterotopic liver transplantation in our understanding of portal flow in the maintenance of hepatic integrity. The studies of Marchioro demonstrated that the organ which has first access to the portal flow retains its size, while the other liver atrophies.[10] These experiments led to further investigations which demonstrated that the hepatotrophic factors are contained in blood derived from the pancreatic-gastroduodenal-splenic venous drainage. Blood from the intestine has little influence on hepatic substance.[46] Insulin was proven to be the primary hepatotrophic substance contained in portal blood by investigations demonstrating that hepatic atrophy is avoided by the administration of insulin.[11]

Others have studied the dog's metabolism after liver transplantation. Faris reported that plasma cholesterol, triglyceride, and glucose levels are normal in dogs with good liver function.[47] Glucose metabolism in the peritransplant period was studied by DeWolf. He reported that the blood glucose progressively decreases in dogs during the anhepatic phase from 135 ± 9 mg/dl to 88 ± 8 mg/dl. The glucose rises to a mean value of 240 mg/dl 30 min after reperfusion.[48]

Tashiro noted that the blood sugar rises to a greater level (935 ± 189 mg/dl) after reperfusion in animals that die than in those that survive (<500 mg/dl).[49] The study of physiology and metabolism after liver transplantation remains a largely unexplored area where important work may be done.

C. REJECTION

In 1960 Moore noted that the bilirubin begins to rise 4 to 5 days after liver transplantation in nonimmunosuppressed dogs. Histologic study of these rejecting livers reveals infiltration of the portal areas with mononuclear cells, lymphocytes, and plasma cells. The parenchymal cells are relatively spared, while the portal tracts and centrilobular veins demonstrate progressive cellular infiltration. The biliary ductal epithelium is damaged by this infiltrate.[17] The features of hepatic allograft rejection described by Moore remain important diagnostic criteria today: lymphocytic infiltration of the portal triads, central vein endothelialitis, and biliary injury.

Starzl provided further details of the course of untreated rejection in dogs. Eighteen dogs that survived at least 4 days after transplantation were studied. The animals begin to eat after the second postoperative day; their physical activity is normal for the first 4 or 5 days. Typically, the urine becomes dark on the fifth or sixth day. Jaundice follows in 12 to 24 hours. The dietary intake is significantly reduced, and death follows in 2 to 4 days. Blood chemistry studies reveal that the bilirubin begins to rise on day 5 to 6. In addition, the animals become hypoglycemic as the liver fails. Gross examination of the rejected liver reveals a large, firm, tan organ. The histologic features of rejection described by Moore were confirmed in these studies although parenchymal destruction was also noted after 6 days.[6] It was later noted that an elevation in the serum transaminases accompanied the increased bilirubin seen in rejection.[16]

Paronetto studied the mechanisms of liver transplant rejection. He noted binding of gamma-globulin and complement to hepatocytes, bile ducts, vessels, and macrophages in the graft.[51] Groth presented data demonstrating that cellular infiltration is the primary event in acute rejection, while the deposition of immunoglobulin and complement noted above are secondary. Using a xenon washout technique, Groth also noted that portal venous and hepatic arterial blood flow are decreased during rejection.[51] The important role played by the dog model in our understanding of liver transplant rejection is evident. Although such studies have contributed greatly to our understanding of acute rejection, much work remains to be done before chronic rejection is well understood.

D. IMMUNOSUPPRESSION

As we have seen in the preceding section, without some method to modify the host's immune response rejection becomes clinically evident after 5 to 6 days and the animals typically die shortly thereafter. Although most new immunosuppressive agents are first tested in kidney transplantation, almost every agent has also been evaluated for use in liver transplantation. One of the first methods tested in the search for clinically useful immunosuppression was total body irradiation. Toxicity is severe with this technique. In one report, six of six recipients died after 1400 rads of total body irradiation and liver transplantation.[52]

In the 1964 Starzl report the use of azathioprine, a derivative of 6-mercaptopurine, in 40 dogs and 5 patients after liver transplantation. Azathioprine is antiproliferative; it inhibits the replication of DNA. The azathioprine was given to the dogs at a dose of 2 to 10 mg/kg/day. The highest dose tolerated without leukopenia was chosen. This regimen delayed the development of hyperbilirubinemia in the dogs until 9 to 10 days posttransplant.[7] Moore also used azathioprine in canine liver transplantation; in some animals the azathioprine was combined with azaserine or cortisone. Despite the reduction in rejection seen, the animals still survived only 3.5 to 12 days.[53] Starzl further studied azathioprine in a large series of canine liver transplants. An effort was made to avoid leukopenia in this series. The azathioprine was noted to increase the perioperative mortality from 1 of 23 dogs not receiving azathioprine to 32 of 116 dogs (28%) treated with azathioprine. The most common cause of death in the animals receiving azathioprine was pneumonitis. Despite this increased perioperative mortality, 44 dogs survived at least 25 days and 24 dogs survived at least 50 days. This clearly demonstrated azathioprine's ability to potentiate allograft survival. It is interesting that 29% of the animals had uncontrolled rejection despite azathioprine, while 23% had no rejection while on the azathioprine. The largest group of dogs (49%) had some biochemical evidence of rejection which spontaneously reversed without increasing the azathioprine.[54]

Antilymphocyte serum and antilymphocyte globulin were used in canine renal and hepatic transplantation in 1967. Control animals survived for an average of 11.3 ± 4.6 days after kidney transplantation and 7.1 ± 2.2 days after liver transplantation. The dogs treated with antilymphocyte serum survived for a mean of 27.2 ± 13 days after kidney transplantation and 26.8 ± 26 days after liver transplantation. Antilymphocyte globulin also significantly improved survival after transplantation, to 22.9 ± 23 days for kidney recipients and 36 ± 30 days for liver recipients.[55] Newer immunosuppressive agents have also been tested in the canine liver transplant model. Cyclosporine, a unique cyclic peptide produced by fungi, has revolutionized liver transplantation. Cyclosporine has resulted in far better survival after liver transplantation than was previously attained; 1-year patient survivals of 70 to 80% are now frequently reported.[32] Cyclosporine's primary side-effect is nephrotoxicity. In a search for a less nephrotoxic form of cyclosporine, an analogue of cyclosporine (Nva^2-cyclosporine) was tested in dogs undergoing liver transplantation.[56] Unfortunately, the purported advantages of Nva^2-cyclosporine have not yet been realized. FK-506 is a another new immunosuppressive agent which has been tested in canine liver transplantation. FK-506 is produced by bacteria normally found in the soil. It exhibits immunosuppressive properties that are similar, but not indentical, to those of cyclosporine. Both drugs inhibit the production of interleukin-2 by T-lymphocytes. FK-506 is much more potent than cyclosporine. Todo reported that 8 of 10 dogs that received FK-506 after liver transplantation survived for at least 1 month.[57] FK-506 has now entered clinical trials in kidney and liver transplantation.

IV. CONCLUSIONS

The use of dogs has been essential in the development of liver transplantation for clinical application. The dog model has obviously been important in the refinement of the surgical techniques used for orthotopic and segmental liver transplantation. The techniques used for living donor liver transplantation were first tested in the dog and then successfully applied, in modified form, to human patients. In addition to providing an arena in which to test new techniques, the dog laboratory provides a training facility in which surgeons and anesthesiologists may develop the teamwork necessary for clinical liver transplantation. The dog model is not only used for technical exercises, but has also been important in the testing of new preservation solutions and immunosuppressive agents. Liver transplant experiments in dogs have also significantly contributed to our understanding of hepatic physiology. The heterotopic liver transplant experiments were helpful in defining the role of insulin as a hepatotrophic factor. Despite these successes, many problems remain. Immunosuppression remains relatively nonspecific and results in too many infections. Chronic rejection is still a significant cause of graft loss. How the function of the liver changes after transplantation is not completely understood. The dog will therefore remain an important experimental model for liver transplantation.

REFERENCES

1. **Welch, C.W.,** A note on the transplantation of the whole liver in dogs, *Transplant. Bull.,* 2, 54, 1955.
2. **Cannon, J.A.,** Brief report, *Transplant. Bull.,* 3, 7, 1956.
3. **Bax, N.M.A., Vermeire, B.M.J., Dubois, N., Madern, G., Meradji, M., and Molenaar, J.C.,** Orthotopic nonauxiliary homotransplantation of part of the liver in dogs, *J. Ped. Surg.,* 17, 906, 1982.
4. **Cherqui, D., Emond, J.C., Pietrabissa, A., Michel, M., Roncella, M., Brown, S.B., Whitington, P.F., and Broelsch, C.E.,** Segmental liver transplantation from living donors: report of the technique and preliminary results in dogs, H.P.B. *Surgery,* 2, 189, 1990.
5. **Denmark, S.W., Shaw, B.W., Starzl, T.E., and Griffith, B.P.,** Veno-venous bypass without systemic anticoagulation in canine and human liver transplantation, *Surg. Forum,* 34, 380, 1983.
6. **Starzl, T.E., Kaupp, H.A., Brock, D.R., and Linman, J.,** Studies on the rejection of the transplanted homologous dog liver, *Surg. Gynecol. Obstet.,* 112, 135, 1961.
7. **Starzl, T.E., Marchioro, T.L., Rowlands, D.T., Kirkpatrick, C.H., Wilson, W.E.C., Rifkind, D., and Waddell, W.R.,** Immunosuppression after experimental and clinical homotransplantation of the liver, *Ann. Surg.,* 160, 411, 1964.
8. **Marchioro, T.L., Huntley, R.T., Waddell, W.R., and Starzl, T.E.,** Extracorporeal perfusion for obtaining postmortem homografts, *Surgery,* 54, 900, 1963.
9. **Jamieson, N.V., Sundberg, R., Lindell, S., Claesson, K., Moen, J., Vreugdenhil, P.K., Wight, D.G.D., Southard, J.H., and Belzer, F.O.,** Preservation of the canine liver for 24-48 hours using simple cold storage with UW solution, *Transplantation,* 46, 517, 1988.
10. **Marchioro, T.L., Porter, K.A., Dickinson, T.C., Faris, T.D., and Starzl, T.E.,** Physiologic requirements for auxiliary liver homotransplantation, *Surg. Gynecol. Obstet.,* 121, 17, 1965.
11. **Serrou, B., Michel, H., Gelis, C., Solassol, C., Pujol, H., and Romieu, C.L.,** Study of the role of the splanchnic organs in hepatic atrophy for the purpose of auxiliary transplantation of the liver, *Ann. Surg.,* 180, 274, 1974.
12. **Howell, W.H.,** *Dissection of the Dog,* Henry Holt, New York, 1889, 21.
13. **Shively, M.J. and Beaver, B.G.,** Dissection of the Dog and Cat, Iowa State University Press, Ames, IA, 1985, 92.
14. **Evans, H.E. and de Lahunta, A.,** *Miller's Guide to the Dissection of the Dog,* W.B. Saunders, Philadelphia, 1988, 191.
15. **Arey, L.B.,** Throttling veins in the livers of certain mammals, *Anat. Rec.,* 81, 21, 1941.
16. **Todo, S., Kam, I., Lynch, S., and Starzl, T.E.,** Animal research in liver transplantation with special reference to the dog, *Semin. Liver Dis.,* 5, 309, 1985.
17. **Moore, F.D., Wheeler, H.B., Demissianos, H.V., Smith, L.L., Balankura, O., Abel, K., Greenberg, J.B., and Dammin, G.J.,** Experimental whole-organ transplantation of the liver and of the spleen, *Ann. Surg.,* 152, 374, 1960.

18. **Starzl, T.E., Kaupp, H.A., Brock, D.R., Lazarus, R.E., and Johnson, R.V.,** Reconstructive problems in canine liver homotransplantation with special reference to the postoperative role of hepatic venous flow, *Surg. Gynecol. Obstet.,* 111, 733, 1960.
19. **Goodrich, E.O., Welch, H.F., Nelson, J.A., Beecher, T.S., and Welch, C.S.,** Homotransplantation of the canine liver, *Surgery,* 39, 244, 1956.
20. **Jamieson, N.V., Sundberg, R., Lindell, S., Kalayoglu, M., Southard, J.H., and Belzer, F.O.,** A simplified technique for transplantation of the canine liver, *Acta Chir. Scan.,* 154, 511, 1988.
21. **Monden, M., Barters, R.H., and Fortner, J.G.,** A simple method of orthotopic liver transplantation in dogs, *Ann. Surg.,* 195, 110, 1982.
22. **Kam, I., Lynch, S., Todo, S., Dewolf, A., McSteen, F., Jakab, F., Ericzon, B., Takaya, S., and Starzl, T.E.,** Low flow venovenous bypass in small dogs and pediatric patients undergoing replacement of the liver, *Surg. Gynecol. Obstet.,* 163, 33, 1986.
23. **Cooperman, A.M., Woods, J.E., and McIlrath, D.C.,** Simplified method of canine orthotopic hepatic tranplantation, *Am. J. Surg.,* 122, 797, 1971.
24. **Cutropia, J.C., Coratolo, F., Spinetta, A., Kei, J., Ribas, A., Assini, L., Delle Donne, G., Bianchi, M., Guinazu, A., and Verdaguer, J.A.,** Transplante hepatico ortotopico experimental, *Rev. Esp. Enferm. Apar. Dig.,* 38, 553, 1972.
25. **Stuart, F.P., Torres, E., and Moore, F.D.,** The association of upper gastrointestinal ulceration and orthotopic hepatic allotransplantation in the dog, *Transplantation,* 5, 804, 1967.
26. **Bengoechea-Gonzalez, E., Awane, Y., and Reemtsma, K.,** Experimental liver homotransplantation, *Arch. Surg.,* 94, 1, 1967.
27. **Mehrez, I.O., Nabseth, D.C., Kekis, B.P., Apostolou, K., Gottlieb, L.S., and Deterling, R.A.,** Homotransplantation of the canine liver: a new technic, *Ann. Surg.,* 159, 416, 1964.
28. **Hagihara, P. and Absolon, K.B.,** Experimental studies on homologous heterotopic liver transplantation, *Surg. Gynecol. Obstet.,* 119, 1297, 1964.
29. **Sigel, B., Baldia, L., and Dunn, M.R.,** Studies of liver lobes autotransplanted ouside of the abdominal cavity, *Surg. Gynecol. Obstet.,* 124, 525, 1967.
30. **Slapak, M., Beaudoin, J.G., Lee, H.M., and Hume, D.M.,** Auxiliary liver homotransplantation: a new technique and evaluation of current techniques, *Arch. Surg.,* 100, 31, 1970.
31. **Jerusalem, C., Van Der Heyde, M.N., Schmidt, W.J., and Tjebbes, F.A.,** Heterotopic liver transplantation. II. Unfavorable outflow conditions as a possible cause for late graft failure, *Eur. Surg. Res.,* 4, 186, 1972.
32. **Broelsch, C.E., Emond, J.C., Whitington, P.F., Thistlethwaite, J.R., Baker, A.L., and Lichtor, J.L.,** Application of reduced size liver transplants as split grafts, auxiliary orthotopic grafts, and living related segmental transplants, *Ann. Surg.,* 212, 368, 1990.
33. **Van Der Heyde, M.N., Cooper J.R., Carter, J.H., and Welch, C.S.,** Survival of hepatectomized dogs with partial liver transplants, *Surgery,* 59, 1079, 1966.
34. **Aoki, H., Wang, T.C., Matsumura, H., Funabiki, T., Kakumoto, Y., and Hatori, T.,** Partial liver transplantation with primary revascularization in dogs: preliminary report on new methods and its theoretical background, *Keio J. Med.,* 16, 205, 1967.
35. **Lygidakis, N.J., Chamuleau, R.A.F.M., Rothuizen, J., Grijm, R., Van Baal, J.G., Kox, K., Van Altena, E., Van Joost, H.E., Van Den Brom, W.E., Jobis, A.C., and Brummelkamp, W.H.,** Segmental auxiliary liver transplantation in dogs: a search for an ideal graft — illusion or reality?, *Eur. Surg. Res.,* 19, 265, 1987.

36. **Taira, N., Kanou, T., Orihara, A., Kawai, M., Uchida, K., Satake, M., Katou, Y., Haba, T., Hayashi, S., Yamada, N., Tominaga, Y., Tanaka, Y., and Takagi, H.,** Partial liver transplantation in dogs: preserving the IVC of the recipient and use of a heparinized catheter (Anthron) for portal vein bypass, *Tranplant. Proc.,* 21, 2362, 1989.
37. **Smith, B.,** Segmental liver transplantation from a living donor, *J. Ped. Surg.,* 4, 126, 1969.
38. **Owens, T.C., Prevedel, A.E., and Swan, H.,** Prolonged experimental occlusion of thoracic aorta during hypothermia, *Arch. Surg.,* 70, 95, 1955.
39. **Borgardus, G.M. and Schlosser, R.J.,** The influence of temperature upon renal ischemic damage, *Surgery,* 39, 970, 1956.
40. **Lillehei, R.C., Goott, B., and Miller, F.B.,** The physiologic response of the small bowel of the dog to ischemia including in vitro preservation of the bowel with successful replacement and survival, *Ann. Surg.,* 150, 543, 1959.
41. **Collins, G.M., Bravo-Shugarman, M., and Terasaki, P.I.,** Kidney preservation for transplantation, *Lancet,* 2, 1219, 1969.
42. **Benichou, J., Halgrimson, C.G., Weil, R., Koep, L.J., and Starzl, T.E.,** Canine and human liver preservation for 6 to 18 hr by cold infusion, *Transplantation,* 24, 407, 1977.
43. **Marchioro, T.L., Huntley, R.T., Waddell, W.R., and Starzl, T.E.,** Extracorporeal perfusion for obtaining postmortem homografts, *Surgery,* 54, 900, 1963.
44. **Slapak, M., Wigmore, R.A., and MacLean, L.D.,** Twenty-four hour liver preservation by the use of continuous pulsatile perfusion and hyperbaric oxygen, *Transplantation,* 5, 1154, 1967.
45. **Brettschneider, L., Daloze, P.M., Huguet, C., Porter, K.A., Groth, C.G., Kashiwagi, N., Hutchison, D.E., and Starzl, T.E.,** The use of combined preservation techniques for extended storage of orthotopic liver homografts, *Surg. Gynecol. Obstet.,* 126, 263, 1968.
46. **Starzl, T.E., Francavilla, A., Halgrimson, C.G., Francavilla, F.R., Porter, K.A., Brown, T.H., and Putnam, C.W.,** The origin, hormonal nature, and action of hepatotrophic substances in portal venous blood, *Surg. Gynecol. Obstet.,* 137, 179, 1973.
47. **Faris, T.D., Marchioro, T.L., Herrmann, T.J., and Starzl, T.E.,** Late function of the orthotopic liver homograft, *Surg. Forum,* 16, 222, 1965.
48. **DeWolf, A.M., Kang, Y.G., Todo, S., Kam. I., Francavilla, A.J., Polimeno, L., Lynch, S., and Starzl, T.E.,** Glucose metabolism during liver transplantation in dogs, *Anesth. Analg.,* 66, 76, 1987.
49. **Tashiro, S., Kawamoto, S., Nakakuma, K., Hayashida, N., Kanemitsu, K., Sugihara, S., Tsuji, T., Uchino, R., Nanakawa, K., Kimura, M., Saito, N., and Miyauchi, Y.,** Evaluation of early graft function in orthotopic liver transplantation in dogs, *Transplant. Proc.,* 21, 1344, 1989.
50. **Paronetto, F., Horowitz, R.E., Sicular, A., Burrows, L., Kark, A.E., and Popper, H.,** Immunologic observations on homografts. I. The canine liver, *Transplantation,* 3, 303, 1965.
51. **Groth, C.G., Porter,K.A, Otte, J.B., Daloze, P.M., Marchioro, T.L., Brettschneider, L., and Starzl, T.E.,** Studies of blood flow and ultrastructural changes in rejecting and nonrejecting canine orthotopic liver homografts, *Surgery,* 63, 658, 1968.
52. **Starzl, T.E., Butz, G.W., Brock, D.R., Linman, J.T., and Moss, W.T.,** Canine liver homotransplants: the effect of host and graft irradiation, *Arch. Surg.,* 85, 460, 1962.
53. **Moore, F.D., Birtch, A.G., Dagher, F., Veith, F., Krisher, J.A., Order, S.E., Shucart, W.A., Dammin, G.J., and Couch, N.P.,** Immunosuppression and vascular insufficiency in liver transplantation, *Ann. N.Y. Acad. Sci.,* 120, 729, 1964.

54. **Starzl, T.E., Marchioro, T.L., Porter, K.A., Taylor, P.D., Faris, T.D., Herrmann, T.J., Hlad, C.J., and Waddell, W.R.,** Factors determining short- and long-term survival after orthotopic liver homotransplantation in the dog, *Surgery,* 58, 131, 1965.
55. **Starzl, T.E., Marchioro, T.L., Porter, K.A., Iwaski, Y., and Cerilli, G.J.,** The use of heterologous antilymphoid agents in canine renal and liver homotransplantation and in human renal homotransplantation, *Surg. Gynecol. Obstet.,* 124, 301, 1967.
56. **Todo, S., Porter, K.A., and Kam, I.,** Canine liver transplantation under Nva^2-cyclosporine versus cyclosporine, *Transplantation,* 41, 296, 1986.
57. **Todo, S., Podesta, L., ChapChap, P., Kahn, D., Pan, C-E., Ueda, Y., Okuda, K., Imventarza, O., Casavilla, A., Demetris, A.J., Makowka, L., and Starzl, T.E.,** Orthotopic liver transplantation in dogs receiving FK-506, *Transplant. Proc.,* 19, 64, 1987.

Chapter 7

Liver Transplantation in the Pig

D. Kahn, Rosemary Hickman, H. Pienaar, and J. Terblanche

CONTENTS

I. INTRODUCTION

The earliest description of experimental liver transplantation appears to be that of Welch, who performed auxiliary grafts in dogs without removing the host liver.[1] Orthotopic liver grafts were performed in dogs by Moore et al. in Boston[2] and Starzl et al.[3] in Chicago, leading to the first phase of clinical transplantation in Denver in 1963.[4] Within a few years, several undesirable characteristics of the dog as an experimental liver transplant model had become apparent (hepatic venous sphincter spasm and intolerance to portal venous clamping[5]), leading Garnier et al.[6] in France and Peacock and Terblanche[7] in England to explore the use of the pig. This animal's gastrointestinal and endocrine

0-8493-3629-5/94/$0.00+$.50

physiology was deemed to be more like a human's, and in addition, the pig did not manifest hepatic venous sphincter spasm when transplanted. It did, however, share with the dog a poor tolerance to portal venous occlusion.

It soon became apparent during those early experiments that the pig appeared to possess another intriguing attribute. The pig was able to accept liver grafts with minimal or no evidence of rejection and was able to overcome mild rejection spontaneously.[8] Subsequent studies have indicated that this "tolerance" may even be encountered in animals with major mixed lymphocyte culture (MLC) incompatibility.[9] Despite intensive work, the reason for this apparently mild response, which seems to be confined to the pig and to some strains of rat,[10] has not been found. Indeed, some workers believe that it may be entirely due to inbreeding of stock.[11] At this time, a further incentive for using pigs in liver transplantation research has been given by the experience gained in many laboratories using isolated porcine liver perfusion in the treatment of fulminant hepatic failure.[12] Recently, another advantage of the pig has become obvious. The use of the pig for research purposes has proven to be ethically better tolerated then the use of dogs. Despite these advantages, most of the long-established laboratories have remained with whatever animal they had originally chosen. Workers in Holland,[13] Cambridge, England,[14] and Cape Town,[15] though, continue to use pigs.

II. TECHNIQUES

A. ANATOMY

The pig liver rests in the right upper quadrant of the abdomen. It is secured by three ligamentous attachments: the left triangular; the falciform; and the posterior ligaments. The right triangular ligament is not present in the pig. The liver is vascularized by: (1) the portal vein, (2) the hepatic artery, and (3) its attachment to the inferior vena cava.

The pig liver has classically been described as being composed of three lobes: right, median, and left. The principal lobe, the median, shows greater development than either the right or the left lobes. The three lobes meet at the posterior third of the median lobe. The median lobe can be further divided into left and right secondary segments as delineated by its superficial intramediate fissure. The next most developed lobe is the right within which the gallbladder is embedded.

1. Hepatic Artery

After entrance into the liver, the hepatic artery follows a posterioinferior concave arc ascending, through the median lobe and arcing into the left lobe with its terminal branches. The artery invariably follows the course of the portal vein and is located ventral to it. Immediately after entrance the artery gives off two to three branches destined for the right lobe. One passes behind the portal vein and supplies the posterior zone of the lobe while the remainder supply the anterior face. The next branch develops as the artery turns left and it ascends toward the gallbladder supplying the right segment of the median lobe. As the artery proceeds to the left it develops branches which supply the left median and left lobes.

2. Portal Vein

The portal vein proceeds as does the artery through a posterioinferior concave arc. With the exception of one small branch given off the concave surface of the arc as it enters the liver the others emerge from the convex anterosuperior surface. The portal vein consistently divides into four segments. The first is the hilar segment, which is a single extraparenchymal vessel approximately 4 cm in length. The second portion is an arc 2 to 3 cm in length that gives off branches to the right lobe and the right segment of the median lobe. The third portion proceeds through a 3- to 4-cm horizontal course emitting a branch to the left lobe. The fourth and final segment supplies the left median and left lobe.

3. Hepatic Veins

The hepatic veins enter the anterior and anterolateral faces of an intraparenchymal inferior vena cava. There are three branches: left, median, and right. The median drains the left median segment, the right hepatic vein drains the right lobe and right median segment. The left lobe is drained by the left hepatic vein.

4. Biliary Tract

The biliary tree is divided into two hepatic ducts that converge to form the common bile duct outside the liver parenchyma. They drain the right and left halves of the liver. The common bile duct is firmly attached to the inferior surface of the liver by a thick connective tissue layer. In general the principal routes follow the portal vessels course.

B. ANESTHESIA

In general, animals 25 to 32 kg are easiest to handle. Pigs require at least 24 hours of food restriction to ensure an empty gastrointestinal tract at the time of surgery. Water is allowed ad lib. If food is present in the stomach or gut, the recipient procedure should be aborted, and the animal closed. If one continued, the result would be significant intestinal distension. As a result, abdominal closure would be difficult to impossible and associated with severe pulmonary restriction.

Anesthesia techniques and monitoring are kept simple in this laboratory because of a shortage of skilled assistants. No premedication is used because previous experience has suggested that administration of atropine does not appear to influence the volume of bronchial and pharyngeal secretions. In cases where animals are liable to develop malignant hyperpyrexia, the administration of a beta-blocking agent can be beneficial.[19] Anesthesia is induced with intravenous thiopentone sodium injected into an ear vein (20 to 30 mg/kg), while the awake animal is restrained. Endotracheal intubation is performed with the animal lying on its back and the head extended over the edge of the trolley which facilitates the introduction of the endotracheal tube. A long-bladed laryngoscope is helpful in reaching the epiglottis, which is 10 to 15 cm from the teeth. The tube must be turned 90° from the midline during insertion to avoid interference from the cricoid cartilage. A large bore stomach tube is routinely passed to ensure decompression of the stomach.

C. THE STANDARD TECHNIQUE OF LIVER ALLOGRAFT

The method of liver transplantation in the pig which is currently used is an adaptation of techniques described by Terblanche et al.,[16] Dent et al.,[17] and Hickman et al.[18]

Once sedated and intubated, the animal is manually ventilated and transported to the operating laboratory, where it is placed supine and secured to the table. The table is equipped with a foam-rubber sheet 4 cm thick, in order to decrease the risk of pressure-related injuries. Anesthesia is maintained with nitrous oxide (4 l/min) and oxygen (2 l/min) given through a Magill-type circuit. A heated humidifier is necessary if the ambient temperature is below 25°C. Hypothermia develops rapidly in an unwarmed animal. A fan heater under the table is also used to warm the surrounding air. The skin is either left unshaved, or shaved very gently with a scalpel blade, avoiding scraping as much as possible. The skin is prepared with an iodine in alcohol solution.

Through a right neck incision, 3 to 4 cm lateral to the trachea, catheters (8 French) are introduced into the right carotid artery for pressure monitoring and blood collection, and into the right internal jugular vein for fluid administration. The right external jugular vein which will be used for the porto-systemic bypass should not be dissected at this time.

The laparotomy incision is made the full length of the abdomen, extending from the xiphoid process of the sternum to the symphysis pubis, using electrocautery. The incision is made in the midline in female animals, but deflects to the right to avoid the penis and urethra in males. This deflection reveals many small vessels which require coagulation. The incision is opened with a large self-retaining retractor and the entire bowel is enclosed in a sterile plastic bag, which is twisted closed loosely around the root of the small and large bowel mesentery. Care should be taken when clamping the bag to prevent excessive occlusion. Securing the bowel in this fashion ensures that the mass of intestine can easily be retracted, as well as conserving humidity and warmth.

1. Dissection of the Donor Liver

The liver is retracted downwards by the surgeon and the falciform ligament divided. The left triangular ligament is also divided. The gastrohepatic omentum is incised and opened for 2 to 3 cm while the assistant retracts the stomach. Insertion of the left forefinger from laterally to medially allows the contents of the porta hepatis to be easily identified and the peritoneum across the entire porta hepatis is incised. The left gastric artery and veins are ligated and divided. Further dissection in this area by blunt dissection with a swab is useful, but in animals from 27 to 32 kg there are many small vessels which require coagulation. The left hand can retract the posterolateral border of the portahepatis revealing the full length of the common hepatic artery to which are applied at least two lymph nodes which must be carefully removed without damage to the artery. It is possible to dissect the full required length of artery cephalad to the pancreas (3 to 4 cm) from this site.

Returning to the portahepatis, the common bile duct is identified and left intact with a good deal of surrounding connective tissue. The portal vein often has several lymph nodes applied to its medial border which should be carefully removed. The pancreaticoduodenal vein must be preserved[20] along with any anastomosis performed cephalad to it. The pancreaticoduodenal artery, which is one of the four to five terminal branches of the common hepatic artery, may be divided. The close application of the medial border of the hepatic artery and the lateral border of the portal vein must be separated. This is aided by the lateral retraction of the portal vein to reveal the proximal section of the common hepatic artery/celiac artery which is surrounded by the many branches of the celiac plexus. The artery should be dissected free to this point. (For dissection of an aortic segment, see below.)

The infrahepatic vena cava is initially approached from the left side. The retrocaval peritoneum is incised revealing the adrenal gland. This dissection should not include the separation of the gland from the vena cava because the adrenal veins are sessile and must be avoided. Dissection of part of the pancreas and the bowel overlying the inflow from the renal veins grants greater mobility to this area. The vena cava should then be retracted to the left and the peritoneal attachment of the liver to the diaphragm and posterior abdominal wall divided. This section of the inferior vena cava is intrahepatic. Cephalad dissection of the peritoneal reflection should be done carefully in order to avoid opening the vena cava where it is "tented" by a strong diaphragmatic reflection into the hepatic venous orifices. It is important, however, to dissect clear the upper vena cava as much as possible. Medial deflection of the liver allows this last part of the dissection to be completed easily.

During this initial phase of the dissection, which should take approximately 30 to 40 min, about 200 to 300 ml of intravenous fluids are infused and mean arterial pressure should be maintained above 70 mm/hg.

2. Preparation of the Recipient

Preparation of the liver for removal in the recipient is similar, but slightly less involved than in the donor. Dissection of the hepatic artery needs only to include enough length for the application of the proximal clamp, and subsequent reanastomosis in the recipient.

In order to prevent potentially lethal gastric mucosal hemorrhage once rejection develops, a highly selective vagotomy is also routinely performed. We use the technique of Terblanche and Hickman,[21] which dissects free the lesser curve of the stomach for some 2 cm but preserves the nerve supply to the antrum.

3. Insertion of the Bypass

The pig tolerates up to 10 min of total portal occlusion without sustaining irreversible hypotension.[22] Thus, a passive bypass is inserted between the splenic vein and the right external jugular vein. After administration of heparin (1 mg/kg) the splenic vein and splenic artery are ligated with a single tie and the vessels clamped proximally. A small nick is made in the vein which is then extended just enough to allow insertion of the bypass catheter. The bypass is composed of two cardiac catheters[22-24] previously primed with saline via a T-connector. Care should be taken not to advance the splenic vein catheter more than 2 to 3 cm or it may lead to obstruction of the portal vein. The second bypass catheter is inserted into the right external jugular vein. This vein may be distended by proximal clamping and manual compression of the neck to increase venous return. The catheter should not be advanced further than 2 to 3 cm. Insertion of the catheters is facilitated by lubrication of the catheter tips with liquid paraffin. The bypass system should be assessed at this time for spontaneous flow without initiating portal venous occlusion, since adequacy of the bypass is critical to the final success of the operation. The bypass is then left open.

4. Harvesting of the Liver

The donor and recipient operations should be scheduled so that there is minimal delay between initiation of bypass in the recipient and harvesting of the donor liver. If the liver is to be perfused with a cooled (4°C) crystalloid solution, a cannula[22-24] should be inserted into the portal vein after proximal clamping. The cannula is loosely secured with a ligature and the perfusion started. Adequate wash-out (1 l) of the portal venous bed appears to be improved by intermittent manual occlusion of the suprahepatic vena cava, with subsequent increased flow velocities. The hepatic artery is not occluded during flushing[23] following the suggestion that continued arterial perfusion during cooling protects mitochondrial function. The arterial blood becomes progressively cooler and hemodiluted with this method. Within 5 to 7 min, 1 l of fluid can be flushed through the liver. The perfused liver is removed by transection of the portal vein, infrahepatic vena cava, and the hepatic artery obliquely. The suprahepatic inferior vena cava is isolated by incising the diaphragm surrounding it; a 3- to 4-cm length is harvested for later trimming.

5. Implantation of the Donor Liver

The recipient liver is removed after clamping the portal vein cephalad to the pancreaticoduodenal vein, the hepatic artery and the infrahepatic vena cava immediately cephalad to the adrenal veins. A toothed vascular clamp is applied across the suprahepatic vena cava and the diaphragm during downward retraction of the liver in order to preserve as much length as possible. Use of the toothed vascular clamp lessens the potential risk of slipping off the smooth thick diaphragm, a problem with the use of a conventional vascular clamp. The common bile duct, hepatic artery, portal vein, infrahepatic vena cava, and suprahepatic vena cava are divided and the liver removed. The shaft of the clamp on the suprahepatic vena cava is secured tightly to the left border of the skin incision with a nylon stitch to reduce movement. Following removal of the liver, a swab is placed in the retrohepatic fossa and the donor liver placed upon it. The suprahepatic vena cava of the donor liver is trimmed to leave a 0.6- to 0.8-cm cuff. The anastomosis of the suprahepatic vena cava is made with 4-0 Ethibond and is commenced by placing two corner sutures. The posterior layer of the anastomosis is performed from within the lumen. After completion of the anterior layer of the anastomosis, the swab behind the liver is removed. The recipient and donor portal veins are aligned and trimmed if necessary. The anastomosis of the portal vein is made with 5-0 Ethibond and is performed by the same technique as for the suprahepatic vena cava. Precaution must be taken not to constrict or twist this anastomosis. The portal clamp is released and placed on the donor infrahepatic vena cava; the liver is filled with blood until it distends, but care should be taken to maintain a mean arterial pressure of greater than 50 mmHg using the infusion of fluids as necessary. Approximately 100 ml of blood is then run to waste from the infrahepatic vena cava and the circulation restored by removal of the suprahepatic vena cava clamp. The bypass may be clamped at this point. The infrahepatic vena caval anastomosis is completed with 5-0 Ethibond using the same technique as for the other vessels; the last two loops of this suture should be left loose and tightened as the proximal clamp is briefly released to free the air from this segment. The anastomosis is then definitively opened.

During the anhepatic phase, which lasts approximately 15 to 20 min, the blood pressure should not be allowed to fall below 60 mmHg. This often occurs with clamping of the infrahepatic vena cava to which pigs appear to be particularly susceptible. In addition, the patency of the bypass should also be repeatedly confirmed by rapid infusion of a saline bolus if the hypotension persists or the bowel appears dusky. Portal venous obstruction is easily assessed in the pig as hypotension supervenes swiftly and the bowel becomes cyanotic. This must be corrected immediately. Failure to do so leads to irreversible hypotension that fails to correct even when portal circulation is subsequently restored.[22]

After reconstitution of the portal and vena caval circulations, a 2- to 5-min break should be taken while the blood pressure and general status of the pig are assessed. The hepatic arterial anastomosis is then completed with a single continuous 7-0 suture. The knot is tied 1 to 2 mm from the edge of the vessel thereby leaving a "growth factor" which is taken up as the vessel distends. It is very important to ensure that the length of the hepatic artery is correct, neither too long resulting in kinking nor too short resulting in stretching. Each situation may lead to complete or partial occlusion of the vessel.

The bypass is removed from the splenic vein which is repaired and unclamped. The bile duct is repaired with a continuous 6-0 polypropylene or polyglycolic acid suture; this is not difficult to perform as the bile duct in the pig is extremely well vascularized[25] and as a result postoperative stricture is extremely uncommon.

6. The Aortic Segment

Where an aortic segment is preferred in the donor graft, the dissection of the hepatic artery must be extended to the celiac artery with ligation of the splenic and gastric branches. This dissection is difficult and is known colloquially as "lion country". It is easier to approach the celiac artery first by identification of its origin from the aorta; thereafter, the vessel should be followed distally with sequential ligation of its branches. Posterior to the pancreas, the dissection becomes especially difficult, but once completed, it is possible to see the cephalad length of the artery already prepared. Returning to the aorta it is helpful to dissect the left lateral border of the vessel free of peritoneum. Slight medial traction will reveal the four to five lumbar branches which need to be divided at the time of final removal. At the time the liver is removed, the diaphragm around the aorta is divided so that a length of 10 to 12 cm of thoracic aorta can be harvested. At the distal end of the aortic segment, it is necessary only to proceed 1 to 2 cm before one encounters another lumbar branch.

7. Postoperative Care

The intraoperative biochemical changes have been described previously.[15] In general, these changes are not marked. Within 1 hour postoperatively, the animal is awake and has stabilized with normal electrolytes, blood gas, and temperature values. The animal is nursed on its side for another one to two hours and is then removed to a warmed cage. The orogastric tube is removed and the ties around the endotracheal tube are loosened so that the animal can eject it spontaneously. The intravenous fluids are continued overnight at an approximate rate of 60 ml/hour. If acid base analysis at 6 hours reveals a metabolic alkalosis, the prognosis for survival is good. This alkalosis should persist for at least 5 days.[15] Intravenous fluids containing 1 gm/1 Penicillin are continued for 2 to 3 days at a rate of 2 l/day. Food is offered the first day and the animal is usually eating normally within 3 days.

D. AUTOGRAFT OF THE LIVER

The technique of transplantation of a liver autograft in the pig is identical to that described above except that the liver is removed from, and replaced into the same animal. It is essentially a model used to assess the effects of its surgery alone on the animal. If the liver is to be perfused, it must be removed prior to perfusion as the animal cannot tolerate the rapid infusion of an additional liter of fluid. Particular care must be taken with division of the vessels, especially the suprahepatic vena cava, which is short. Preservation of the pancreaticoduodenal vein is also important.

E. AUXILIARY LIVER GRAFT

The donor liver is prepared with an aortic segment as described above.[26] In the recipient, the right kidney is removed and particular care is paid to preparing a long length of infrahepatic vena cava. Also, in the recipient, a segment of abdominal aorta is prepared 3 to 4 cm caudal to the renal arteries. At this site there are several large lymphatic channels which leak considerably if damaged, but this is almost unavoidable.

Upon implantation, the liver is placed into the right renal fossa and the donor suprahepatic vena cava anastomosed end-to-side to the recipient infrahepatic vena cava. The portal vein is disconnected from the host liver, swung laterally, and anastomosed end-to-end to the donor portal vein. The donor infrahepatic vena cava is used to vent waste blood upon reperfusion and is then ligated or oversewn. The donor aortic segment is anastomosed end-to-side to the infrarenal recipient aorta using 4-0 Ethibond; the recipient aorta is partially occluded with a Satinsky vascular clamp and a 1-cm incision made. The anastomosis can be performed by suturing one side from the external surface and then, by deflecting the clamp laterally, reveal the opposite side for completion of the anastomosis from outside the lumen. The celiac artery is occluded with a bulldog clamp and great care is taken to ensure that all air is expelled from the segment prior to the release of the clamps.

The biliary anastomosis is performed between the fundi of the gallbladders using 3-0 chromic catgut.[27]

F. REDUCED LIVER GRAFT

The anatomy of the blood supply to the liver of the pig has been described in detail.[28,29] The common hepatic artery divides variably into three to four branches which supply the right lobe, the right part of the middle lobe, the left part of the middle lobe, and the left lobe sequentially as the vessel traverses a posterioinferior concave arc. This precludes use of the right part of the liver for reduced grafting, but use of the left part is possible. On the bench, the capsule is incised 1 cm lateral to the diaphragmatic reflection off the suprahepatic inferior vena cava and the incision continues as a curve to the medial border of the gallbladder. The parenchyma is easily separated off the major branches using a finger fracture technique and these may be identified, ligated, and divided. The remainder of the vessels are much smaller and ligation may not be possible. The cut surface of the liver is oversewn first with a layer of 2-0 chromic catgut 1 cm parallel to the cut edge and then with 2-0 chromic catgut over and over the cut surface behind the first layer. Implantation of the liver is performed as described for the allograft.

G. STORAGE

The porcine liver may be cold-stored adequately for 4 hours in either lactated Ringer's solution, Eurocollins, or University of Wisconsin (UW) solution. Further extension to 6 or more hours has proven less successful, and it has not been possible in this laboratory to use UW solution for the extended periods tolerated by human[30] or dog[31] livers. An unexpected finding has been that the simple storage without any flushing at all is tolerated for 6 hours. In this technique, the liver is removed, placed in one plastic bag within another containing lactated Ringer's solution and kept submerged in crushed ice. Recipients of these livers recover swiftly and may be standing and foraging within 4 hours of the operation.[15]

III. APPLICATIONS OF EXPERIMENTAL LIVER TRANSPLANTATION

The technique of liver transplantation in the pig can be applied to a number of experimental protocols allowing for the development of a better understanding of the transplantation process. Application to (1) autografts, (2) transhepatic sampling, (3) the anhepatic model, (4) the brain-dead donor, (5) the evaluation of immunosuppressive therapies, and (6) effects of portal diversion will be discussed in the following sections.

A. AUTOGRAFT

This technique has several advantages: (1) it fulfills all the requirements of a true control, and thus, may be used to define many effects of the procedure where rejection is not relevant, (2) it is economical, and (3) it provides an excellent method of teaching the technique of liver transplantation. This is a technically more difficult surgery because there is no "space" length to any of the vessels as in the allografts.

B. TRANSHEPATIC SAMPLING

Methods of transhepatic sampling have been described previously.[32,33] These techniques are simple, easily performed, and allow for direct access to blood samples that indicate hepatic handling of substrates. Briefly, the procedure entails the placement under general anesthesia of small flexible catheters into the portal vein via the splenic vein and the hepatic vein via the jugular vein. The positioning of the catheter tips allows for the sampling of blood both before and after it perfuses the liver.

C. ANHEPATIC MODEL

The technique of rendering the pig permanently anhepatic with minimal disturbance has recently been described.[34,35] The animals in this study survived up to 32 hours and were awake and biochemically stable for at least 26 hours, provided that the glucose infusion was restricted. This provides a valuable model for assessment of the anhepatic period.

D. BRAIN-DEAD DONOR

Several important questions exist regarding the physiological effects of brain death on the functioning of transplanted organs. A simple model of experimentally induced brain death has been described by Pienaar et al.[36] They found that brain death could be consistently induced in normal pigs with the placement, and the inflation of an 18 French foley catheter subdurally. Such animals could be maintained for approximately 16 hours with rigorous monitoring and prompt intervention. The use of this model of donor will more closely resemble the model used in human liver transplantation and thus may provide significant information.

E. THE EFFECTS OF PORTAL DIVERSION

During the anhepatic phase of transplantation there is portal diversion in effect. The immediate effects of this are not known. Preliminary studies of glucose homeostasis during the first 2 hours after portal diversion indicate remarkably rapid changes in glucoregulatory hormones; however, further investigations are necessary before generalizations may be made.[37]

F. IMMUNOSUPPRESSIVE THERAPIES

Because the porcine liver displays such similarity to the human's in terms of anatomy and function, liver transplantation in this model can prove to be a significant model for the evaluation of new and standard immunosuppressive therapies. From the standard, Cyclosporin A, to newer agents yet to be established, this model can serve to help delineate immunosuppressive activity as well as any physiologic change induced by the agent's administration.

IV. CONCLUSIONS

Liver transplantation in swine provides a great opportunity to study the techniques and immunologic phenomena involved in transplantation. Pigs can be used to study orthotopic, auxiliary, and partial liver grafts. When normal donors and recipients are used the effects of operative intervention alone can be assessed. The operation also serves to provide valuable technical experience in vascular and transplant surgery. Overall, the swine is an important model in the development of clinical liver transplants in humans because of its similar physiologic and metabolic characteristics.

REFERENCES

1. **Welch, C.S.,** A note on transplantation of the whole liver in dogs, *Transplant. Bull.,* 2, 54, 1955.
2. **Moore, F.D., Wheeler, H.B., Demissianos, H.V., Smith, L.L., Balankura, O., Abel, K., Greenberg, J.B., and Dammin, G.J.,** Experimental whole organ transplantation of the liver and of the spleen, *Ann. Surg.,* 152, 374, 1960.
3. **Starzl, T.E., Kaupp, H.A., Brock, D.R., Lazarus, R.E., and Johnson, R.V.,** Reconstructive problems in canine liver homotransplantation with special reference to the post-operative role of hepatic venous flow, *Surg. Gynecol. Obstet.,* 111, 733, 1960.
4. **Starzl, T.E., Marchioro, T.L., van Kaulla, K.N., Hermann, G., Brittain, R.S., and Waddell, W.R.,** Homotransplantation of the liver in humans, *Surg. Gynecol. Obstet.,* 117, 659, 1963.
5. **Eiseman, B. and Spencer, F.C.,** Man's best friend?, *Ann. Surg.,* 159, 159, 1964.
6. **Garnier, H., Clot, J.P., Bertrand, M., Camplez, P., Kunlin, A., Gorin, J.P., Le'Goaziou, F., Levy, R., and Cordier, G.,** Biologie experimentale: greffe de foie chez le porc: approche chirurgicale, *C.R. Acad. Sci. Paris,* 260, 5621, 1965.
7. **Peacock, J.H. and Terblanche, J.,** Orthotopic homotransplantation of the liver in the pig, in *The Liver,* Read, A.E., Ed., Butterworths, London, 1967, 333.
8. **Garnier, H., Clot, J.P., and Chomette, S.,** Orthotopic transplantation of the porcine liver, *Surg. Gynecol. Obstet.,* 130, 105, 1970.

9. **Parker, J.R., Hickman, R., and Terblanche, J.,** Mixed lymphocyte culture studies in pigs which survive transplantation without immunosuppression, *Transplantation,* 19, 276, 1976.
10. **Houssin, D., Charpentier, B., Gugenheim, J., Baudot, P., Tamisier, D., Lang, P., Gigou, M., and Bismuth, H.,** Spontaneous long-term acceptance of RT-1-incompatible liver allografts in inbred rats, *Transplantation,* 36, 615, 1983.
11. **Calne, R.Y., White, H.J.O., and Yoffa, D.E.,** Observations of orthotopic liver transplantation in the pig, *Br. Med. J.,* 2, 478, 1967.
12. **Hickman, R., Saunders, S.J., King, J.B., Harrison, G.G., and Terblanche, J.,** Pig liver perfusion in the treatment of fulminant hepatic necrosis, *Scand. J. Gastroenterol.,* 6, 563, 1971.
13. **Reuvers, C.B., Terpstra, O.T., and Balis, A.L.,** Auxiliary transplantation of part of the liver improves survival and provides metabolic support in pigs with acute liver failure, *Surgery,* 98, 914, 1985.
14. **Calne, R.Y., Ed.,** *Liver Transplantation: The Cambridge Kings College Hospital Experience,* Calne, R.Y., Ed., Grune & Stratton, New York, 1987.
15. **Pienaar, H., Stapleton, G.N., Bracher, M., Lotz, Z., Rose-Innes, C., Fourie, J., and Hickman, R.,** Six hour porcine liver storage without flushing or perfusion, *Transplantation,* 52, 38, 1991.
16. **Terblanche, J., Peacock, J.H., Hobbs, K.E.F., Hunt, A.C., Bowes, T., Tierris, E.J., Palmer, D.B., and Blecher, T.E.,** Orthotopic liver homotransplantation: experimental study in the unmodified pig, *S. Afr. Med. J.,* 42, 486, 1967.
17. **Dent, D.M., Hickman, R., Uys, C.J., Saunders, S.J., and Terblanche, J.,** Natural history of liver allo and auto transplantation in the pig, *Br. J. Surg.,* 58, 407, 1971.
18. **Hickman, R., van Hoorn, W.A., and Terblanche, J.,** Exchange transplantation of the liver in the pig, *Transplantation,* 24, 237, 1971.
19. **Harrison, G.G., Biebuyck, J.F., Terblanche, J., Dent, D.M., Hickman, R., and Saunder, S.J.,** Hyperpyrexia during anaesthesia, *Br. Med. J.,* 3, 594, 1968
20. **Cuschieri, A., Baker, P.R., and Holley, M.P.,** Porta-caval shunt in the pig, *J. Surg. Res.,* 17, 387, 1974
21. **Terblanche, J. and Hickman, R.,** The prevention of gastric ulceration by highly selective vagotomy in a new peptic ulcer experimental model, the bile duct ligated pig, *Surgery,* 84, 206, 1978.
22. **Battersby, C., Hickman, R., Saunders, S.J., and Terblanche, J.,** Liver function in the pig, the effects of 30 minutes normothermic ischaemia, *Br. J. Surg.,* 61, 27, 1974.
23. **Otto, G., Wolff, H., Nerlings, I., and Gellert, K.,** Preservation damage in liver transplantation, *Transplantation,* 42, 122, 1986.
24. **Starzl, T.E., Iwatsuki, S., and Shaw, B.W.,** A growth factor for fine vascular anastomoses, *Surg. Gynecol. Obstet.,* 159, 164, 1984.
25. **Northover, J. and Terblanche, J.,** Bile duct blood supply: its importance in human liver transplantation, *Transplantation,* 26, 67, 1978.
26. **Kahn, D., Hickman, R., McLeod, H., Terblanche, J., et. al.,** The stimulatory effect of a partially hepatectomized auxiliary graft upon the host liver, *S. Afr. Med. J.,* 61, 362, 1982.
27. **Crosier, J.H., Immelman, E.J., Hickman, R., Uys, C.J., van den Ende, J., Terblanche, J., and van Schalkwyk, D.J.,** Cholecystojejunocholecystostomy: a new method of biliary drainage in auxiliary liver allotransplantation, *Surgery,* 87, 514, 1980.
28. **Roncone, A., Pienaar, H., Mahlati, G., Kahn, D., and Hickman, R.,** Ex vivo versus in situ resection of segmental liver autografts in pigs, *Transplantation,* 53, 1020, 1992.

29. **Camprodon, R., Solsona, J., Guerrero, J.A., Mendoza, C.G., Serpura, J., and Fabregat, J.,** Intrahepatic vascular division in the pig — basis for partial hepatectomies, *Arch. Surg.,* 112, 38, 1977.
30. **Kalayoglu, M., Sollinger, W.H., and Stratta, R.J.,** Extended preservation of the liver for clinical transplantation, *Lancet,* 261, 711, 1988.
31. **Moen, J., Claesson, K., Pienaar, H., Lindell, S., Ploeg, R.J., McAnutty, J.F., Vreugdenhil, P., Southard, J.H., and Belzer, F.O.,** Preservation of the dog liver, kidney and pancreas using the Belzer UW solution with a high Na^+ and low K^+ content, *Transplantation,* 47, 940, 1989.
32. **Hickman, R., Vinik, A.I., and van Hoorn, W.A.,** Transhepatic hormone levels in the portacaval-shunted pig — the effects of arginine upon gastrin and glucagon release, *Am. J. Clin. Nutr.,* 32, 2009, 1979.
33. **Hickman, R., Tyler, M., McLeod, H., and Fourie, J.,** Transhepatic sampling during experimental porcine liver autotransplantation — its application to measurements of insulin, glucagon and glucose, *J. Surg. Res.,* 49, 1, 1990.
34. **Hickman, R., Dent, D.M., and Terblanche, J.,** The anhepatic model in a pig, *S. Afr. Med. J.,* 48, 263, 1974.
35. **Hickman, R., Bracher, M., Tyler, M., Lotz, Z., and Fourie, J.,** The effect of total hepatectomy upon coagulation and glucose homeostasis in the pig, *Dig. Dis. Sci.,* 37, 328, 1991.
36. **Pienaar, H., Schwartz, I., Roncone, L.Z., and Hickman, R.,** Function of kidney grafts from brain dead donor pigs — the influence of dopamine and tri iodo-thyronine, *Transplantation,* 50, 580, 1990.
37. Hickman, R., Tyler, M., Rose-Innes, C., Lotz, Z., and Fourie, J., How rapidly do hyperinsulinaemia and hyperglucagonaemia develop after porta-caval shunting? *J. Surg. Res.,* 53, 20, 1992.

Chapter 8

Liver Transplantation in Primates

Oscar Imventarza, Horacio L. Rodriguez Rilo, Alejandra Oks, John J. Fung, Thomas E. Starzl

CONTENTS

I. INTRODUCTION

The first description of experimental liver transplantation was by J. Cannon in 1956, although few details about the operation were supplied.[1] Concurrent developments of orthotopic liver transplantation were begun by Moore et al.[2] and Starzl et al.[3] By 1960, successful dog liver transplantation was achieved. This model allowed for technical and immunosuppressive developments which were essential to the eventual application in humans.[4] Nevertheless, a number of differences have been identified between canine and primate hepatic physiology, such as the presence of hepatic vein musculature in the dog which may act as a "throttle mechanism",[5] as well as the sensitivity of the canine liver to histamine-mediated vasoconstriction.

Myburgh and co-workers first described the use of non-human primates for experimental liver transplantation,[6] and independently by Fortner.[7] Non-human primates offer a number of advantages in the study of liver transplantation. Unlike the canine liver, the liver anatomy of non-human primates is similar to humans. Amongst the higher order primates, similarities exist between the major histocompatibility complex, as well as the cellular markers found in the immune system.[8] The blood groups are similar to the A and B blood types, although O blood types have only been reported in man. In addition, a number of human pathogenic virus, such as Hepatitis B, can be studied in certain of these primates.

Primates are comprised of two suborders, Prosimii and Anthropoidea.[9] Prosimian primates resemble squirrels or rats more than true monkeys. The Anthropoidea suborder can be further subdivided into five different families: new world monkeys, old world monkeys, lesser apes, great apes, and man. From an investigational standpoint, the most frequently used species are the old world monkeys. This family includes Rhesus monkeys *(Macaca mulata),* Cynomologus monkeys *(Macaca fascicularis)* and baboons *(Papio cynocephalus).* Great apes include the chimpanzee *(Pan troglodytes)* and gorillas *(Gorilla gorilla),* but are not used in great numbers because of their endangered status.

The purpose of this chapter is to describe the techniques of liver procurement and transplantation in non-human primates, for purposes of experimental and perhaps eventual clinical applications.

0-8493-3629-5/94/$0.00+$.50

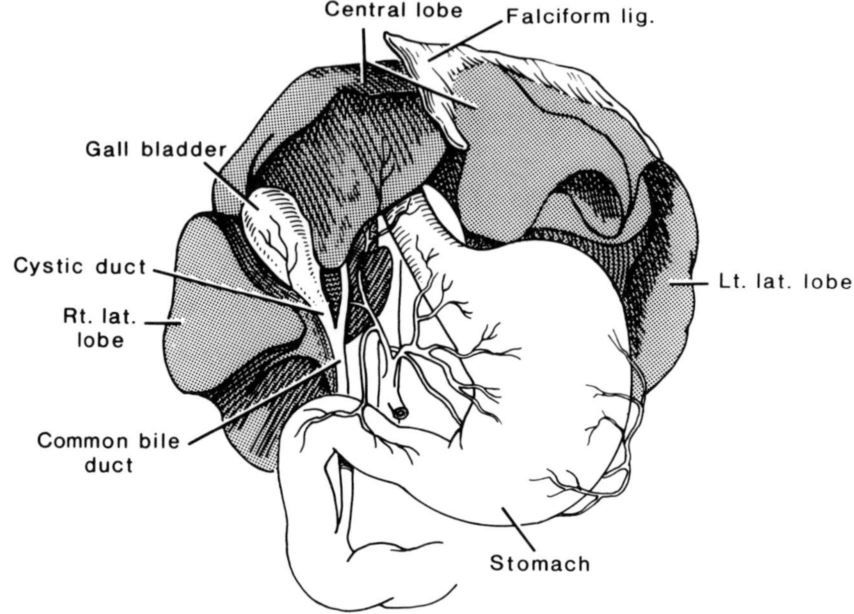

Figure 1 Non-human primate liver anatomy.

II. TECHNIQUE

A. ANATOMY

Liver anatomy among primates is similar, with a right and left lateral lobe placed dorsally and a single large ventral central lobe.[10,11] The liver of the *Macaca* and *Papio* is notable for lobation, with four identifiable lobes. In the higher order primates, the central lobe fuses with the right and left lateral lobes. The quadrate lobe is much more narrow in non-human primates than in man, and the caudate lobe may encircle the circumference of the inferior vena cava. The ligamentous attachments are similar to those described in man (Figure 1).

The blood supply to the liver is similar in the Anthropoidea, with a great variation in the arterial blood supply, as noted by several authors.[7-13] The portal venous system is essentially identical with that of the higher order primates. The hepatic venous drainage is similar to man, with small short hepatic venous tributaries draining the right and central lobes, and two large hepatic venous branches, one right and one left.

In all primates, the gallbladder lies closely attached to the right or central lobe. The arterial supply is usually from a branch from the right hepatic artery. The cystic duct joins the common hepatic duct a variable distance to form the common bile duct, before emptying into the duodenum.

B. GENERAL CONSIDERATION

Generally, the donor should be smaller in size than the recipient. Blood type compatibility should be established and the animals should be fasted the evening prior to the operation.

Animals are tranquilized by the use of intramuscular injection of ketamine (10 mg/kg). Using intravenous induction with 25 to 35 mg/kg sodium thiopental, a cuffed endotracheal tube is inserted and attached to a respirator. Ventilation is with a mixture of oxygen and nitrous oxide, with inhalation anesthesia using isofluorane at a concentration between 0.5 and 2.0%.

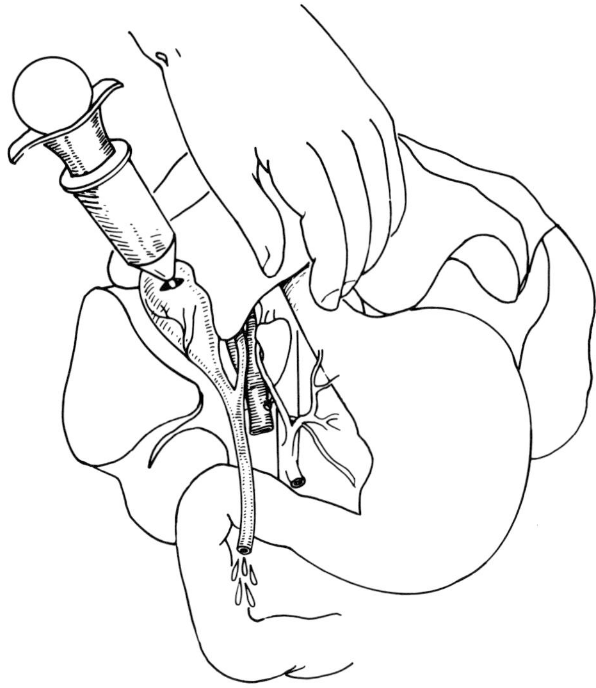

Figure 2 Procurement technique. Common bile duct is being flushed through gallbladder incision.

Arterial pressure monitoring is accomplished by canalization of the radial artery and central venous monitoring utilizes a catheter in the jugular vein. Low-dose dopamine or phenylephrine can be used to maintain adequate systolic blood pressure. Measurement of blood gas and electrolytes should be done frequently, and correction of abnormalities of ionized calcium and in acid-base balance should utilize calcium chloride and sodium bicarbonate.

Blood can be taken from the donor during the organ procurement, prior to heparinization, for immunologic studies or for transfusion into the recipient to maintain adequate oxygen carrying capacity.

C. DONOR PROCUREMENT

Little has changed in the procurement technique since described in detail by Starzl for human organ procurement[14] and applied to baboons by Myburgh and co-workers.[15]

Allowing adequate time for hydration and stabilization of donor blood pressure and oxygenation, the abdomen and chest are thoroughly prepped and draped. A long midline incision from the sternal notch to the pubis is made, and adequate exposure maintained. Care must be taken in handling the liver, since it is very soft and more friable than that of man. The ligamentous attachments of the liver are incised, first cutting the ligamentum teres, followed by the falciform and left triangular ligaments. The left phrenic vein can now be visualized and ligated. The hepatogastric ligament is examined for the presence of an accessory left hepatic artery from the left gastric artery. The hilar structures are identified, starting in the right side of the porta hepatis. The common bile duct is ligated as inferior as possible. The gallbladder is incised and flushed through the cut common bile duct (Figure 2).

The hepatic artery is identified and traced back, ligating the right gastric and gastroduodenal artery. An aberrant right hepatic artery, if present, can be felt as a pulsatile structure posterior to the portal vein. Care must be taken during the arterial dissection, since the vessels are fragile, and overmanipulation may result in intimal dissection and subsequent thrombosis of the artery. The celiac axis is mobilized and the celiac ganglion is divided. The portal vein can be seen at this time. Careful mobilization to the junction of the splenic and superior mesenteric vein allows the portal vein to be cannulated, either through the superior mesenteric vein or the splenic vein. The aorta is mobilized at the aortic bifurcation and cannulated.

The connective tissue attached to the retrohepatic vena cava is bluntly dissected by lifting the caudate lobe. The liver is retracted to the left and the right triangular ligament is divided and the vena cava is freed circumferentially. The pleural spaces are opened and the intrathoracic aorta is mobilized and clamped. The preservation solution is allowed to perfuse the aortic and portal vein cannulas. The intrathoracic vena cava is incised to allow venting of the flush solution. The liver is allowed to cool *in situ* with 750 ml aortic flush and 500 ml portal vein flush. The liver is then removed and taken to the back table in an ice-filled basin.

Stitches are applied to the corners of both the suprahepatic and infrahepatic vena cava. A tongue of hepatic tissue is carefully dissected from the lower vena cava to obtain an adequate length of lower vena cava. The right adrenal vein is ligated. The diaphragm surrounding the suprahepatic vena cava is trimmed. A Carrell patch around the celiac axis is then prepared and the liver is ready for implantation.

D. RECIPIENT OPERATION

The initial steps in the dissection of the recipient liver are identical to those described in the donor operation, except for the incision. The porta hepatis is approached by ligating the common hepatic duct as high as possible. The hepatic artery is ligated at the bifurcation into the right and left hepatic arteries and mobilized to the level of the gastroduodenal artery. The portal vein is skeletonized from the connective tissue. The suprahepatic and infrahepatic vena cava are mobilized circumferentially and Potts clamps are applied. A vascular clamp is applied to the portal vein and the liver is removed. While some investigators have used the veno-venous bypass described by Shaw and co-workers,[16] other have not found this necessary, especially if the venous anastomosis can be completed in a short time frame (30 min). The vascular anastomoses are performed in an end-to-end manner using continuous non-absorbable monofilament polypropylene sutures. The sequence of anastomosis is generally the upper vena cava followed by the lower vena caval anastomosis. The portal vein is usually connected prior to the hepatic artery. The technique used for all vascular anastomoses utilizes corner stitches and a continuous everting over and over suture, completing the posterior wall prior to starting the anterior wall (Figure 3).

In smaller caliber vessels, such as the portal vein, but also with small vena cavas, a "growth factor" is utilized to prevent stenosis following expansion of the circumference with reperfusion (Figure 4). The growth factor is usually the length of one diameter of the portal vein, while it is much shorter for the vena cava. Once the venous anastomosis are completed, the clamps are released to allow for timely decompression of the mesentery.

The hepatic artery anastomosis is performed between the proper hepatic artery with a Carrell patch from the donor, and the common hepatic artery at the level of the takeoff of the gastroduodenal artery of the recipient (Figure 5).

There are a number of techniques available for the biliary reconstruction. The preferred method for reconstruction is with an end-to-end choledochocholedochostomy using a silastic stent (Figure 6), although the side-to-side choledochocholedochostomy (Figure 7) described by Neuhaus can be utilized. Choledochojejunostomy, choledochoduodenostomy, or the Weidell-Caine gallbladder conduit (Figure 8) can also be utilized.

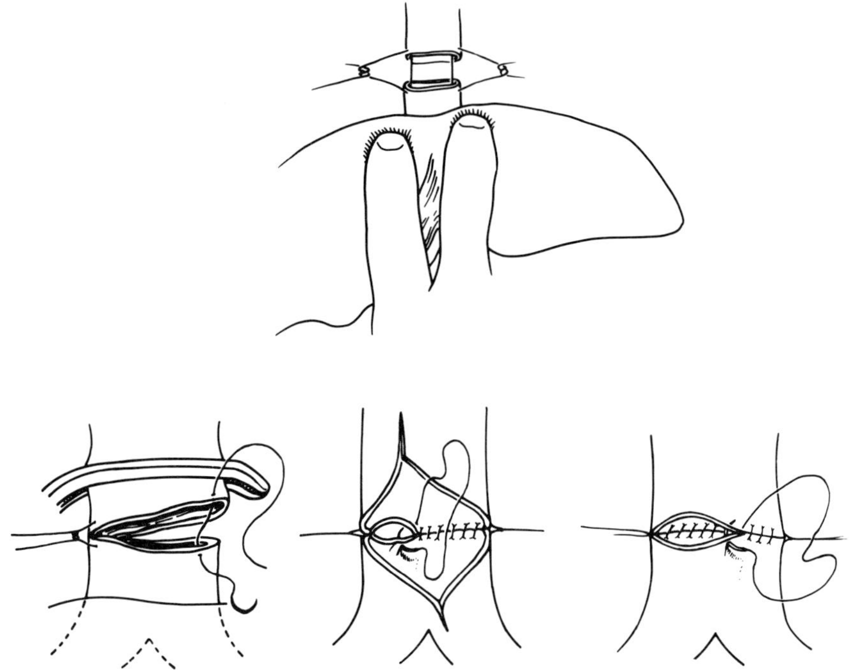

Figure 3 Vascular anastomosis: (A) both ends are approximated with corner stitches; (B) corner stitches have been placed; (C) continuous everting suture of posterior wall; and (D) continuous everting suture of anterior wall.

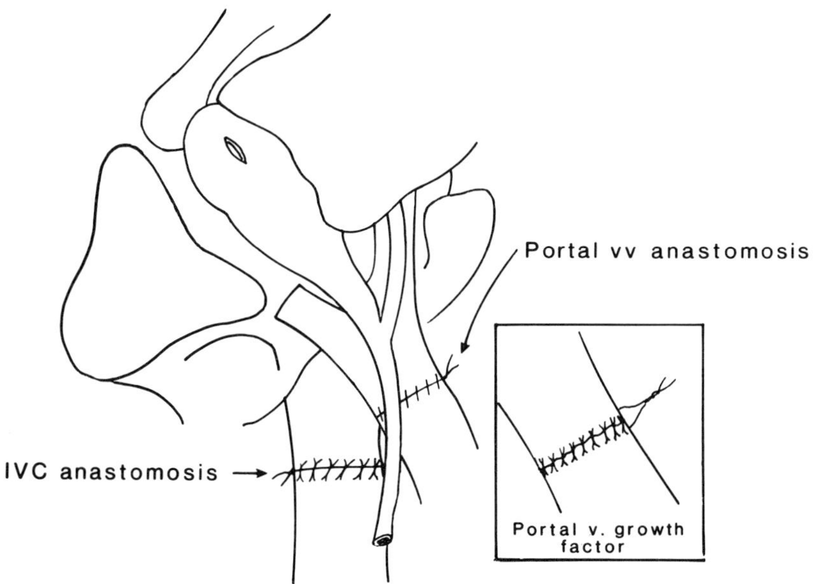

Figure 4 Growth factor before unclamping (picture), and final result after expansion of the circumference.

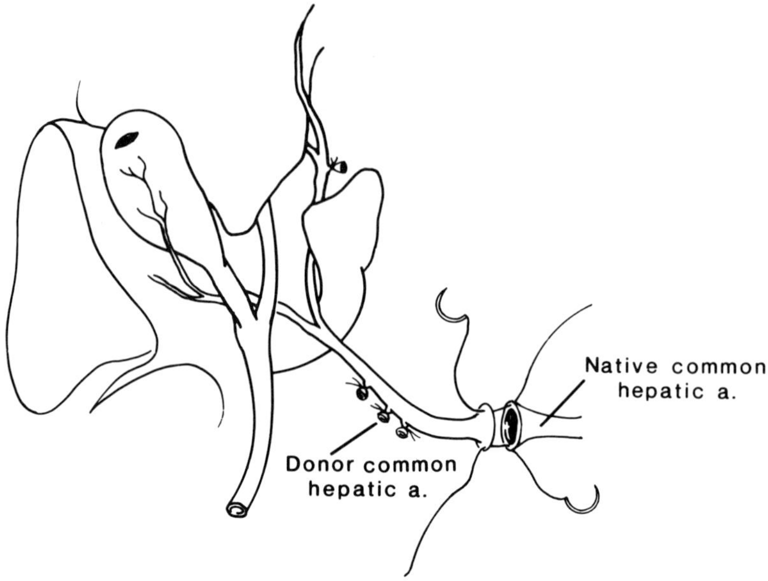

Figure 5 Hepatic artery anastomosis (see text).

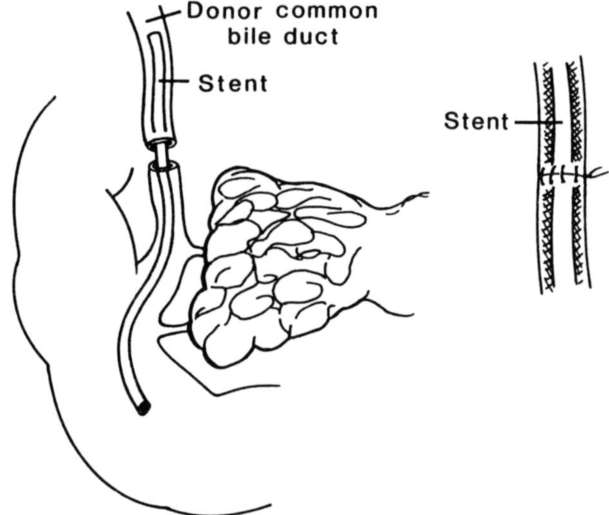

Figure 6 End-to-end choledochocholedochostomy using a silastic stent as a factor.

Incisions in the skin must be closed by an intradermal suture, otherwise metal clips are apt to be scratched off.

E. POSTOPERATIVE CONSIDERATIONS

In a number of non-human primates, including the rhesus monkey and baboon, the males have large canine teeth. These can be vicious weapons, and it is recommended to either extract them or to shorten them prior to, or at the time of surgery.

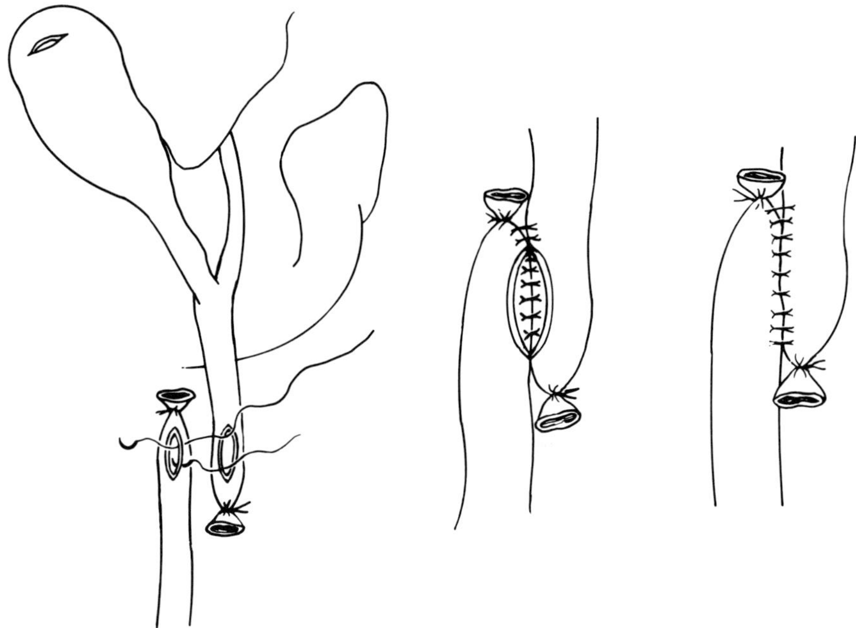

Figure 7 Side-to-side choledochocholedochostomy.

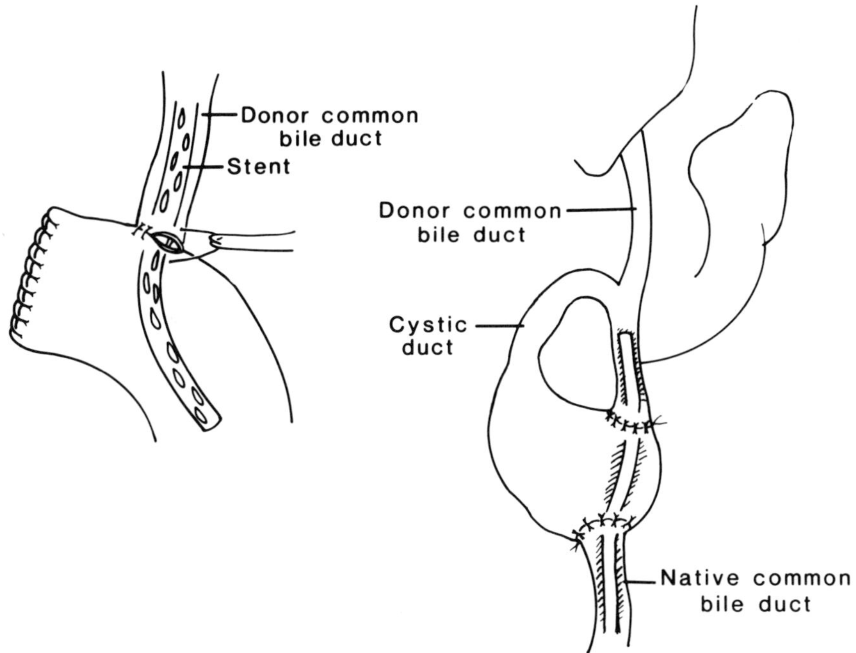

Figure 8 (A) Choledochojejunostomy. (B) Biliary reconstruction using the gallbladder as a conduit (Weidell-Calne).

Primates are quite strong. For animals weighing less than 15 lb, manual restraint is often satisfactory, while larger animals should be sedated with ketamine. Administration of sedatives is greatly facilitated by keeping the animals in a "squeeze-back" type cage.

Shortly after transplantation, animals should be monitored in a recovery setting, with warming lamps. Animals may be extubated shortly after transplantation. Intravenous lines cannot be used once the animals are awake, and any intravenous medications must be completed, or the animal will require further sedation. Antibiotics are given to the animals at the initiation of surgery and once again at the completion of the transplant.

Animals are allowed to drink ad libitum for the first day, and a solid diet is given by the third day following transplantation. Immunosuppressive drugs can be given either as an intramuscular injection or by mixing it with food.

Tests, such as phlebotomies and liver biopsies, require sedation. These studies should be done with a brief fasting period, to avoid aspiration during sedation.

III. CONCLUSIONS

Non-human primates are an important resource in biomedical research. Because of their similarity to humans anatomically and immunologically, refinements in the understanding of liver transplantation may well be derived from studying non-human primates. In addition, this resource may also prove to be an answer to the growing problem of organ shortage. Concordant xenotransplantation has been shown to be successful, both in animal models and in a few human cases.[19-22] While ethical issues surround the application of concordant xenotransplantation, responsible investigations using non-human primates must continue.

REFERENCES

1. **Cannon, J.A.,** Organs, *Transplant. Bull.,* 3, 7, 1956.
2. **Moore, F.D., Wheeler, H.B., Demissianos, H.V., Smith, L.L., Baldnkura, O., Abel, K., Greenberg, J.B., and Dammin, G.J.,** Experimental whole-organ transplantation of the liver and of the spleen, *Ann. Surg.,* 152, 374, 1960.
3. **Starzl, T.E., Kaupp, H.A., Jr., Brock, D.R., Lazarus, R.E., and Johnson, R.U.,** Reconstructive problems in canine liver homotransplantation with special reference to the postoperative role of hepatic venous flow, *Surg. Gynecol. Obstet.,* 111, 733, 1960.
4. **Starzl, T.E., Marchioro, T.L., Von Kaulla, K.N., Hermon, G., Brittain, R.S., and Waddell, W.R.,** Homotransplantation of the liver in humans, *Surg. Gynecol. Obstet.,* 117, 659 1963.
5. **Neill, S.A., Gaisford, W.D., and Zuidema, G.D.,** A comparative anatomic study of the veins in the dog, monkey and human, *Surg. Gynecol. Obstet.,* 117, 451, 1963.
6. **Myburgh, J.A., Smit, J.A., Mieny, C.J., and Mason, J.A.,** Hepatic allotransplantation in the baboon. The effects of immunosuppression and administration of donor-specific antigen after transplantation, *Transplantation,* 12, 202, 1971.
7. **Fortner, J.G. and Shiu, M.H.,** Primate liver transplantation, *Primates Med.,* 7, 64, 1972.
8. **Socha, W.W. and Moor-Jankowski, J.,** Primate animal model for xenotransplantation: serological criteria of donor-recipient selection, in *Xenograft 25,* Hardy, M.D., Ed., Elsevier, Amsterdam, 1989, chap. 10.
9. **Holmes, D.D.,** Nonhuman primates, in *Clinical Laboratory Animal Medicine,* Iowa State University Press, Ames, IA, 1984, 67.
10. **Lineback, P.,** The respiratory, digestive and urinary systems, in *Anatomy of the Rhesus Monkey (Macaca mulatta),* Bast, T.H., et al., Williams & Wilkins, Baltimore, MD, 1933, 210.

11. **Swindler, D.R. and Wood, C.D.,** *An Atlas of Primate Gross Anatomy, Baboon, Chimpanzee and Man,* R.E. Kreiger, Seattle, WA, 1982, 379.
12. **Bourne, G.H.,** *The Rhesus Monkey,* Academic Press, New York, 1975.
13. **Lineback, P.,** The vascular system, in *Anatomy of the Rhesus Monkey (Macaca mulatta),* Bast, T.H., et al., Williams & WIlkins, Baltimore, MD, 1933, 248.
14. **Starzl, T.E., Makola, T.R., Shaw, B.W., Hardesty, R.L., Rosendhol, R.L., Griffith, B.P., Iwodsuki, S., and Boulison, H.T.,** A flexible procedure for multiple organ procurement, *Surg. Gynecol. Obstet.,* 158, 223, 1984.
15. **Myburgh, J.A., Mieny, C.J., Vetten, B., Isaacs, F., Morake, J., Nemudzivhadi, A., and Neube, L.,** The technique of orthotopic hepatic allotransplantation in the baboon, *S. Afr. J. Surg.,* 9, 81, 1971.
16. **Shaw, B.W., Jr., Martin, D.J., Marquez, J.M., Kang, Y.G., Bugbee, A.C., Jr., Iwatsuki, S., Griffith, B.P., Hardesty, R.L., Bahnson, H.T., and Starzl, T.E.,** Venous bypass in clinical liver transplantation, *Ann. Surg.,* 200, 524, 1984.
17. **Neuhaus, P., Neuhaus, R., Piehimayr, R., and Vonnahme, F.,** Technique of biliary reconstruction after liver transplantation, *Res. Exp. Med.,* 180, 239, 1982.
18. **Mybrugh, J.A., Smit, J.A., Stark, J. H., and Browde, S.,** Total lymphoid irradiation in kidney and liver transplantation in the baboon: prolonged graft survival and alterations in T cell subsets with low cumlative dose regimens, *J. Immunol.,* 132, 1019, 1984.
19. **Leger, L., Chapuis, Y., Lenriot, J.P., Deloche, A., and Frenoy, P.,** Transplantation heterotopique d'un foie de babouin a l'homme, *Presse Med.,* 78, 429, 1970.
20. **Pouyet, M. and Berard, Ph.,** Deux cas de transplantation heterotopique vraie de foie de babouin, au cours d'hepatites aigues malignes, *Lyon Chir.,* 67, 288, 1971.
21. **Caine, R.Y., Davis, D.R., Pena, J.R., Bainer, H., Vries, M.D., Herbertson, B. M., Joysey, V.C., Millard, P.R., Seaman, M.J., Samuel, J.R., Stibbe, J., and Westbroedk, D.L.,** Hepatic allografts and xenografts in primates, *Lancet,* 1, 103 1970.
22. **Starzl, T.E., Marchioro, T.L., Peters, G.N., Kirkpatrick, C.H., Wilson, W.E.C., Porter, K.A., Rif kind, D., Ogden, D.A., Hitchcock, C.R., and Waddell, W.R.,** Renal heterotransplantation from baboon to man: experience with 6 cases, *Transplantation,* 2, 752, 1964.

Chapter 9

Transplantation of Hepatocytes*

Achilles A. Demetriou, Jacek Rozga, and Albert D. Moscioni

CONTENTS

I. INTRODUCTION

Whole organ transplantation remains the only clinically effective method of treating both acute liver failure and specific genetic defects of liver function which result in liver failure. However, there are several factors which limit the application of liver transplantation in the treatment of liver insufficiency. These include: limited donor availability, inability to obtain an organ at the time when it is urgently needed, and need for life-long immunosuppression and high cost. There is, thus, a need for developing alternate therapeutic methods. Hepatocyte transplantation has been evaluated over a long period of time, primarily experimentally, as a possible therapeutic modality for acute and chronic liver insufficiency and for the treatment of specific genetic liver defects. In the latter instance, correction of the defect can result in normal liver function. Potential advantages of the use of isolated hepatocyte transplantation include: (1) cell cryopreservation, (2) *in vitro* cell manipulation to either decrease immunogenicity or correct genetic defects using gene transfer technology, (3) cells from a single donor can be used to treat multiple recipients, and (4) simplicity of the technique, at a potentially low cost. However, experiments to date have yielded disappointing results due to lack of long-term cell function and need for immunosuppression.

* Supported by N.I.H. Grant DK 38763 and a VA Merit Review Award.

0-8493-3629-5/94/$0.00+$.50

II. TECHNIQUES

A. BACKGROUND

Transplantation of hepatocytes,[1-5] injection of various hepatocyte extracts[1] and hepatocyte culture supernatants,[1] have been shown to prolong survival in animals with D(+)galactosamine-induced liver injury and animals with acute liver ischemia.[6-8] Hepatocyte transplantation into the spleen,[9,10] fat pads,[11] dorsal fascia,[12] pancreas,[13] and portal vein[14] has been carried out resulting in partial correction of specific liver metabolic defects, i.e., unconjugated hyperbilirubinemia due to lack of UDP-glucuronyl transferase for bilirubin (Gunn rat) and/or demonstration of viable differentiated hepatocytes morphologically at the various ectopic sites. Several studies attempted to provide a matrix and solid supports for transplanted hepatocytes[15-17] to facilitate cell engraftment and enhance neovascularization.

The hepatocyte transplantation literature is very extensive but very often contradictory and incomplete. In general, there are several problems with most reported hepatocyte transplantation studies: (1) the use of experimental animal models that are not standardized or validated and thus difficult to reproduce, (2) the use of indirect assays of transplanted cell function, (3) the reliance on morphologic evidence of presence of cells in the animal at ectopic sites without demonstration of function, (4) the lack of long-term success; most reported studies demonstrate only a transient effect (several weeks) of transplanted hepatocytes even in congeneic and syngeneic strains of animals, and (5) several groups report potentially significant preliminary observations without further follow-up studies. For a hepatocyte transplantation method to be useful, the technique must be simple, it should allow early engraftment, it should allow transplantation of an adequate number of hepatocytes, and transplanted cells should be able to express differentiated functions *in vivo*.

With current advances in gene transfer technology, there is a need for developing methods for delivering genetically altered cells to individuals with specific genetic liver defects. The technical problems we face in developing methods which will allow successful long-term hepatocyte transplantation are not trivial. It has been shown by several investigators that efficient gene transfer into primary hepatocytes is possible either using retroviral vectors[15,18-23] or DNA-mediated gene transfer techniques.[24,25]

For the past 5 years, our laboratory has been working towards developing techniques of experimental isolated hepatocyte transplantation. Our aim has been to collect data which will serve as a useful information base for future efforts to develop methods of clinical hepatocyte transplantation. Specifically, we described techniques of intraperitoneal cell transplantation utilizing a "physiologic" cell matrix and reliable animal models to test transplanted cell function, methods of immunomodulation of cells prior to transplantation resulting in prolongation of their *in vivo* survival, methods of studying transplanted cell proliferation, immunohistochemical techniques for examining transplanted cell morphology, and methods of harvesting hepatocytes from large animals.

B. PREPARATION AND CHARACTERIZATION OF HEPATOCYTES

Tissue culture studies demonstrated that hepatocyte survival and differentiation *in vitro* was enhanced by utilizing a collagen matrix.[26] We applied similar principles in developing *in vivo* hepatocyte transplantation techniques. We hypothesized that attachment of isolated hepatocytes to a collagen matrix prior to transplantation will improve transplanted cell survival and expression of differentiated hepatocyte function. Normal rat hepatocytes were harvested by collagenase portal vein perfusion and were attached to type-I collagen-coated dextran (Cytodex-3, Pharmacia) microcarriers.[27-29]

Microcarrier-attached hepatocytes *in vitro* have been shown to synthesize and conjugate bilirubin and synthesize liver specific proteins.[29] In addition, the ability of microcarrier-attached rat hepatocytes to carry out cytochrome P-450-dependent functions was examined. The *in vitro* metabolism of Cyclosporin A, a widely used immunosuppressant drug, and 19-*nor*testosterone, an anabolic steroid, were investigated in a continuously perfused culture system.[30] Hepatocytes were attached to collagen-coated dextran microcarriers. In rat hepatocyte cultures, Cyclosporin A was metabolized to most major first-generation mono-hydroxylation and N-demethylation products within 8 hours of culture. 19-Nortestosterone metabolism produced the major isomerization products associated with cytochrome P-450 related liver enzymes, in addition to glucuronidation of the parent substrate. Primary human hepatocytes exhibited a similar spectrum of Cyclosporin A and 19-Nortestosterone metabolism as was observed with the rat hepatocytes. Similar data were obtained with fresh and 6-month cryopreserved hepatocytes.[30]

We have developed a simple technique for cryopreserving hepatocytes from different species. We hypothesized that cryopreservation of hepatocytes attached on microcarriers results in a more uniform exposure to the cryoprotectant of a single layer of cells. Microcarrier-attached hepatocytes were cryopreserved in DMEM, 10% FCS, and 5% DMSO, at –70°C, for periods of up to 9 months. Viability was greater than 80%. Cryopreserved hepatocytes were used for most of the studies described below. Our laboratory has carried out a series of studies aimed at developing techniques for isolating and functionally testing adult human hepatocytes. Human hepatocytes were harvested by collagenase (and/or EDTA) perfusion from portions of human liver obtained from organ donors. Catheters were introduced into all visible vessels on the cut surface of liver fragments and sutured in place. Perfusion was carried out in a humidified chamber maintained at 37°C with oxygenated perfusion buffer. The technique has been described in detail elsewhere.[31] Harvested human hepatocytes were attached to collagen-coated dextran microcarriers and cryopreserved using 5% dimethylsulfoxide as cryoprotectant, at –80°C for 4 to 8 weeks.

Our laboratory has been working towards developing techniques of experimental isolated hepatocyte transplantation. We have described techniques of intraperitoneal adult rat hepatocyte transplantation utilizing a “physiologic” cell matrix and reliable animal models to test transplanted cell function.[1] In addition, we have carried out a series of studies aimed at developing techniques for isolating and functionally testing adult human hepatocytes.[2] Our ultimate objective is to develop methods of human hepatocyte transplantation.

C. THE SURGICAL TECHNIQUE

Approximately 1 to 1.5×10^6 microcarrier-attached normal hepatocytes were transplanted intraperitoneally into adult rats. This was done under light, ether anesthesia. Injected microcarrier-attached cells form three-dimensional aggregates on the anterior surface of the pancreas in rats.

III. APPLICATIONS

A. CORRECTION OF SPECIFIC GENETIC DEFECTS

Microcarrier-attached normal rat hepatocytes were transplanted into animals with specific genetic defects of liver function:[27-29] rats unable to conjugate bilirubin (Gunn) or synthesize albumin (Nagase analbuminemic rats; NAR). We were able to demonstrate survival and function of transplanted normal hepatocytes by: (1) the appearance of bilirubin conjugates, increase in total bilirubin in the bile and decrease in serum bilirubin levels in transplanted Gunn rats (Figures 1, 2, and 3a) and (2) an increase in plasma albumin levels in Cyclosporin-immunosuppressed allogeneic NAR recipients (Figure 4).

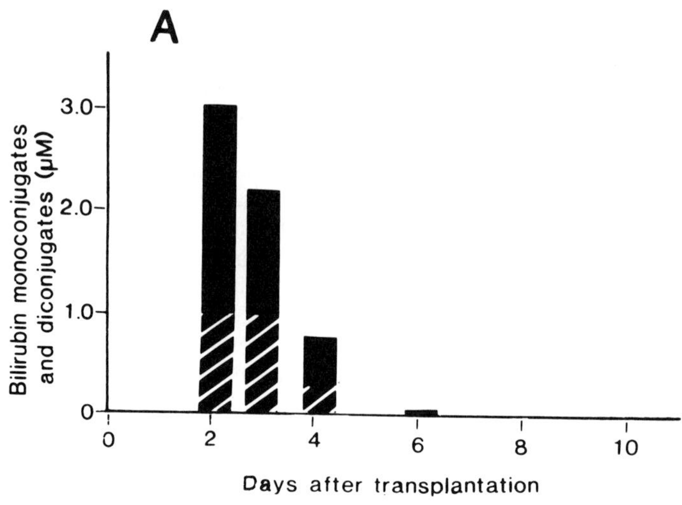

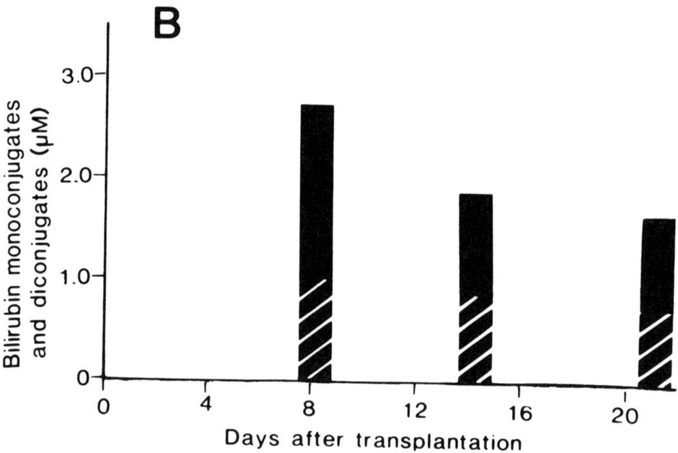

Figure 1 Bilirubin conjugates in the bile of allogeneic (A) and Congeneic (B) Gunn rat recipients at various time points following intraperitoneal transplantation with microcarrier-attached normal rat hepatocytes. Each time point represents an individual animal.

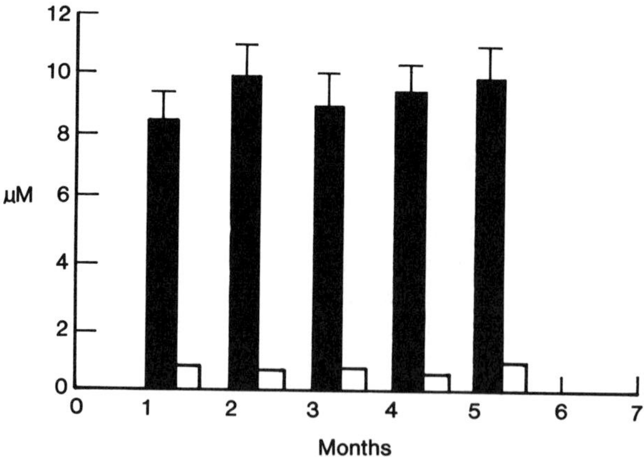

Figure 2 Total bile bilirubin ($\mu M \pm$ SEM) in neonate Gunn rats transplanted with congeneic microcarrier-attached normal rat hepatocytes (shaded bars; n = 4 for each time point) or microcarriers alone (open bars; n = 2 for each time point).

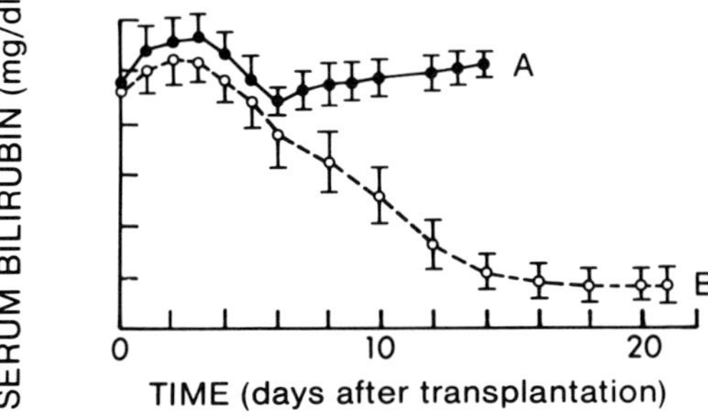

Figure 3 Serum bilirubin concentrations following intraperitoneal transplantation of normal microcarrier-attached rat hepatocytes into allogeneic (A) or congeneic (B) Gunn rat recipients. Data represent mean ± SEM, n = 5 rats for each time point.

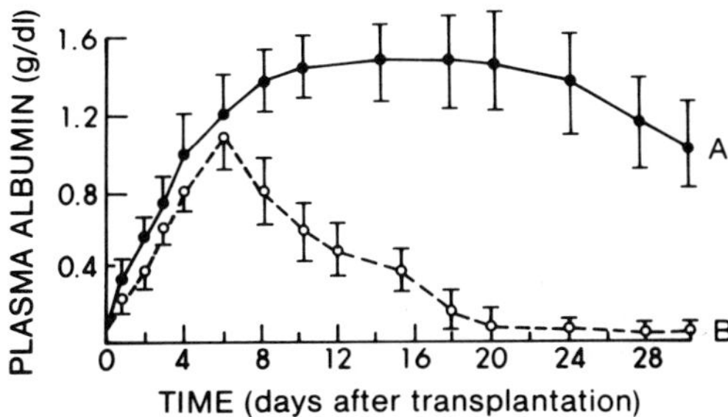

Figure 4 Plasma albumin levels following intraperitoneal transplantation of normal microcarrier-attached rat hepatocytes into allogeneic analbuminemic rat recipients with (A) or without (B) cyclosporin immunosuppression. Data represent means ± SEM, n = 5 rats for each time point.

B. TREATMENT OF ACUTE LIVER INSUFFICIENCY

Microcarrier-attached normal rat hepatocytes were transplanted into rats with acute liver insufficiency following 90% partial hepatectomy.[32] A dramatic improvement in recipient survival and prevention of postoperative hypoglycemia in rats transplanted 4 days prior to 90% partial hepatectomy (Figures 5 and 6) was noted. There were no differences in animal survival between rats given allogeneic vs. syngeneic hepatocytes. It takes about 5 to 6 days for intraperitoneal transplanted allogeneic hepatocyte rejection; we thus conclude that only early hepatocyte survival is important in prolonging animal survival

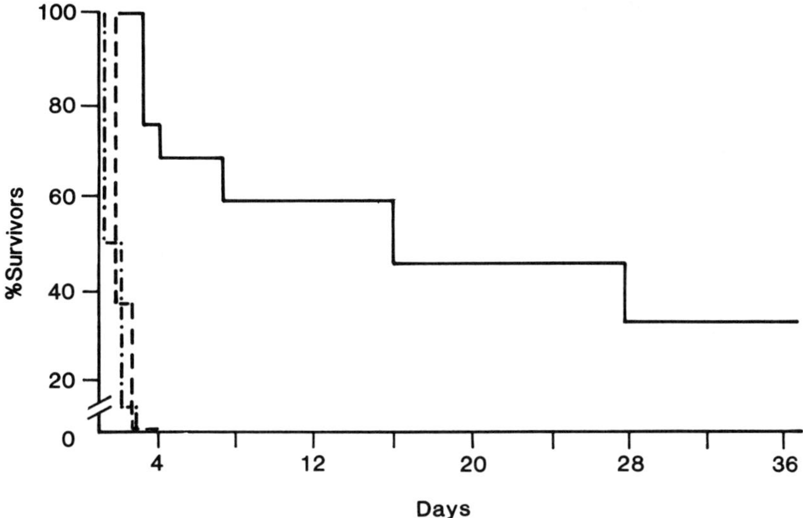

Figure 5 The percentage of animals surviving following 90% partial hepatectomy is shown at various time intervals postoperatively (n = 20 to 30 animals/group). Animals were transplanted with allogeneic hepatocytes without immunosuppression. (—) Microcarrier-attached hepatocytes, intraperitoneal; (– –) hepatocytes alone, intraperitoneal; (– ·) microcarriers alone, intraperitoneal.

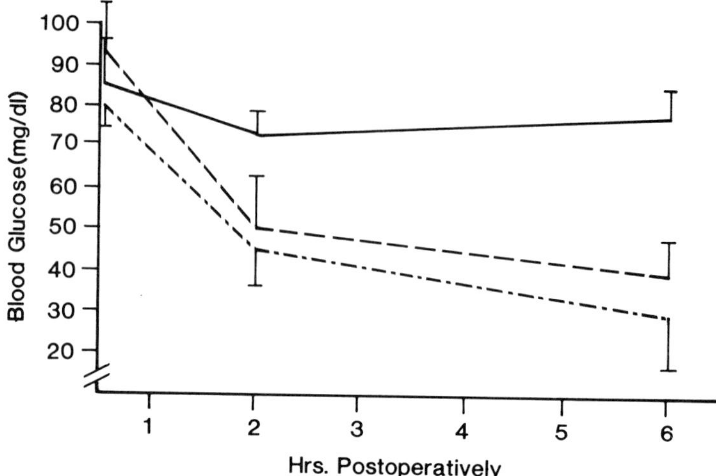

Figure 6 Blood glucose levels (mean ± SEM, n = 12/time point/group) in the early period following 90% partial hepatectomy. Rats were transplanted with allogeneic hepatocytes: (—) microcarrier-attached hepatocytes, intraperitoneal; (– –) hepatocytes alone, intraperitoneal; (– ·) microcarriers alone, intraperitoneal.

in this animal model by providing metabolic support during a time of acute liver insufficiency until the native liver regenerates. These data demonstrate survival and function of transplanted hepatocytes *in vivo,* resulting in partial correction of specific genetic liver function defects and metabolic support in animals with acute liver insufficiency.

C. TRANSPLANTED HEPATOCYTE PROLIFERATION

In the experiments in which animals undergoing 90% partial hepatectomy demonstrated improved survival following hepatocyte transplantation, we noted that transplantation of only a small number of liver cells (1 to 2% of the total liver mass), resulted in improvement in animal survival. One possible explanation is that transplanted hepatocytes proliferated at their ectopic site. Studies were thus undertaken to determine whether transplanted microcarrier-attached hepatocytes proliferate in the peritoneal cavity of syngeneic rat recipients.[33] Microcarrier-attached normal hepatocytes were transplanted intraperitoneally into syngeneic Lewis rats. Transplanted rats underwent 70% partial hepatectomy 3 days later, and 24 hours later, they were injected with [^{3}H]thymidine intravenously. All rats were killed with ether 1 hour later and their livers and peritoneal cavity aggregates were excised and assayed for total DNA and [^{3}H] thymidine incorporation into DNA. There was a significant ($p < 0.01$) increase in [^{3}H]thymidine incorporation into DNA in the regenerating liver remnants when compared to sham-operated animal livers (13,150 ± 1,080 vs. 860 ± 119 dpm/µg DNA); similarly, there was a significant ($p < 0.02$) increase in [^{3}H]thymidine uptake by peritoneal aggregates in partially hepatectomized rats when compared to uptake in sham-operated rats (4,500 ± 610 vs. 729 ± 165 dpm/µg DNA). Data from this and other similar experiments carried out in our laboratory suggested that transplanted liver cells proliferate *in vivo.* Experiments were then carried out to determine which cells in the peritoneal cavity aggregates proliferate. This was done by labeling dividing cells using either [^{3}H]thymidine autoradiography[34] or staining for bromodeoxyuridine.[35] Sections were also stained with a specific hepatocyte probe (anti-albumin antibody) in an attempt to identify proliferating hepatocytes. Using such a "double label" technique we were able to identify a large number of proliferating cells in

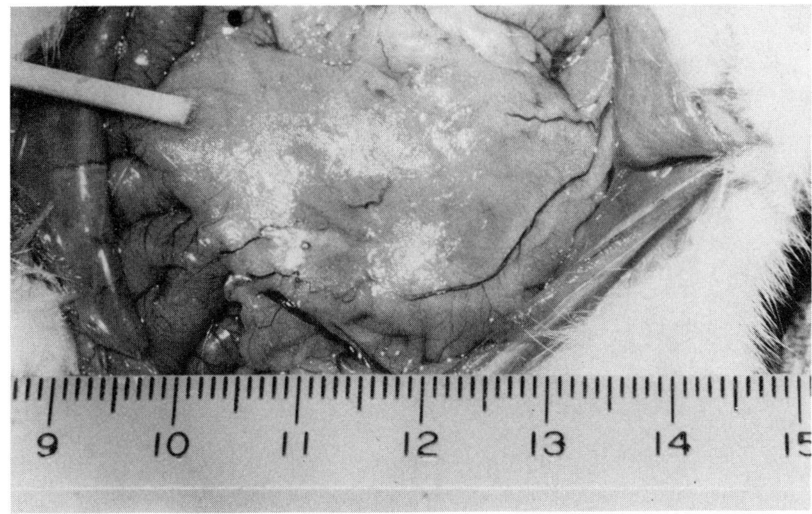

Figure 7 Gross appearance of the peritoneal cavity aggregate in a rat transplanted intraperitoneally with microcarrier-attached hepatocytes. A well-vascularized aggregate is formed in the upper abdomen under the liver on the anterior surface of the pancreas and stomach.

the aggregates; however, only a rare hepatocyte was shown to proliferate. Our data suggest that although cells within the peritoneal cavity aggregates proliferate in response to partial hepatectomy, most proliferating cells were not differentiated hepatocytes; we were thus unable to demonstrate hepatocyte proliferation to a significant extent using immunohistochemical/nuclear labeling techniques.

D. MORPHOLOGIC STUDIES

Intraperitoneally transplanted microcarrier-attached hepatocytes in rats form well-vascularized aggregates on the anterior surface of the pancreas (Figure 7). We examined the morphology of the aggregates using light and electron microscopy and immunohistochemical techniques.[28] In the aggregates, microcarriers are surrounded by "epithelioid" cells. Using standard indirect immunoperoxidase methods, formalin fixed, paraffin embedded sections of NAR rat peritoneal cavity aggregates were stained with anti-rat albumin antibody. A large number of hepatocytes were identified staining for albumin up to 28 days following transplantation in Cyclosporin-treated NAR rats (Figure 8). Most hepatocytes were found in clusters embedded in the connective tissue matrix among microcarriers, indicating that cell-cell contact may be important for transplanted cell survival. Use of the NAR rat recipient allows better definition of the albumin-stained hepatocytes due to low background staining. Electron microscopic examination of the aggregates revealed the presence of giant nucleoid containing peroxisomes characteristic of normal rat hepatocytes[28] and formation of microvilli and canaliculi suggesting a high level of morphologic differentiation. Scanning electron microscopy revealed clusters of cells and microcarriers and newly formed collagen on the surface of the microcarriers.

E. USE OF BIODEGRADABLE COLLAGEN MATRIX

To develop a clinical method of intraperitoneal hepatocyte transplantation, it would be advantageous to utilize a totally biodegradable matrix as a vehicle for cell transplantation to eliminate foreign body reaction in the recipients. We thus attached normal isolated hepatocytes to thin (2 mm) slices of type-I collagen sponges and transplanted 1×10^7 cells

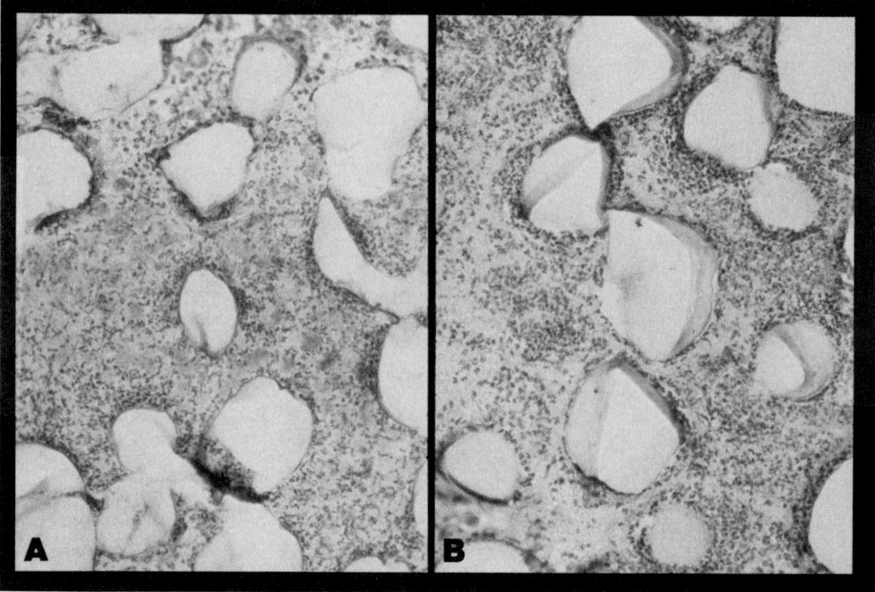

Figure 8 (A) A large number of albumin-stained hepatocytes (dark brown) are seen in the peritoneal cavity aggregate among and on the surface of microcarriers. The clear spaces represent microcarriers. (B) Aggregate from animal transplanted with microcarriers alone.

into NAR rats by inserting several sponge slices (total: 140 mg) into the peritoneal cavity through a small abdominal incision.[37] This resulted in a significant increase in serum albumin levels. Rats with collagen sponge slices alone had no changes in serum albumin (0.05 ± 0.01 g/dl); 16 rats (50%) transplanted with hepatocyte-sponge slices, demonstrated a significant linear (peak level day 6; 2.5 ± 1.1 g/dl, $p < 0.01$) increase in serum albumin. Western Blot analysis confirmed the SDS-PAGE data. This method does not appear to be as effective as the microcarrier technique, however. Most cells in the center of the sponges were not viable and only a relatively small number of viable cells were found on the surface of the sponge.

F. MODULATION OF TRANSPLANTED CELL ANTIGENICITY

1. Ultraviolet Irradiation

Exposure to ultraviolet (UV) irradiation has been shown to prolong pancreatic islet allograft survival. The purpose of this study was to examine the effect of UV irradiation on the survival and function of normal hepatocytes *in vivo*. Microcarrier-attached normal rat hepatocytes were pretreated with UV irradiation (600 J/m^2) prior to transplantation into allogeneic NAR recipients.[37] Transplanted hepatocyte survival and function was assessed by serial measurements of plasma albumin levels. There was a significantly prolonged elevation of plasma albumin levels in rats transplanted with UV-treated cells when compared to rats transplanted with sham-irradiated cells (Figure 9). Our results demonstrate that pretreatment with UV irradiation prolongs survival and function of allogeneic, intraperitoneally transplanted microcarrier-attached adult normal rat hepatocytes.

2. Encapsulation

Cell encapsulation has been shown to prolong transplanted cell function because the semipermeable membrane reportedly immunoisolates the cells. Transplantation of encapsulated cells has been carried out by Cai et al.[38] employing 25,000 MW alginate/poly-L-

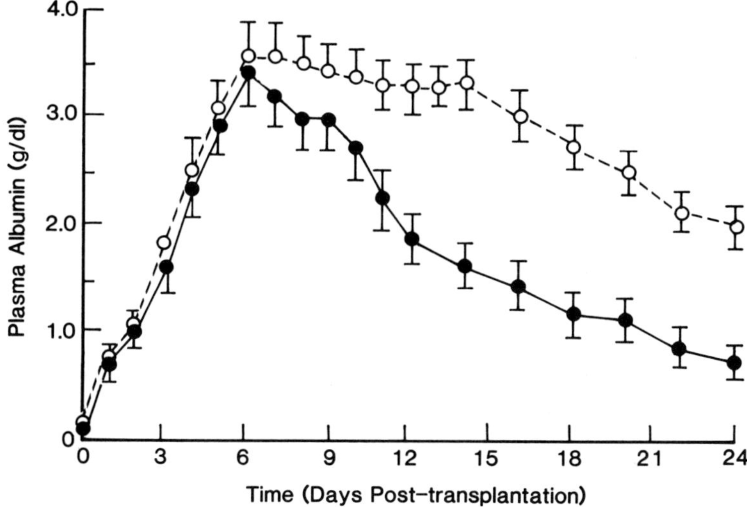

Figure 9 Effect of pretreatment of normal rat hepatocytes with UV irradiation on their function following transplantation into allogeneic NAR rats. Plasma albumin levels were measured at various time points following transplantation and expressed as means ±SEM. There was prolonged elevation of plasma albumin levels in rats transplanted with UV-treated microcarrier-attached hepatocytes (open circles, n = 8) when compared to rats transplanted with sham-irradiated microcarrier-attached hepatocytes (closed circles, n = 7).

lysine (ALP). We carried out a series of experiments using ALP-encapsulated microcarrier-attached hepatocytes. We hypothesized that combining microcarrier and encapsulation techniques would allow cell attachment to a matrix and additionally, immunoisolate the cells. A series of *in vitro* experiments were first carried out to examine transport of macromolecules across the ALP capsule. Encapsulated microcarrier-attached hepatocytes were cultured under standard conditions in defined serum-free media. Albumin release in the media was measured by ELISA. At 24 hours, nonencapsulated, microcarrier-attached hepatocytes released more albumin than encapsulated ones into the medium (25.0 ± 2.0 vs. 4.1 ± 1.2 ng/ml; $p < 0.02$). Subsequently, in a series of *in vivo* experiments, NAR rats were transplanted either with microcarrier-attached allogeneic hepatocytes or with encapsulated microcarrier-attached hepatocytes, both groups without Cyclosporin immunosuppression. No statistically significant differences were noted. A very strong inflammatory response was seen in the peritoneal cavity of animals transplanted with encapsulated microcarrier-attached hepatocytes. No such reaction was seen in controls injected with encapsulated microcarriers alone or animals transplanted with syngeneic microcarrier-attached encapsulated hepatocytes. Encapsulated allogeneic hepatocytes induce a significant pericapsular inflammatory reaction probably due to release of cytokines.

G. HUMAN HEPATOCYTE STUDIES

Isolated normal human hepatocyte function was tested by transplanting isolated, cryopreserved, microcarrier-attached human hepatocytes intraperitoneally into nude Gunn rats, which are unable to conjugate bilirubin due to lack of the enzyme UDP glucuronyl transferase and into nude analbuminemic rats (NAR) which only have trace levels of plasma albumin due to a genetic defect in albumin synthesis. Nude Gunn rats demonstrated excretion of bilirubin glucuronides in their bile for up to 30 days posttransplantation and significant reduction of their serum bilirubin levels.[31] Similarly, recipient nude NAR rats demonstrated a significant increase in plasma albumin levels 8 days posttransplantation and remained at that level throughout the duration of the experiment.[32]

Transplanted human hepatocytes formed three-dimensional aggregates in the peritoneal cavity in close proximity to the pancreas. Studies were carried out to identify transplanted hepatocytes using hepatocyte-specific probes. Using standard indirect immunoperoxidase methods, formalin fixed, paraffin-embedded sections of NAR rat peritoneal cavity microcarrier-cell aggregates were stained with anti-human albumin antibody. Up to 2 weeks posttransplantation, a large number of hepatocytes could be identified staining for albumin within the aggregates, primarily in clusters, among the microcarriers. Subsequently, the number of albumin-stained cells progressively decreased.

H. CELL HARVESTING FROM LARGE ANIMALS

The techniques developed for harvesting, cryopreserving, and testing human hepatocytes[31] were utilized to harvest dog hepatocytes. Dogs underwent partial hepatic resection (left lobe), the liver segment removed was perfused as described above,[31] and harvested hepatocytes were attached to microcarriers and cryopreserved. Approximately 1.5×10^8 microcarrier-attached hepatocytes were re-injected intraperitoneally into their original donor 48 hours following resection. Animals underwent exploratory laparotomy 2, 4, and 8 weeks later and the peritoneal cavity was examined. Microcarrier-cell aggregates were detected on the surface of the pancreas and along the small bowel mesentery. Excisional biopsies revealed the presence of microcarrier-attached hepatocytes within the aggregates. No functional studies were carried out. Morphologically viable hepatocytes were present in the aggregates up to 8 weeks posttransplantation.

IV. CONCLUSION

The ultimate objective of cell transplantation for the treatment of liver and other diseases is to harvest a relatively small number of cells from a patient, correct the genetic defect *in vitro,* select and expand the transfected cell population, and subsequently retransplant the cells in the original donor. This could potentially eliminate the problem of tissue rejection and avoid the need for chronic immunosuppression. However, a number of formidable problems remain to be resolved at each step of this approach. These include difficulty of transfecting large numbers of primary normal adult hepatocytes in culture, difficulty in expanding transfected hepatocyte numbers *in vitro,* inability to obtain long-term survival of transfected hepatocytes *in vivo* and the potential for transformation of transplanted transfected cells.

Our laboratory has been involved in the development of techniques of isolated hepatocyte transplantation. The aim of many investigators, including ourselves, is to develop *in vitro* hepatocyte transfection techniques and examine the survival, function, and stability of product expression of transplanted cells *in vivo.* Previous work by Pasco and Fagan has shown successful but transient transfection of rat hepatocytes using a DNA mediated system.[39] Other work by Anderson et al.[15] using retro–viral mediated gene transfer has been successful for longer periods of time. In these studies, a retroviral gene was incorporated into adult rat hepatocytes and was present in intraperitoneal transplants as demonstrated by Southern blot analysis and NPT activity assays. In a rabbit model, Wilson[40] has shown retroviral transfection of rabbit hepatocytes. In this model, heritable low-density hyperlipidemic rabbit hepatocytes were transfected with virus encoding for low density lipoprotein receptor. After transfection, success was determined by Southern blot analysis of the integrated sequences. There were also tests of expression using *in vitro* receptor precipitation analysis. As in other areas of transplantation, the two major remaining unresolved problems are transplanted cell rejection and limited donor tissue availability. Additionally, long-term or indefinite survival of large numbers of transplanted normal hepatocytes, at any site, remains elusive even in syngeneic recipients. This suggests that the purely technical aspects of isolated hepatocyte transplantation have not been resolved. Absence of indefinite transplanted cell survival remains the major limitation of all currently used experimental techniques of isolated cell transplantation. Further studies are needed to develop improved methods of hepatocyte transplantation to: (1) enhance engraftment, (2) decrease the antigenicity of donor cells (by using highly purified liver cell populations, manipulating *in vitro* culture conditions, selecting specific liver cell subpopulations from harvested cells, using novel encapsulation techniques, and pretreating transplanted cells with specific antisera or UV irradiation), and (3) identify specific hepatocyte subpopulations which may respond to mitogens and growth factors *in vitro* or *in vivo,* thus expanding the available supply/mass of normal hepatocytes.

REFERENCES

1. **Baumgartner, D., LaPlante-O'Neill, P.M., Sutherland, D.E.R., and Najarian, J.S.,** Effects of intrasplenic injection of hepatocytes, hepatocyte fragments and hepatocyte culture supernatants on D-galactosamine induced liver failure in rats, *Eur. Surg. Res.,* 15, 129, 1983.
2. **Sommer, B.G., Sutherland, D.E.R., Matas, A.J., Simmons, R.L., and Najarian, J.S.,** Hepatocellular transplantation for treatment of D-galactosamine-induced acute liver failure in rats, *Transplant. Proc.,* 11, 578, 1979.
3. **Makowka, L., Rotstein, L.E., Falk, R.E., and Falk, J.A.,** Reversal of toxic and anoxic induced hepatic failure by syngenic, allogenic and xenogenic hepatocyte transplantation, *Surgery,* 88, 244, 1980.
4. **Grundman, R., Koebe, H.G., and Waters ,W.,** *Res. Exp. Med.,* 186, 141, 1986.
5. **Makowka, L., Rotstein, L.E., Falk, R.E., and Falk, J.A.,** Studies into the mechanism of reversal of experimental acute hepatic failure by hepatocyte transplantation, *Can. J. Surg.,* 24, 39, 1981.
6. **Fasola, C.G., Matas, A.J., Payne, W.D., and et al.,** Reversal of acute hepatic failure in a large animal by autologous hepatocyte transplantation, *Surg. Forum,* 40, 347, 1989.
7. **Fasola, C.G., Kim, Y.S., Manivel, C.J., and et al.,** Reversal of acute hepatic failure in pigs by xenogeneic hepatocyte (HC) transplantation and by injection of supernatant from xenogeneic HC cultures, *Surg. Forum,* 41, 374, 1990.
8. **Toledo-Pereira, L.H., Gordon, D.A., and MacKenzie, G.H.,** Immunologic response to liver cell allografts, *Am. Surg.,* 48, 28, 1982.
9. **Vroemen, J.P., Buurman, W.A., Schutte, B., and Maessen, J.G.,** The cytokinetic behavior of donor hepatocytes after syngenic hepatocyte transplantation into the spleen, *Transplantation,* 45, 600, 1988.
10. **Vroemen, J.P., Blanckaert, N., Buurman, W.A., Heirwegh, K.P., and Koostra, G.,** Treatment of enzymic deficiency by hepatocyte transplantation in rats, *J. Surg. Res.,* 39, 267, 1985.
11. **Jirtle, R.L. and Michalopoulos, G.,** Effects of partial hepatectomy on transplanted hepatocytes, *Cancer Res.,* 42, 3000, 1982.
12. **Jirtle, R.L., Biles, C., and Michalopoulos, G.,** Morphologic and histochemical analysis of hepatocytes transplanted into syngeneic hosts, *Am. J. Pathol.,* 101, 115, 1980.
13. **Vroemen, J.P., Buurman, W.A., van der Linden, C.J., and Visser, R.,** Transplantation of isolated hepatocytes into the pancreas, *Eur. Surg. Res.,* 20, 1, 1988.
14. **Sutherland, D.E.R., Numata, M., Matas, A.J., and Najarian, J.S.,** Hepatocellular transplantation in acute liver failure, *Surgery,* 82, 124, 1977.
15. **Anderson, K.D., Thompson, J.A., DiPietro, J.M., and Montegomery, K.T.,** Gene expression in implanted rat hepatocytes following retroviral mediated gene transfer, *Somat. Cell Mol. Genet.,* 15, 215, 1989.

16. **Thompson, J.A., Anderson, K.D., DiPietro, J.M., and Zwiebl, J.A.,** Site directed neovessel formation in vivo, *Science,* 241, 1349, 1988.
17. **Vacanti, J.P., Morse, M.A., Saltzman, W.M., and Domb, A.J.,** Selective cell transplantation using bioabsorbable artificial polymers as matrices, *J. Ped. Surg.,* 23, 3, 1988.
18. **Ledley, F.D., Darlington, G.J., Hahn, T., and Woo, S.L.C.,** Retroviral gene transfer into primary hepatocytes: Implications for genetic therapy of liver-specific functions, *Proc. Natl. Acad. Sci. U.S.A.,* 84, 5335, 1987.
19. **Peng, H., Armentano, D., Graham, L., Mackenzie-Graham, L., and Shen, R.F.,** Retroviral-mediated gene transfer and expression of human phenylalanine hydroxglase in primary mouse hepatocytes, *Proc. Natl. Acad. Sci. U.S.A.,* 85, 8146, 1988.
20. **Wolff, J.A., Yee, J.K., Skelly, H.F., and Moores, J.,** Genes in primary cultures of adult rat hepatocytes, *Proc. Natl. Acad. Sci. U.S.A.,* 84, 3344, 1987.
21. **Wolff, J.A., Yee, J.K., Skelly, H.F., and Moores, J.,** Adult mammalian hepatocyte as target cell for retroviral gene transfer: a model for gene therapy, *Somat. Cell Mol. Gent.,* 13, 423, 1987.
22. **Wilson, J.M., Jefferson, D.M., Chowdhury, J.R., and Novikoff, P.M.,** Retroviral-mediated transduction of adult hepatocytes, *Proc. Natl. Acad. Sci. U.S.A.,* 85, 3014, 1988.
23. **Wilson, J.M., Johnston, D.E., Jefferson, D.M., and Mulligan, R.C.,** Correction of the genetic defect in hepatocytes from the Watanabe heritable hyperlipidemic rabbit, *Proc. Natl. Acad. Sci. U.S.A.,* 85, 4421, 1988.
24. **Dubensky, T.W., Cambell, B.A., and Villarreal, L.P.,** Direct transfection of viral and plasmid DNA into the liver or spleen of mice, *Proc. Natl. Acad. Sci. U.S.A.,* 81, 7529, 1984.
25. **Wu, G.Y. and Wu, C.H.,** Receptor-mediated in vitro gene transformation by a soluble DNA carrier system, *J. Biol. Chem.,* 262, 4429, 1987.
26. **Reid, L.M. and Rojkind, M.,** New technique of re?culturing differentiated cells; reconstituted basement membrane grafts, *Meth. Enzymol.,* 58, 263, 1979. title ok??
27. **Demetriou, A.A., Whiting, J., Feldman, D., and Levenson, S.M.,** Replacement of liver function in rats by transplantation of microcarrier-attached hepatocytes, *Science,* 233, 1190, 1986.
28. **Demetriou, A.A., Levenson, S.M., Novikoff, P.M., and Novikoff, A.B.,** Survival, organization, and function of microcarrier-attached hepatocytes transplanted in rats, *Proc. Natl. Acad. Sci. U.S.A.,* 83, 7475, 1986.
29. **Demetriou, A.A., Whiting, J., Levenson, S.M., and Chowdhury, N.R.,** New method of hepatocyte transplantation and extracorporeal liver support, *Ann. Surg.,* 204, 259, 1986.
30. **Moscioni, A.D., Backfisch, G., Black, D., and Demetriou, A.A.,** Long-term cryopreserved human hepatocytes maintain drug metabolizing ability, *Surg. Forum,* 41, 3, 1990.
31. **Moscioni, A.D., Chowdhury, J.R., Barbour, R., Brown, L., Chowdhury, N.R., Competiello, L.S., Lahiri, P., and Demetriou, A.A.,** Human liver cell transplantation. Prolonged function in athymic-Gunn and athymic-analbuminemic hybrid rats, *Gastroenterology,* 96, 1546, 1989.
32. **Demetriou, A.A., Reisner, A., Sanchez, J., and Levenson, S.M.,** Transplantation of microcarrier-attached hepatocytes into 90% partially hepatectomized rats, *Hepatology,* 8, 1006, 1988.
33. **Arepally, G., Felcher, A., Moscioni, A.D., and Demetriou, AA.,** Proliferation of intraperitoneally transplanted microcarrier-attached rat liver cells, *Surg. Forum,* 39, 396, 1988.

34. **Grisham, J.W.,** A morphologic study of deoxyribonucleic acid synthesis and cell proliferation in regenerating rat liver: autoradiography with thymidine-^{3}H, *Cancer Res.,* 22, 842, 1962.
35. **Shimizu, A., Tarao, K., Takemiya, S., and Harada, M.,** S-phase cells in diseased human liver determind by an in vitro BrdU anti-BrdU method, *Hepatology,* 8, 1535, 1988.
36. **Lau, H., Reemtsma, K., and Hardy, M.A.,** Prolongation of rat islet allograft survival by direct ultraviolet irradiation of the graft, *Science,* 233, 607, 1984.
37. **Phay, J., Felcher, A., Levenson, S.M., and Demetriou, A.A.,** Prolongation of function of transplanted microcarrier-attached hepatocytes by ultraviolet irradiation, *Surg. Forum,* 39, 398, 1988.
38. **Cai, Z., Shi, Z., O-Shea, G.M., and Sun, A.M.,** Microencapsulated hepatocytes for bioartificial liver support, *Artif. Organs,* 12, 388, 1988.
39. **Pasco, D.S. and Fagan, J.B.,** Efficient DNA-mediated gene transfer into primary cultures of adult rat hepatocytes, *DNA,* 8 & 7, 335, 1989.
40. **Wilson, J.M.,** Retrovirus-mediated transduction of adult hepatocytes, *Proc. Natl. Acad. Sci. U.S.A.,* 88, 3014, 1988.

Section III

Pancreas Transplantation

Chapter 10

Whole Pancreas Transplantation in the Rat

Marshall J. Orloff and Mark S. Orloff

CONTENTS

I. INTRODUCTION

The technique of whole pancreas transplantation was developed in the experimental laboratory in animals, particularly in rats. It has been used for studies of endocrine and exocrine function of the pancreas, pancreatic cancer, pancreatitis, liver regeneration, and diabetes mellitus. Unquestionably, the widest and most important use of whole pancreas transplantation has been in animal studies of the pathophysiology, pathology, metabolic derangements, and treatment of diabetes.

Diabetes mellitus is a leading cause of morbidity, mortality, and economic cost. In 1987 it was estimated that a total of 14 million persons in the U.S. had diabetes mellitus, and that annual health care costs for diabetes were $20.4 billion. Diabetes is responsible for 10% of all days of hospitalization yearly in the U.S., and is currently the sixth ranking

0-8493-629-5/94/$0.00+$.50

primary cause of death, not including deaths of diabetics attributed to heart disease and stroke. For these reasons, research on all aspects of diabetes is very important.

During the past 15 years, a great deal has been learned about the etiology of the various forms of diabetes. The weight of evidence indicates that destruction of the beta cells of the pancreatic islets of Langerhans is the ultimate cause of Type I or insulin-dependent diabetes mellitus (IDDM), and that dysfunction of the beta cells plays an important role in Type II or non-insulin dependent diabetes (NIDDM). Although conventional treatment of diabetes with insulin injections and diet has reduced the risk of death from ketoacidosis, and increased the life span of the diabetic population, it has not prevented the development or progression of the debilitating secondary complications that are responsible for its morbidity and mortality. This failure of conventional treatment has been attributed to imperfect, less-than-physiologic control of metabolic abnormalities such as hyperglycemia. The main stimulus to research in pancreas transplantation and other forms of endocrine pancreas replacement therapy is the hope that such treatment will provide more precise and even physiologic beta cell function.

There are enormous difficulties associated with efforts to study the living pathology of diabetes in humans because of ethical considerations and the long periods of observation required to determine the effectiveness of any form of treatment. Therefore, diabetes research requires studies in animal models with diabetes that is similar to the human disease. The NIH-sponsored Task Force on Animals Appropriate for Studying Diabetes Mellitus and Its Complications concluded that animals "...have proven invaluable for elucidating the etiology and pathophysiology of diabetes and have contributed to the development of therapy...". Two chemical agents, alloxan and streptozotocin, have been used most widely to induce diabetes in animals. Both agents selectively destroy the beta cells of the islets of Langerhans and produce a marked decrease in insulin secretion, persistent hyperglycemia, the full range of metabolic abnormalities, including ketoacidosis, and the common clinical manifestations of diabetes. Both agents are associated with pathologic lesions that closely mimic those found in human IDDM if the animals are observed for a sufficiently long period of time. The authors and many other investigators have documented severe nephropathy, neuropathy, eye lesions, cardiovascular lesions, gonadal lesions, and microangiopathy that are strikingly similar to the pathologic abnormalities found in patients with Type I diabetes. The NIH Task Force stated: "At present the optimal experimental model in which all these facets can reasonably be studied and integrated seems to be the chemically induced diabetic rat."

Early studies of pancreas transplantation involved the placement of fragments of embryonic, fetal, neonatal, or adult pancreatic tissue into various locations, such as the subcutaneous tissues, testes, and anterior chamber of the eye, in pancreatectomized or alloxan-diabetic recipient animals and, in a few instances, in diabetic humans. Although some of these allograft fragments appeared to function for brief periods, none produced persistent lowering of blood sugar, and all were rapidly rejected. Transplantation of the pancreas with vascular connections was initially accomplished in the 1920s in dogs by the use of cannulas to connect the blood vessels. In 1957, Lichtenstein and Barshack first performed pancreas allografting with direct vascular anastomoses. In 1962 and 1963, Reemtsma and his colleagues and De Jode and Howard were the first to obtain significant short-term survival with vascularized heterotopic allografts, and to demonstrate lowering of the blood sugar for up to 16 days in pancreatectomized dogs. Reemtsma et al. were unsuccessful in their attempts to prolong graft survival with immunosuppression therapy. In 1967, Largiader et al. were the first to perform orthotopic allotransplantation of the pancreas and duodenum. They demonstrated both endocrine and exocrine function of the pancreaticoduodenal grafts for up to 9 days. Since then, a number of investigators have successfully transplanted the pancreas in dogs and pigs, usually in a heterotopic location, and have demonstrated graft function for short periods of time. However, all of these transplants have undergone immunologic rejection, and survival for more than two months has been rare despite immunosuppression therapy. As a result, it has not been possible to determine the value of pancreas transplantation in the treatment of diabetes mellitus from studies in dogs or pigs.

In 1970, the first microvascular surgical technique of heterotopic pancreaticoduodenal transplantation in rats was developed in the senior author's laboratory.[1-6] Use of this technique of transplantation of isografts between highly inbred rats has provided the opportunity to perform long-term studies of the function of the transplanted pancreas without the complications of immunologic rejection or immunosuppression therapy. Initial studies demonstrated that pancreas transplants placed in alloxan-diabetic rats shortly after induction of diabetes consistently maintained good endocrine function and relieved the metabolic abnormalities of diabetes for the full life span of the rat.[4] Generally, three techniques of heterotopic whole pancreas transplantation have been used in both experimental studies and clinical trials, namely, transplantation of the pancreas alone with its exocrine ducts ligated or occluded, transplantation of the pancreas and duodenum together so as to avoid any complications of exocrine duct obstruction, and transplantation of a segment, usually the body and tail, of the pancreas, with or without ductal drainage. Drainage of the exocrine pancreatic duct has been accomplished by direct anastomosis to the ureter, bladder, or small bowel. It has been suggested that ligation of the pancreatic ducts might cause autodigestion of the gland and ultimate destruction of the islets of Langerhans, but our studies have demonstrated that the duct-ligated whole pancreas transplant has good endocrine function for the life span of the rat, and our findings have been confirmed in the rat and dog by others.[2] Nevertheless, it is technically more difficult to harvest a pancreas transplant without the duodenum than with the duodenum, and we have used the pancreaticoduodenal transplant as our standard model.

Since the treatment of diabetes would not be enhanced by removal of the diseased pancreas, transplantation of the pancreas as an auxiliary (heterotopic) graft connected to the systemic vasculature has been used in almost all experimental studies and clinical trials. Heterotopic transplantation has the advantage of being far simpler technically than orthotopic transplantation. Despite earlier suggestions to the contrary, substantial evidence indicates that portal venous drainage of the transplant is not essential for satisfactory endocrine function. Our long-term studies have shown no difference in satisfactory endocrine function between systemic and portal venous drainage of pancreas isografts.

II. TECHNIQUES

A. ANATOMY

The pancreas of the rat is more diffuse, softer, and less clearly delineated than the human pancreas. Its location is approximately the same as that of the pancreas in humans, although it is more spread out. It is darker in color than the adjacent adipose tissue. The right lobe and body of the pancreas are embedded in the mesoduodenum and beginning of the mesojejunum. The left lobe is a branched, flattened part that runs along the dorsal aspect of the stomach embedded in the dorsal leaf of the greater omentum, and along the splenic artery toward the hilum of the spleen. The bile duct is surrounded by pancreatic tissue along most of its length. In a 300-g rat, the pancreas weighs about 1 g.

The pancreatic acini are drained by 15 to 40 excretory ducts that fuse to form at least two and sometimes five to eight main ducts that open into the ductus choledochus or bile duct. Often, small ducts open directly into the duodenum. The largest collecting duct is from the left lobe (splenic duct) and it is the first duct to enter the bile duct. Figure 1 shows the gross anatomy of the pancreas.

The arterial blood supply to the pancreas is provided by two main arteries, the cranial and caudal pancreaticoduodenal arteries. The cranial pancreaticoduodenal artery runs

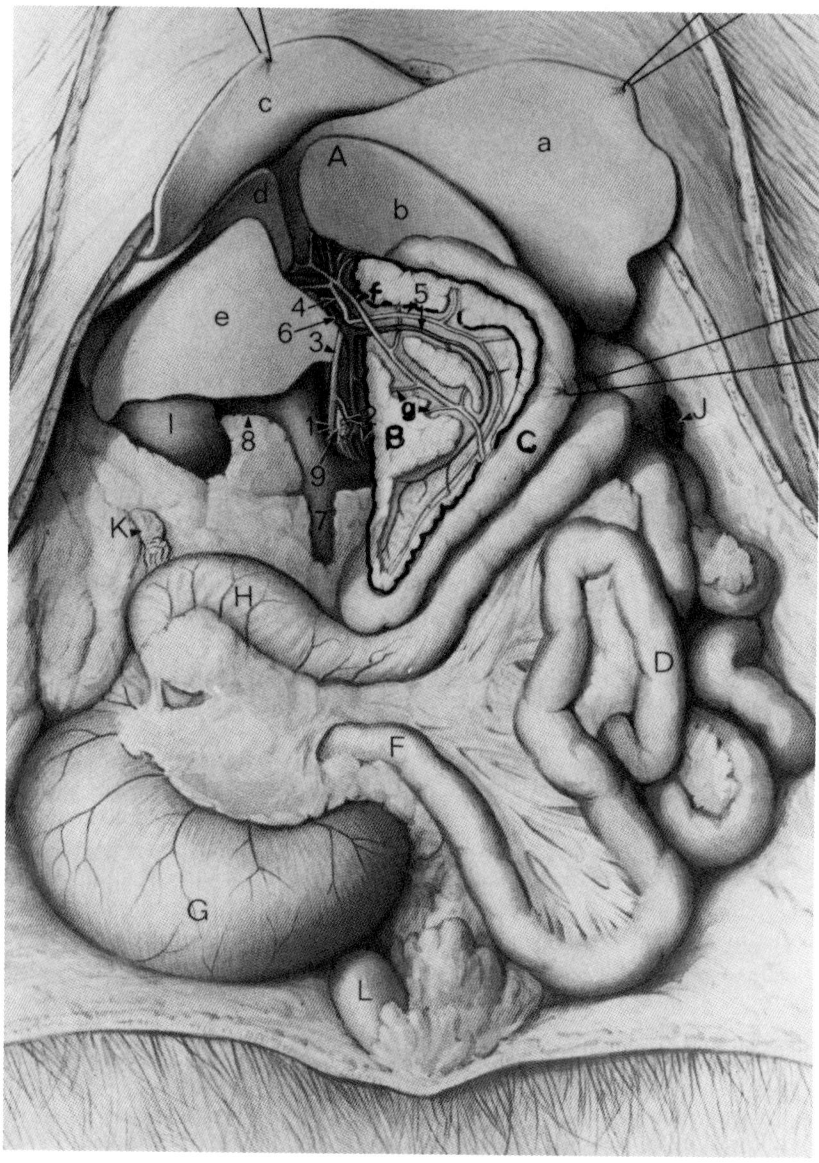

Figure 1 Gross anatomy of the pancreas showing the right lobe and body in the mesoduodenum. The left lobe is not shown. A: liver; B: pancreas; C: duodenum; f: bile duct; g: pancreatic duct emptying into the bile duct; 4: hepatic artery; 5: gastroduodenal artery and vein; 6: portal vein.

caudally in the mesoduodenum and is a branch of the gastroduodenal artery which, in turn, is a branch of the hepatic artery or, sometimes, a direct branch of the celiac artery (Figure 2). The celiac artery in the rat is similar to the human celiac artery, arising from the abdominal aorta as the first unpaired branch, and branching into a splenic artery, left gastric artery, and hepatic artery which, in turn, gives off the gastroduodenal artery and then the cranial pancreaticoduodenal artery. The caudal pancreaticoduodenal artery is the second branch of the cranial (superior) mesenteric artery which, in turn, arises from the

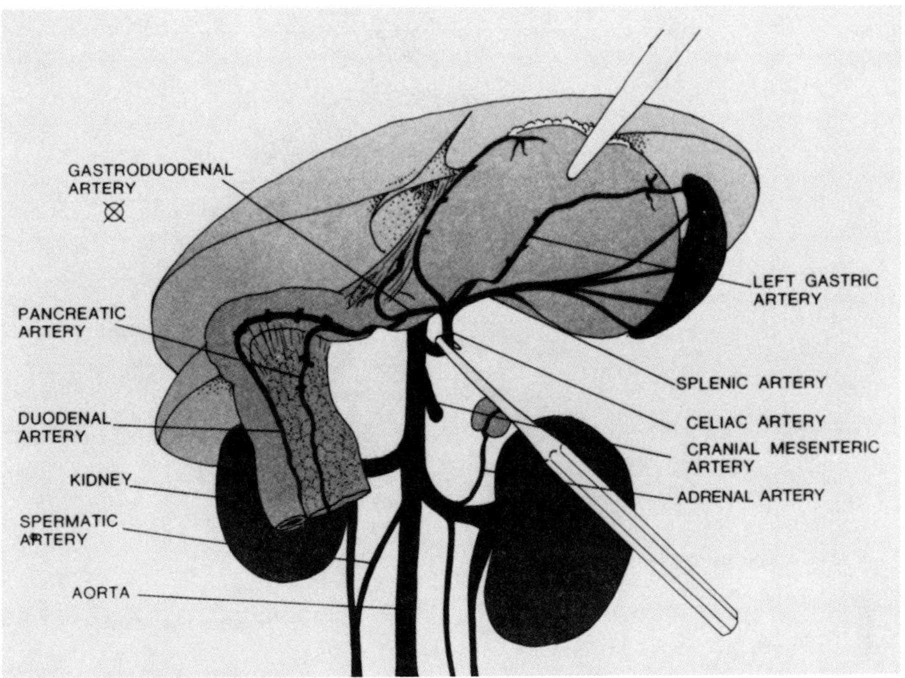

Figure 2 Pancreatic arterial vascular supply. The right lobe and body of the pancreas are supplied by a branch of the gastroduodenal artery and a branch of the cranial (superior) mesenteric artery (not shown). The left lobe of the pancreas is supplied by branches of the splenic artery that are not shown.

aorta as the second unpaired branch 3 to 5 mm caudal to the celiac artery. The caudal and cranial pancreaticoduodenal arteries connect with each other. The splenic artery gives a number of tiny branches to the left lobe of the pancreas.

The venous drainage of the pancreas is into the hepatic portal vein (Figure 3). The cranial (superior) pancreaticoduodenal vein drains the right lobe and body of the pancreas and is the last tributary (i.e., closest to the liver) to enter the portal vein directly. The caudal (inferior) pancreatic vein drains the left lobe of the pancreas into one of the splenic veins which, in turn, drain into the portal vein. The portal vein in the rat is similar to the human portal vein, draining the cranial (superior) mesenteric vein, gastrosplenic vein, caudal (inferior) mesenteric vein, right gastroepiploic vein, and veins from the pancreas.

The rat duodenum is relatively longer than the human duodenum, and is more mobile since it is suspended by a mesentery, the mesoduodenum. It arises in the midline from the distal end of the pylorus, runs transversely toward the right abdominal wall, rises dorsally after an initial curve along the right margin of the liver toward the right kidney, curves as the transverse duodenum toward the midsagittal plane and after another right-angled turn runs cranially as the ascending duodenum. The jejunum is the longest part of the intestine and its loops fill the right ventral part of the abdomen. The small intestine is six times the length of the body, from snout to anus.

In our technique of pancreaticoduodenal transplantation, the vasculature of the graft consists of a segment of the aorta containing the arteries that ultimately supply the pancreas and duodenum, and a segment of the hepatic portal vein containing the venous tributaries that drain the graft. In the recipient, the graft aortic segment and graft portal vein are anastomosed to the host aorta and caudal (inferior) vena cava, respectively,

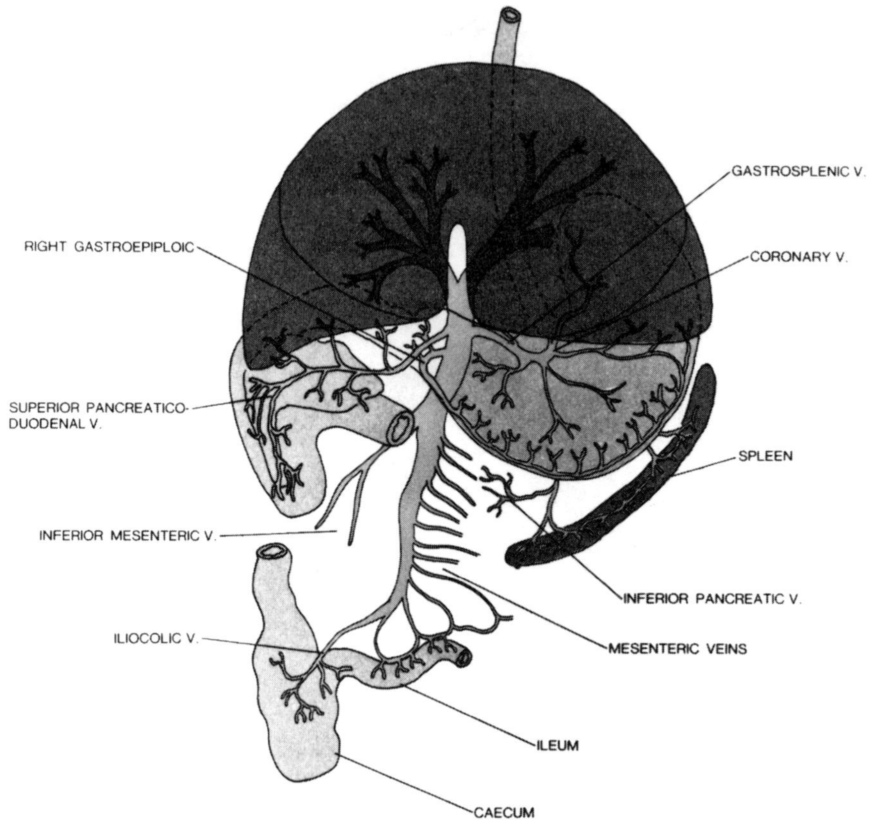

Figure 3 Venous drainage of the pancreas. The right lobe and body drain directly into the portal vein by way of the superior pancreaticoduodenal vein. The left lobe is drained by the inferior pancreatic vein, which drains into one of the splenic veins and then into the portal vein.

below the level of the renal vessels. Therefore, the anatomy of the abdominal aorta and inferior vena cava is germaine.

The abdominal aorta begins at the aortic hiatus between the left and right medial crus of the diaphragm and descends through the abdominal cavity in the midline until it bifurcates into the common iliac arteries at the level of the first sacral vertebra (Figure 4). The first branches of the abdominal aorta are the phrenic arteries that supply the diaphragm. The cranial adrenal arteries may arise from the phrenic arteries, or directly from the aorta or, on the right side, from the right renal artery. The second branch of the abdominal aorta, and the first unpaired branch, is the celiac artery which ultimately supplies part of the pancreas. It arises at the level of the third lumbar vertebra (upper pole of right kidney) and, after about 1 cm, branches into the splenic, left gastric, and hepatic arteries. The third major branch of the abdominal aorta, also an unpaired artery, is the cranial (superior) mesenteric artery which arises 3 to 5 mm caudal to the celiac artery and gives off branches to the small and large intestine and pancreas. The fourth major branches of the abdominal aorta are the renal arteries. The right renal artery arises immediately behind the cranial (superior) mesenteric artery and about 5 mm cranial to the left renal artery. The right renal artery passes behind (dorsal to) the inferior vena cava on its way to the right kidney. The fifth branches of the abdominal aorta are the paired testicular (spermatic) or ovarian (uterian) arteries. The sixth branches of the abdominal aorta are the paired deep circumflex iliac (iliolumbar) arteries. The left and right branches arise at different levels.

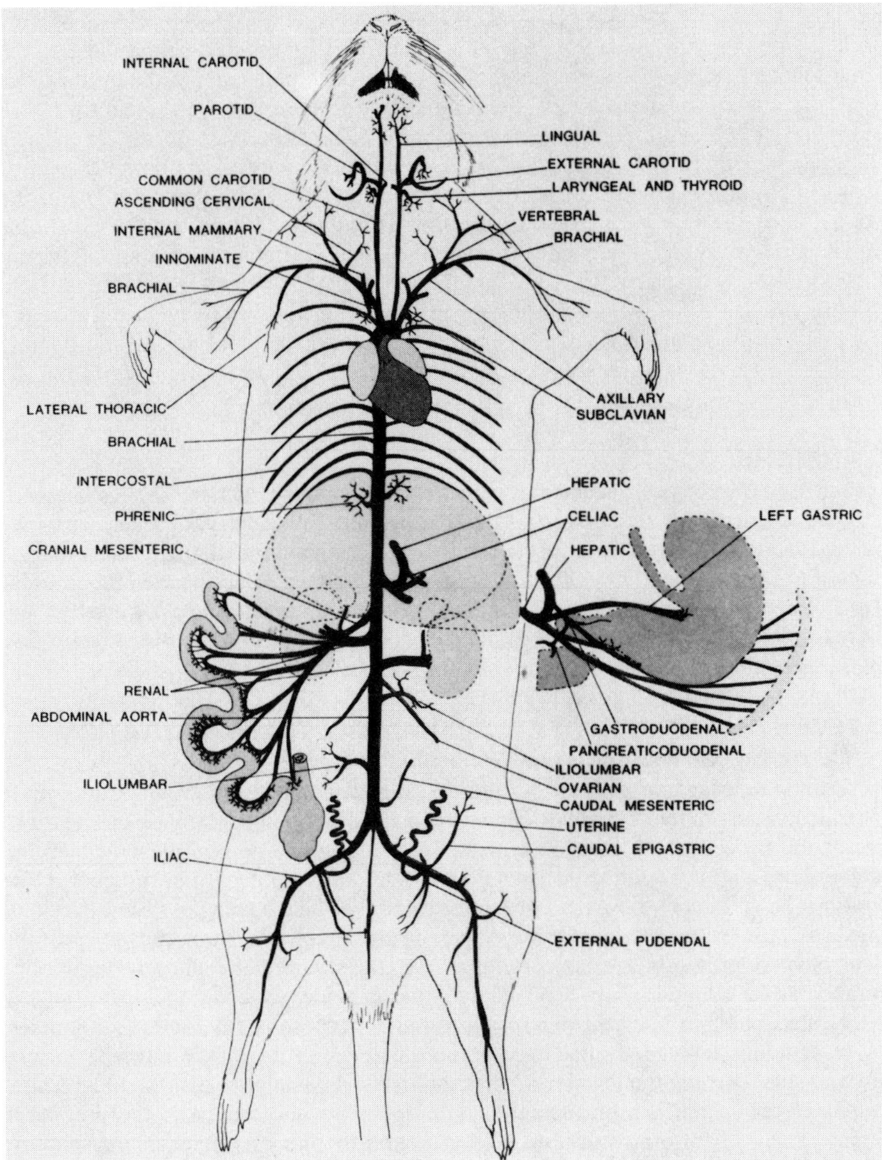

Figure 4 Major arteries of the rat as seen from the ventral surface. The important branches of the abdominal aorta involved in the whole pancreas transplant are the celiac artery and cranial (superior) mesenteric artery. In the recipient, the graft is anastomosed to the abdominal aorta below the level of the renal arteries.

The left deep circumflex iliac artery arises about 10 to 15 mm caudal to the left renal artery, while the right deep circumflex iliac artery takes off about 15 mm distally. The seventh branch of the abdominal aorta is the unpaired caudal (inferior) mesenteric artery that supplies the left colon and rectum and is much smaller than its cranial (superior) mesenteric sister. At the level of the last lumbar or first sacral vertebra, the abdominal aorta divides into the paired common iliac arteries and the single middle sacral artery. Finally, in its course through the abdomen the aorta gives rise to five unpaired lumbar arteries from its dorsal surfaces, each of which promptly divides into a right and left branch. Sometimes the branches originate separately from the aorta. The first lumbar artery arises close to the origin of the diaphragm, the second at the level of the right renal artery, the third cranial to the testicular (or ovarian) artery, the fourth close to the right deep circumflex iliac (iliolumbar) artery, and the last one near the aortic bifurcation.

The caudal (inferior) vena cava is formed by the fusion of the common iliac veins, dorsal and slightly caudal to the aortic bifurcation. It runs along the right side of the aorta as far as the visceral surface of the liver (Figure 5). Within the liver it drains seven hepatic veins and then passes through the caval foramen of the diaphragm into the thorax where it joins the right cranial vena cava to enter the right atrium. Beginning at its caudal origin, the tributaries that enter the vena cava are the paired iliolumbar veins, right spermatic (or ovarian) vein, paired renal veins, and hepatic veins. The left renal vein crosses the ventral surface of the aorta on its way from the left kidney.

B. ANESTHESIA

We have used a variety of injectable anesthetic agents and inhalant anesthetics in the rat over the years and have found most of them to be quite satisfactory. In recent years we have used an injectable anesthetic cocktail consisting of ketamine (50 mg/kg), acepromazine (1 mg/kg), and xylazine (5 mg/kg) mixed together in a single solution given intramuscularly. The depth of anesthesia is controlled by repeatedly monitoring the lid reflex and muscle movement. Operations are performed in a constant temperature room to minimize hypothermia and dehydration.

C. TECHNIQUE OF PANCREATICODUODENAL TRANSPLANTATION

1. Microsurgical Technique — General Comments

Pancreaticoduodenal transplantation in the rat is a demanding operation that requires microsurgical technique. However, any surgeon and many nonsurgeons can learn to perform the procedure if they are patient, have a genuine commitment to learning microsurgery, and are willing to spend a substantial amount of time practicing the operation. The relatively new field of microsurgery has been made possible by technological advances in the design of optical magnifying equipment. Numerous magnifying instruments are presently available, ranging from the simple binocular jeweler's loupe (a binocular lens mounted on a headband) that provides 2 to 4× magnification, to sophisticated, complex operating microscopes with foot controls, fiberoptic light sources, built-in cameras, television attachments, and observation arms. We have performed many transplants using the jeweler's loupe, but its disadvantage is that the focal length shortens as the magnification increases. The operating microscope is the preferable instrument for performing microsurgery in rats, although establishing coordination between the eyes and hands while looking through the microscope requires substantial patience and practice.

Because of the restrictive size of the operative field and the vertical plane of vision, instruments used in microsurgery must have extremely fine tips with limited movement and multiple configurations. Such instruments are readily available and some of them are shown in Figure 6A. Standard ophthalmic surgical instruments, jeweler's forceps, spring-handled scissors, a Castroviejo needle holder, retractor for rat surgery, and a Lee portacaval shunt clamp are very useful. The essential manipulation in performing microsurgery is a precise, graded pinch-closure between index finger and thumb, with the instrument supported on the first web space.

Nylon is the most widely used suture material for microsurgery, and it is fixed by hand to a channel made in the needle. Suture sizes range from 8-0 to 11-0 (18 μm), approximately half the size of a human hair (Figure 6B). Needles come in several configurations and sizes, and selection is based on the type of tissue being sutured. We

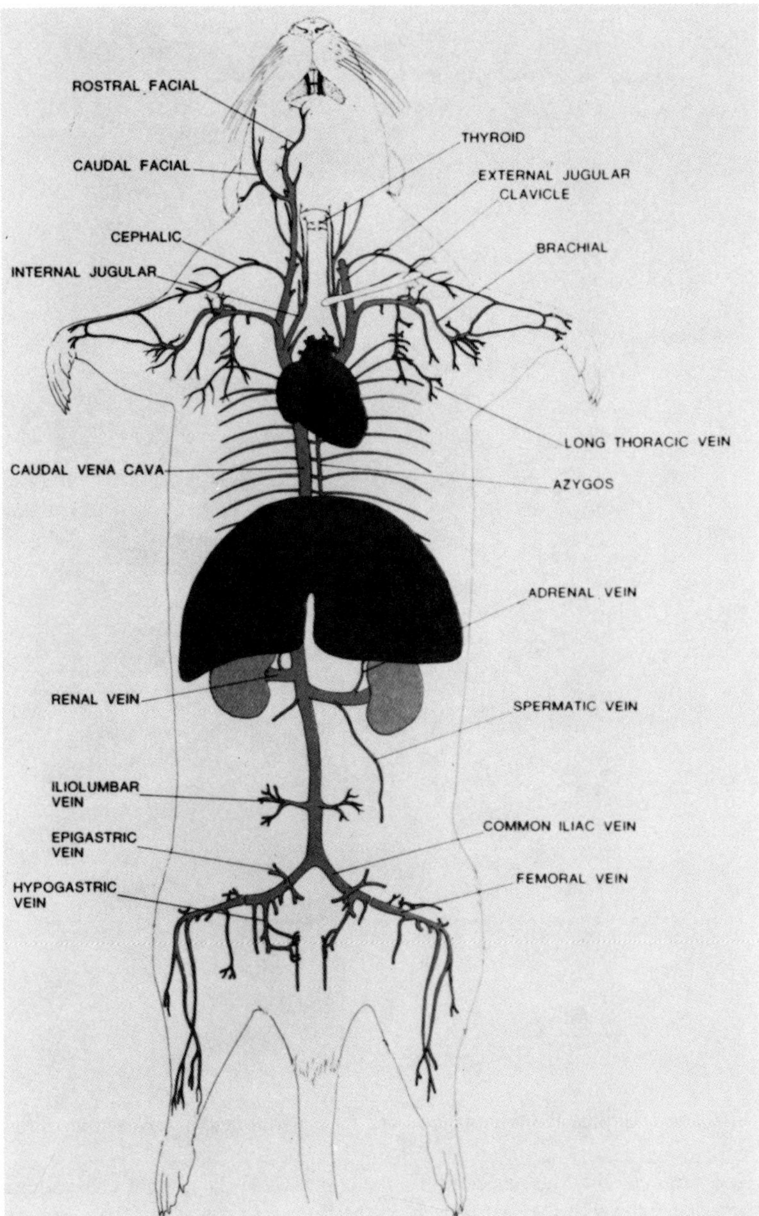

Figure 5 Major systemic veins of the rat as seen from the ventral surface. In whole pancreas transplantation, the portal vein of the graft is anastomosed to the caudal (inferior) vena cava below the level of the renal veins.

use 9-0 nylon sutures for the vascular anastomoses in the rat pancreaticoduodenal transplant, and 7-0 silk for the duodenal anastomosis.

The transition from performing surgery with the naked eye to operating with a microscope is large, and considerable practice is required to limit hand movements to a millimeter range. Certain steps can be taken to make the transition easier, as follows:

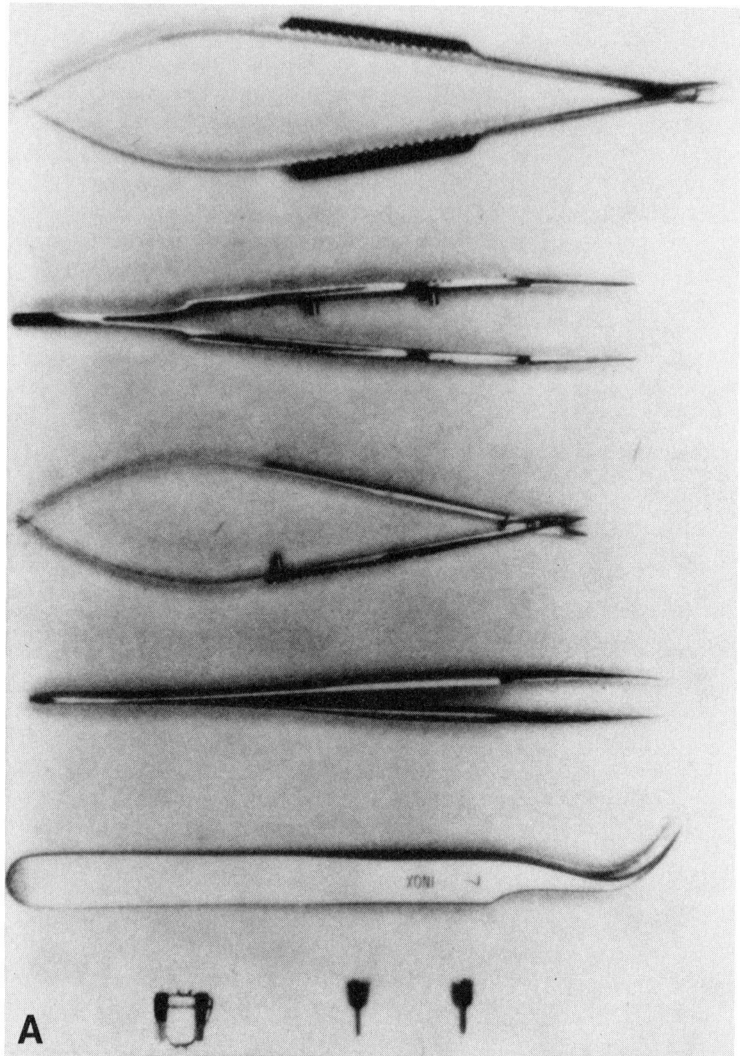

Figure 6A Microsurgical instrumentation used for organ transplantation procedures.

1. The surgeon should be as comfortable as possible, seated on a chair adjusted to the proper height with the feet resting on the floor or a foot rest. The chair should slide easily under the table permitting the surgeon's arms to rest comfortably on the table. Arm support is essential for hand stability when performing microsurgery.
2. The eyepiece of the operating microscope should be at a level that makes it unnecessary for the surgeon to bend his or her neck into an unnatural, uncomfortable position.
3. All supplies and equipment should be arranged so that the operator can get at ligatures, sutures, and instruments easily, with no need to look away from the operative field. Microsurgical sutures are so fine that the ends can be lost just from the air turbulence created by movements of the hands over the field. To avoid this problem, the abdominal incision should be draped with moist gauze which secures the ends of sutures and ligatures being used in the operative field.

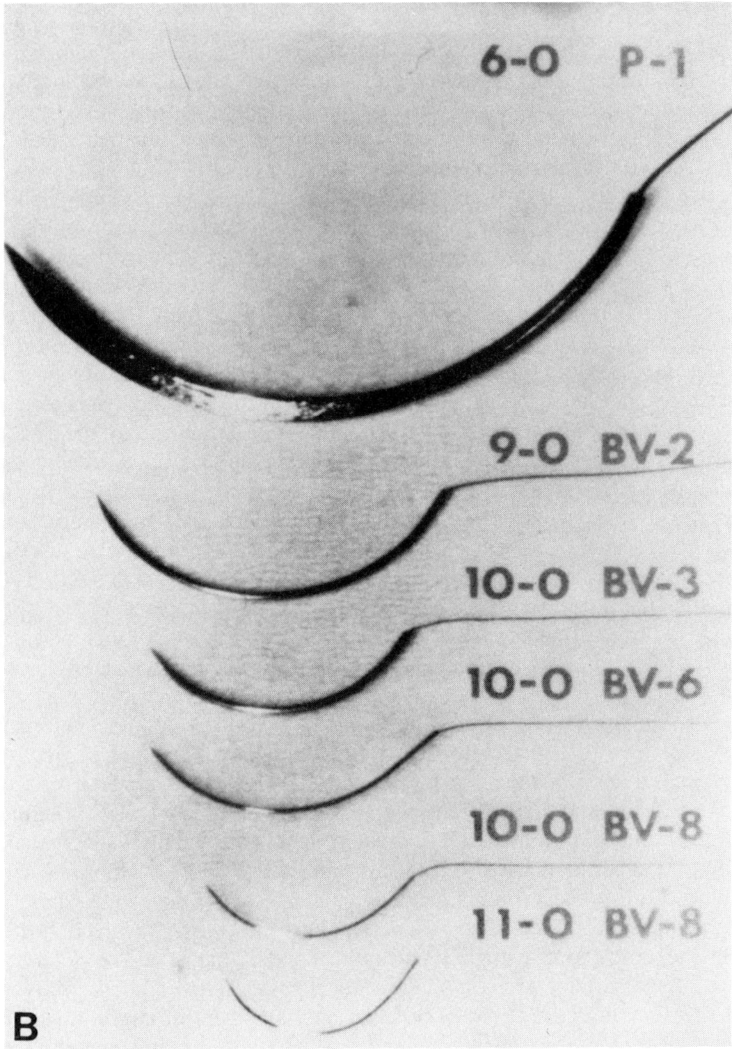

Figure 6B Microvascular suture material.

4. When not in use, the needles of vascular sutures should be hooked into a white styrofoam block taped to the operating table. Vascular sutures with swedged-on needles are very fine and nearly invisible on a dark surface.
5. Needles should never be locked in a needle holder when suturing a blood vessel. The inevitable jerk that occurs when the lock is released is enough to displace a carefully placed suture joining two fragile vessels. When the surgeon finishes suturing, the needle should be left in the same place each time so that the surgeon knows exactly where it is. Much time is saved by not having to search for a needle that is outside of the field of magnification. Needles should never by left attached to needle holders.
6. All ties are done with instrument. Hand tying fine sutures or ligatures without breaking them is nearly impossible.
7. Sponging and some blunt dissection is done with cotton-tipped applicator sticks. Substantial bleeding can often be controlled by pressure with the cotton applicator.

D. THE DONOR OPERATION

We have performed whole pancreas transplants in animals ranging in weight from the tiny 25-g Chinese hamster to the 35-kg mongrel dog.[1-15] We have used essentially the same surgical technique in each species. Most of our studies in rodents have involved male Lewis-strain rats that weighed 300 to 350 g and had blood vessels that were 2 to 3 mm in diameter for the vascular anastomoses.

In all operations the anesthetized rat is fastened with pins in the supine position to a rat operation board that is 32 × 32 cm in dimensions and made of 6-mm-thick wood covered with 6-mm-thick cork, and sealed with a water repellent spray. The lower chest and abdomen are shaved and cleansed with alcohol or some other antiseptic solution.

The pancreaticoduodenal graft is harvested from the donor rat through a midline abdominal incision. The intestines are retracted to the left and the ligament of Treitz is divided. The duodenum is then gently lifted with the left thumb and index finger, and blunt dissection is used to separate the duodenum and attached pancreas from the ascending and transverse colon to expose the portal vein. Tributaries of the portal vein are ligated with 7-0 to 9-0 silk ligatures and divided. Next the stomach is lifted and retracted upward and the intestines are shifted to the right, exposing the aorta and its two major anterior branches, the celiac artery and the superior (cranial) mesenteric artery. The peritoneum is dissected off of the aorta, and the second and third lumbar arteries arising from the dorsal aspect of the aorta are ligated with 7-0 silk and divided. The splenic vessels in the hilum are ligated with 7-0 silk and divided close to the spleen, and splenectomy is done. The left gastric artery is ligated with 7-0 silk behind the stomach and divided, and the greater omentum is dissected off of the stomach and pylorus by blunt dissection.

A 6-0 silk ligature is tied around the duodenum 5 mm below the pylorus, and the duodenum is divided proximal to the ligature. The intestines are then shifted to the left, where they are retracted, and the bile duct and hepatic artery are isolated close to the liver, ligated with 7-0 silk, and divided. The portal vein is now ligated with 7-0 silk ligatures at the lower margin of the pancreas and divided. A 6-0 silk ligature is passed around the aorta below the superior (cranial) mesenteric artery and it is left loose. A right-angle clamp is placed across the aorta 1.5 cm above the take-off of the celiac artery. The graft is perfused with 7 to 8 ml iced 0.85% saline maintained at 4°C by injection through the aorta below the cross clamp, using a 10-ml syringe and 25-gauge needle. As the perfusion is concluded, the distal ligature around the aorta is pulled tight and tied, following which the portal vein is transected at the liver hilum and left open for anastomosis. The aorta is transected just below the cross clamp and just below the ligature, thereby leaving the proximal end open for anastomosis. The duodenum is divided distally at the caudal margin of the pancreas and left open for anastomosis. The pancreaticoduodenal graft is finally removed and placed in 0.85% saline at 4°C (Figure 7).

When transplantation of the duct-ligated pancreas without the duodenum is performed, the pancreatic ducts that enter the bile duct and duodenum are meticulously isolated, ligated with 9-0 ligatures, and divided. The tiny arteries and veins between the duodenum and pancreas are ligated and divided, and the pancreas is carefully separated from the duodenum. The pancreatic blood vessels in the mesoduodenum are carefully preserved.

E. THE RECIPIENT OPERATION

The abdominal cavity is entered through a midline incision. The intestines are retracted to the left. The peritoneum and adjacent tissues are dissected off of the aorta and inferior (caudal) vena cava from just below the level of the left renal vein distally to the take off of the inferior mesenteric artery. The aorta and inferior (caudal) vena cava are carefully

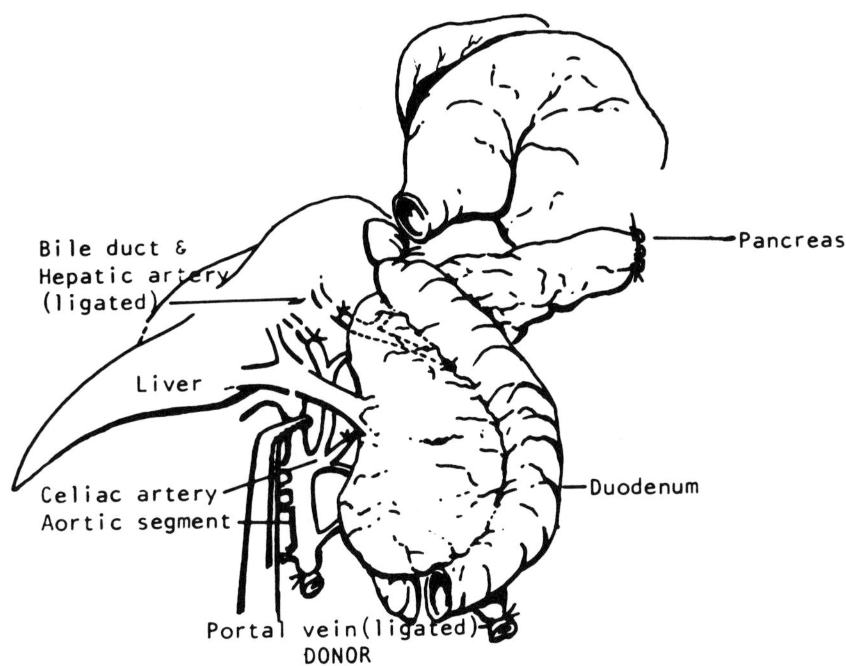

Figure 7 Removal of the pancreaticoduodenal graft from the donor. The duodenum has been ligated and divided at the pylorus and the distal end has been divided but left open for anastomosis. Splenectomy has been done. The portal vein has been ligated and divided below the pancreas and remains to be divided at the liver hilum. The bile duct and hepatic artery have been ligated and remain to be transected. The aortic segment with the attached celiac and superior mesenteric arteries has been ligated and divided inferiorly and cross-clamped superiorly in preparation for transection.

separated from each other. The testicular (or ovarian) artery is ligated and divided if it is in the way.

The aorta and inferior vena cava are occluded together with a curved Lee portacaval shunt clamp, which is similar to a Satinsky clamp. The pancreaticoduodenal graft is positioned in the abdominal cavity for anastomosis of the graft blood vessels to the aorta and inferior vena cava. An ellipse is cut out of the anterior wall of the recipient aorta similar in length to the open end of the graft aortic segment. It is important to make a shallow cut into the aorta, since a deep U or V incision will result in narrowing of the lumen when the anastomosis is sutured. Also adventitia should not be removed beyond the edge of the incision or the aorta will be easily torn when the anastomosis is attempted. A venotomy is performed in the anterior wall of the vena cava similar in length to the open end of the graft portal vein. A narrow ellipse is cut out of the vena cava wall. End-to-side anastomoses are performed with continuous 9-0 nylon sutures between the graft aortic segment and the recipient aorta, and the graft portal vein and the recipient inferior (caudal) vena cava. The aortic anastomosis is performed first (Figure 8). Corner stay sutures are placed beginning and ending on the outside of each vessel. One stay suture is tied and the needle is passed into the lumen (inside) of one vessel. A continuous posterior suture is then run from inside until the opposite end is reached, when it is brought to the outside and tied to the corner stay suture at that end. The anterior row of the anastomosis is performed easily from outside with a continuous

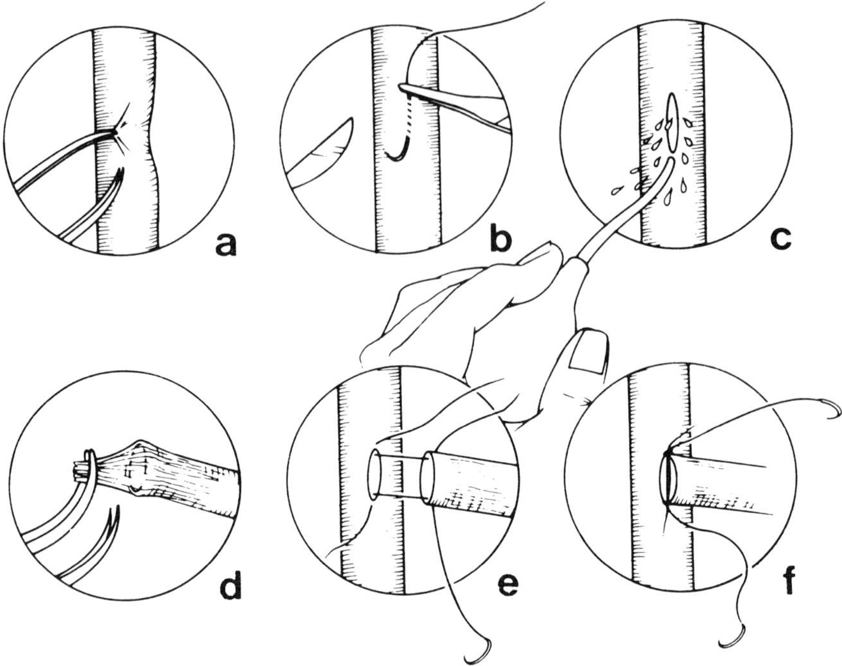

Figure 8 Anastomosis of the open end of the graft aortic segment to the side of the recipient aorta: (a) an ellipse of the aortic wall is excised or (b) alternatively, a curved needle is inserted in the long axis of the aorta and a scalpel incision is made in the vessel as the needle is lifted; (c) the lumen is flushed with saline; (d) the open end of the donor aortic segment is trimmed; (e) stay sutures are placed in the corner of each vessel and one of these is tied; (f) the needle of the tied stay suture is brought inside the lumen and the posterior continuous row of the anastomosis is performed from inside. When the opposite corner is reached, the needle is brought outside the lumen and the suture is tied to the stay suture. The anterior continuous row of sutures is then placed (not shown).

over-and-over stitch. To allow visualization and avoid catching the back of the aortic wall with the suture, the last three or four stitches in the anterior anastomosis are left loose until the last stitch is placed. Then, gentle traction is made on each loop until satisfactory approximation is achieved, and the suture is tied to the corner stay suture. Anastomosis of the graft portal vein to the recipient vena cava is performed in a manner identical to the aortic anastomosis. The clamp occluding the recipient aorta and inferior vena cava is removed and gentle pressure is applied to the anastomoses with a gauze sponge for 1 to 2 min. If there is bleeding from either anastomosis, the field is irrigated with saline, and the clamp is reapplied. The point of bleeding is then sutured and the occluding clamp is again removed. The open (distal) end of the graft duodenum is anastomosed end-to-side to the recipient duodenum with 7-0 silk sutures. In some of our studies we have used a conventional two-layer anastomosis, and in others we have used a single-layer anastomosis with an inverting Connell stitch. Both types of anastomotic techniques have been satisfactory and we have not had any duodenal leaks. In the two-layer anastomosis, the outer layer consists of interrupted seromuscular sutures, the inner posterior layer is done with a continuous through-and through stitch, and the inner anterior layer is accomplished with a continuous inverting Connell stitch. Sometimes it is necessary to trim the open end of the graft duodenum to obtain satisfactory tissue for the anastomosis.

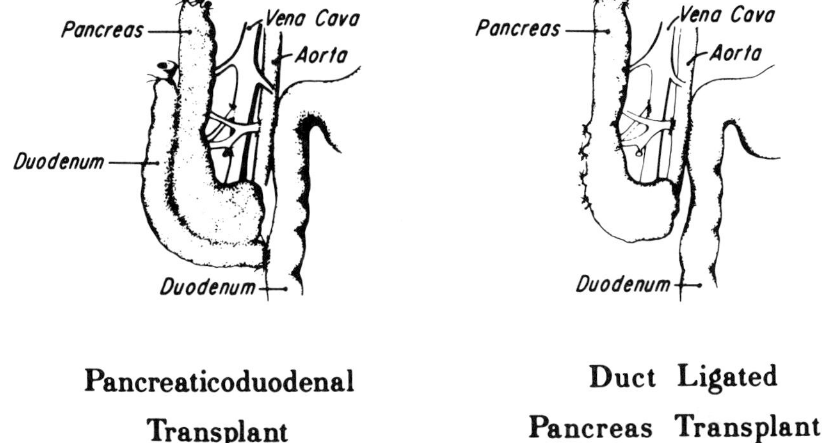

Figure 9 Two methods of whole pancreas transplantation — pancreaticoduodenal or duct ligated pancreas transplantation.

The duct-ligated pancreas graft without the duodenum is transplanted to the recipient by exactly the same technique as is used in the pancreaticoduodenal transplant except that no duodenal anastomosis is involved.

The two types of whole pancreas transplants are diagrammed in Figure 9.

In our laboratory, the mortality rate of pancreaticoduodenal transplantation in nondiabetic rats over a period of 10 years has been about 10%. In rats with well-established diabetes, the mortality rate has been as high as 30%, depending upon the duration and severity of the diabetes. Such rats are much less healthy and much more fragile than nondiabetic or early diabetic animals.

Graft ischemia time, from the moment when the blood supply is cut off in the donor to the time when the graft is completely revascularized in the recipient, has averaged 25 ± 5 minutes in our laboratory.

III. APPLICATION

The development in our laboratory of a microvascular technique of heterotopic transplantation of the whole pancreas, with and without the duodenum, in highly inbred rats has made it possible for us and for others to perform extensive studies of the effect of pancreas transplants on the metabolic abnormalities and pathologic lesions of experimental diabetes mellitus.[1,2] After demonstrating that pancreas transplants functioned vigorously for the life of the rat,[3,4] our research has been directed at answering the crucial question: Can whole pancreas transplantation prevent, stabilize, and, most importantly, reverse the metabolic derangements and renal, ophthalmic, neural, and vascular complications of diabetes? In attempting to answer this question, we have performed long-term studies using highly accurate quantitative techniques.[7-14] The results of some of our studies are summarized below.

A. COMPARISON OF THE METABOLIC CONTROL ACHIEVED BY WHOLE PANCREAS AND PANCREATIC ISLET TRANSPLANTS

To compare the long-term effectiveness of whole pancreas transplantation and pancreatic islet transplantation in controlling the metabolic disorders of alloxan diabetes, metabolic studies were performed monthly for 2 years in four groups of highly inbred rats: (1) NC-116 nondiabetic controls; (2) DC-273 untreated alloxan-diabetic controls; (3) PDT-182 rats that received syngeneic pancreaticoduodenal transplants shortly after induction of diabetes with alloxan;[10,13] and (4) IT-92 rats that received an intraportal injection of at least 1500, but usually 2000, syngeneic pancreatic islets shortly after induction of diabetes with alloxan. Whole pancreas transplantation maintained strict metabolic control throughout the 2 years of study. In contrast, pancreatic islet transplantation failed to maintained precise metabolic control. In group IT, plasma glucose concentration initially fell to normal but then was elevated significantly above normal beginning with the third posttransplant month; plasma insulin levels declined progressively after the sixth posttransplant month; glucose tolerance tests had a diabetic glucose tolerance curve as a result of a markedly deficient plasma insulin response; and body weight gain and growth were significantly less than in group PDT. The results of these long-term metabolic studies may explain the effectiveness of whole pancreas transplantation and the ineffectiveness of pancreatic islet transplantation in preventing diabetic nephropathy (see below).

B. STUDIES OF PREVENTION OF DIABETIC LESIONS BY EARLY PANCREAS TRANSPLANTATION

1. Prevention of Diabetic Nephropathy by Whole Pancreas and Isolated Islet Transplants

This study was undertaken to determine if pancreas transplantation performed early in the course of diabetes is capable of **preventing** all of the lesions of diabetic nephropathy, and to compare the long-term preventive effectiveness of whole pancreas and pancreatic islet transplants.[10,12] Metabolic and morphologic studies were performed in four groups of rats: (1) NC-116 nondiabetic controls; (2) DC-273 untreated alloxan-diabetic controls; (3) PDT-182 rats that received syngeneic pancreaticoduodenal transplants not long after induction of diabetes with alloxan; and (4) IT-92 rats that received an intraportal injection of at least 1500 and usually 2000 syngeneic pancreatic islets soon after induction of diabetes with alloxan. Each month for 24 months after diabetes was well established five lesions were scored by light microscopy in 50 glomeruli and related tubules in each kidney by a "blind" protocol: glomerular basement membrane thickening, mesangial enlargement, Bowman's capsule thickening, Armanni-Ebstein lesions of the tubules, and tubular protein casts. There were progressive and highly significant increases in the incidence and severity of all five kidney lesions in the diabetic control rats compared with the nondiabetic control rats. No significant differences were found between the kidneys of Group PDT and those of Group NC, demonstrating that whole pancreas transplantation prevented all of the diabetic kidney lesions throughout the 2-year study period. In contrast, within 3 to 9 months after pancreatic islet transplantation and thereafter, the incidence and severity of the five diabetic kidney lesions were similar in Group IT and Group DC. Whole pancreas transplantation produced precise metabolic control of diabetes throughout the 24 months of study, whereas pancreatic islet transplantation did not accomplish complete metabolic control, particularly beyond the first several months after transplantation. The difference in the completeness of metabolic control achieved by the two types of transplants is the most likely explanation for their sharp difference in effectiveness in preventing diabetic nephropathy.

2. Long-Term Prevention of Mesangial Enlargement by Whole Pancreas Transplantation

Mesangial enlargement (ME) is one of the hallmark lesions of diabetic glomerulosclerosis and is strongly related to the clinical manifestations of diabetic nephropathy in patients. To determine if whole pancreas transplantation can **prevent** ME, we measured the kidney mesangium and all of its components for 28 months after induction of diabetes with alloxan in four groups of rats: (1) NC-55 nondiabetic controls; (2) DC-57 untreated diabetic controls; (3) PDT-97 rats that received a pancreaticoduodenal isograft 1 week after induction of diabetes; and (4) DLPT-126 rats that received a duct-ligated pancreas isograft 1 week after induction of diabetes.[10,12] The mesangium was measured by quantitative morphologic technique in which camera lucida tracing of the mesangium were made at × 1250, and were analyzed using an electronic planimeter connected to a calculator-computer. Diabetic control (DC) rats developed progressive and highly significant increases in total mesangial area, nuclear-free mesangial area (absolute mesangial area), and percentage of glomerular area occupied by nuclear-free mesangial area (relative mesangial area). Whole pancreas transplants (PDT and DLPT) **prevented** the abnormal increases in all indices of ME, and yielded results that were similar to those in the nondiabetic controls (NC). These results provide convincing evidence that whole pancreas transplantation performed early in the course of diabetes is capable of preventing one of the major lesions of diabetic nephropathy for life.

3. Prevention of Glomerular Basement Membrane Thickening by Whole Pancreas Transplantation

Quantitative measurements of glomerular basement membrane thickness (GBMT) were made over a 2-year period in unaltered control rats, diabetic controls, and diabetic rats with whole pancreas transplants performed early in the course of diabetes.[7] The diabetic controls had significantly greater GBMT than the unaltered controls at all time intervals. GBMT in the rats with pancreas transplants was not significantly different from the GBMT in unaltered controls. Pancreas transplantation prevented the increase in GBMT, measured by a highly accurate technique, for the life span of the diabetic rat.

4. Prevention of Diabetic Somatic Neuropathy by Whole Pancreas Transplantation

To determine whether pancreas transplantation is capable of **preventing** diabetic somatic neuropathy, metabolic studies and electron microscopic morphometry of the sciatic nerve were performed monthly for 2 years in four groups of highly inbred rats: (1) NC-28 nondiabetic controls; (2) DC-82 untreated alloxan-diabetic controls; (3) WPT-122 diabetic rats that received a syngeneic whole-pancreas transplant; and (4) IT-90 diabetic rats that received intraportal injections of 1500 to 2000 syngeneic pancreatic islets.[11] Five diabetic nerve lesions were quantitated by a "blind" protocol: intra-axonal glycogen deposits, axons with glycogen deposits, demyelinated axons, intact axoglial junctions in paranodal terminal myelin loops, and basal lamina thickness of vaso nervorum. Untreated diabetic control animals had significant and progressive increases in all five nerve lesions compared to nondiabetic controls ($p < 0.01$). Whole pancreas transplants produced precise metabolic control of diabetes and prevented development and progression of all five diabetic nerve lesions throughout the 2-year study period. Pancreatic islet transplantation produced strict metabolic control and prevented diabetic neuropathy for the first 6 months, but then diabetes recurred and nerve lesions that were similar in severity to those in untreated diabetic reacts developed. The finding that whole pancreas transplantation prevents diabetic somatic neuropathy adds to and extends our previous studies showing that whole-pancreas transplants prevent diabetic nephropathy.

C. STUDIES OF REVERSAL OF ESTABLISHED DIABETIC LESIONS BY WHOLE PANCREAS TRANSPLANTATION

1. Reversal of Mesangial Enlargement by Whole Pancreas Transplantation

To determine if whole pancreas transplants can **stabilize or reverse** the important diabetic kidney lesion of ME, we measured the mesangium for 24 months in three groups of rats: (1) NC-55 nondiabetic controls; (2) DC-57 untreated alloxan diabetic controls;

and (3) DT-190 diabetic rats that received syngeneic whole pancreas transplants following an open kidney biopsy 6, 9, 12, 15, 18, and 21 months after induction of diabetes with alloxan.[8] Kidney sections obtained for 2 years from diabetic controls and diabetic rats prior to transplantation showed progressive and highly significant increases in total mesangial area, nuclear-free mesangial area, and percentage of glomerular area occupied by nuclear-free mesangial area. Pancreas transplantation consistently produced a highly significant **reversal** of well-established ME, regardless of when it was performed. In most instances, the mesangial area was reduced to a size smaller than that measured at 6 months. This study represents the first quantitative demonstration of **reversal** of a major lesion of diabetic nephropathy by whole pancreas transplants.

2. Reversal of Diabetic Somatic Neuropathy by Whole Pancreas Transplantation

To answer the crucial question regarding reversibility of diabetic somatic neuropathy by whole pancreas transplantation, metabolic studies and electron microscopic morphometry of the sciatic and testicular nerves were performed monthly for 2 years in three groups of highly inbred rats: (1) NC-47 nondiabetic controls; (2) DC-90 untreated alloxan diabetic controls; and (3) DC-230 diabetic rats given syngeneic pancreaticoduodenal transplants 6, 9, 12, 15, 18, and 21 months after induction of diabetes.[14] Six diabetic nerve lesions were quantitated by a "blind" protocol: (1) loss of myelinated axons, (2) intra-axonal glycogen deposits, (3) axons with glycogen deposits, (4) demyelinated axons, (5) degenerating axons, and (6) loss of intact axoglial junctions in paranodal terminal myelin loops. In the DT group, testicular nerve specimens were obtained just before transplantation and at death so that each animal served as its own control. As we have observed previously in untreated diabetic controls, all six nerve lesions progressed relentlessly for 2 years, in contrast to nondiabetic controls ($p < 0.01$). Whole pancreas transplants produced complete metabolic control of diabetes for life, and reversed all six lesions in both sciatic and testicular nerves, even when done late in the course of diabetes. There was complete **reversal** of the nerve lesions when pancreas transplantation was done within 15 months of the onset of diabetes. These results provide the first demonstration of reversal of diabetic somatic neuropathy by any form of diabetes therapy, and extend our previous work in which whole-pancreas transplants were found to **prevent** both diabetic neuropathy and nephropathy and reverse mesangial enlargement in the kidney.

REFERENCES

1. **Lee, S., Tung, S.K., Koopmans, H., Chandler, J.G., and Orloff, M.J.,** Pancreaticoduodenal transplantation in the rat, *Transplantation,* 13, 421-425, 1972.
2. **Grambort, D.E., Lee, S., Storck, L.G., Chandler, J.G., Charters, A.C., Elliott, M.K., and Orloff, M.J.,** Pancreatic duct ligation and long-term endocrine function of pancreas transplants, *Surg. Forum,* 24, 299-302, 1973.
3. **Charters, A.C., Lee, S., Storck, L.G., Chandler, J.G., and Orloff, M.J.,** Long-term exocrine function of the pancreas transplant, *Am. J. Surg.,* 129, 16-22, 1975.
4. **Orloff, M.J., Lee, S., Charters, A.C., Grambort, D.E., Storck, L.G., and Knox, D.,** Long-term studies of pancreas transplantation in experimental diabetes mellitus, *Ann. Surg.,* 182, 198, 1975.
5. **Sayers, H.J., Lee, S., and Orloff, M.J.,** Pancreaticoduodenal transplantation in juvenile Syrian and Chinese hamsters, *Surg. Forum,* 26, 454-456, 1975.
6. **Lee, S. and Orloff, M.J.,** Techniques of vascular anastomoses and organ transplantation in the rat, *Proc. Int. Microsurg. Soc.,* 1, 131-152, 1975.

7. **Bell, R.H., Fernandez-Cruz, L., Brimm, J.E., Sayers, H.J., Lee, S., and Orloff, M.J.,** Prevention by whole pancreas transplantation of glomerular basement membrane thickening in alloxan diabetes, *Surgery,* 88, 31-40, 1980.
8. **Orloff, M.J., Yamanaka, N., Greeleaf, G.E., Huang, Y-T., Huang, D-G., and Leng, X-S.,** Reversal of mesangial enlargement in rats with long-standing diabetes by whole pancreas transplantation, *Diabetes,* 35, 347-354, 1986.
9. **Orloff, M.J., Greeleaf, G.E., Urban, P, and Girard, B.,** Lifelong reversal of the metabolic abnormalities of advanced diabetes in rats by whole pancreas transplantation, *Transplantation,* 41, 556-564, 1986.
10. **Orloff, M.J., Macedo, C., Macedo, A., and Greeleaf, G.E.,** Comparison of whole pancreas and pancreatic islet transplantation in controlling nephropathy and metabolic disorders of diabetes, *Ann. Surg.,* 206, 324-334, 1987.
11. **Orloff, M.J., Macedo, A., and Greenleaf, G.E.,** Effect of pancreas transplantation on diabetic somatic neuropathy, *Surgery,* 104, 437-444, 1988.
12. **Orloff, M.J., Macedo, A., Macedo, C., Yamanaka, N., Huang, Y-T., Huang, D-G., Leng, X-S., Stieber, A., Kreidieh, I., and Greeleaf, G.E.,** Prevention, stabilization, and reversal of the metabolic disorders and secondary complications of diabetes by pancreas transplantation, *Transplant. Proc.,* 30(Suppl. 1), 868-873, 1988.
13. **Orloff, M.J., Macedo, A., Greeleaf, G.E., and Girard, B.,** Comparison of the metabolic control of diabetes achieved by whole pancreas transplantation and pancreatic islet transplantation in rats, *Transplantation,* 45, 307-312, 1988.
14. **Orloff, M.J., Greenleaf, G., and Girard, B.,** Reversal of diabetic somatic neuropathy by whole-pancreas transplantation, *Surgery,* 108, 179-190, 1990.
15. **Lee, S., Mazzoni, G., DeMartino, C., and Orloff, M.J.,** *Textbook of Microsurgery,* Antonio Delfino Editore, Rome, 1982.

Chapter 11

Isolation of Pancreatic Islets for Transplantation Studies

Camillo Ricordi

CONTENTS

I. INTRODUCTION

Despite encouraging preliminary results in recent clinical trials, islet transplantation is still an experimental procedure. Potential indications for an islet transplant are: (1) chronic pancreatitis requiring total or near-total pancreatectomy (autotransplantation); (2) patients with insulin-dependent diabetes mellitus already requiring immunosuppression for a previous or simultaneous organ transplant (allotransplantation); (3) advanced malignancies in which a cluster of abdominal organs including the liver and the entire pancreas are resected (liver-islet allotransplantation).

Of the indications mentioned above, type I diabetes is the most significant. Diabetes mellitus is the most common endocrine disease, and as the fourth leading cause of death by disease in western countries, a worldwide public health problem.[1] Estimates for insulin-dependent diabetes (type 1) indicate a prevalence of 0.26% by 20 years of age in the U.S.[2] There is evidence that the incidence of this disease is increasing in several world populations.[3] Prolongation of life is currently achieved by current maintenance therapy with insulin, but an increasing number of diabetic patients are now being treated for complications[4] including end stage renal failure, which represents 10 to 40% of new patients on dialysis.[5] Diabetes is also the leading cause of new cases of blindness in patients over the age of 20.[1]

In patients with type 1 diabetes mellitus, insulin production by pancreatic islets progressively declines and finally disappears, as the beta cells within the islets are destroyed by an autoimmune process resulting from a complex interplay between genetic and unknown environmental factors.[6] Replacement therapy with exogenous insulin is imperfect and has been ineffective in preventing the chronic complications of this disease. Thus, alternative methods for total endocrine replacement have been explored, including transplantation of isolated islets as free grafts.[7]

0-8493-3629-5/94/$0.00+$.50

There are two main problems related to transplantation of islets as a free graft. The first is the need to physically separate and purify the islets (1 to 2 million in a human pancreas) from the non-endocrine tissue that constitutes 98 to 99% of the pancreas. The second problem is immune rejection. In fact, the small total volume of transplanted tissue makes it difficult to monitor episodes of rejection. When rejection is diagnosed, a significant fraction of the transplanted islets is already irreversibly damaged.

Recent reviews on the objectives, the concepts, the problems, and the progress of islet cell transplantation are available.[8-12] The idea of transplanting pancreatic tissue to reverse diabetes is a century old. In fact, in 1891 Minkowski demonstrated for the first time that it was possible to reverse diabetes in a dog after transplantation of pancreatic fragments.[13] The pioneering work of Minkowski was followed by the first attempt to reverse diabetes by transplanting pancreatic tissue to man.[14] In this case, Williams transplanted fragments of a pancreas from sheep into a diabetic child. This attempt, therefore, constituted additionally, the first islet xenograft. It was almost 30 years before the discovery of insulin. In those years, it was still controversial as to whether a sugar-destroying substance was produced by the pancreas. The procedures for isolation of the islets from the exocrine pancreatic tissue have been pioneered by the work of Moskalewski[15] and Lacy and Kostianovsky.[16]

Recent reports of short-term[17] and prolonged[18-22] insulin independence following human islet allotransplantation have indicated that it is possible to replace the endocrine function of the pancreas by an islet transplant to man. Even though rejection remains the major factor limiting clinical trials of islet transplantation, recent improvements in isolation technology and immunosuppressive therapy resulted in encouraging preliminary results. In fact, the procedures developed for the isolation[16] of rodent islets were ineffective in separating islets from the pancreas of larger mammals, including man. The development of more effective procedures for islet isolation and purification from large animals[11,24-27] and human[28-33] pancreata have resulted in significant increases in both number and purity of the separated islets. In addition, the use of more powerful immunosuppressive agents such as Cyclosporine A[17,19,32] or FK-506[18,22,23] has resulted in prolonged human islet allograft survival in some cases.

The purpose of this chapter is to provide a detailed description of the procedures for isolating and purifyng animal and human pancreatic islets.

II. EQUIPMENT AND MATERIALS FOR ISLET ISOLATION AND PURIFICATION

The islet isolation and purification procedure can be performed in a standard tissue culture laboratory equipped with at least one double biological safety cabinet (Class 100). If possible, it is preferable to have two biological safety cabinets to perform the islet separation and the islet purification in separate sterile environments. In addition, at least one refrigerated centrifuge (e.g., IEC PR7000, Fischer Scientific, Pittsburgh, PA) is required. Two refrigerated centrifuges are preferable for large animal isolation procedures, in order to collect larger volumes of pancreatic tissue in a shorter time.

The isolation chamber can be built as indicated in Figure 1 (isolation chamber for large animals, 500-ml volume) and Figure 2 (mini-isolation chamber of 50-ml volume for rodents or small specimens). The chambers can be built of any biocompatible material that is nontoxic and easily sterilized. A transparent chamber may be preferable in laboratories at the beginning of their experience in islet isolation, since it allows direct observation of the pancreatic tissues during the digestion process. The use of either glass or plastic material is acceptable for this purpose. For logistical reasons, we prefer a stainless steel digestion chamber, since it is easily sterilized by autoclaving and more durable than other models. An acceptable compromise is the placement of two transparent

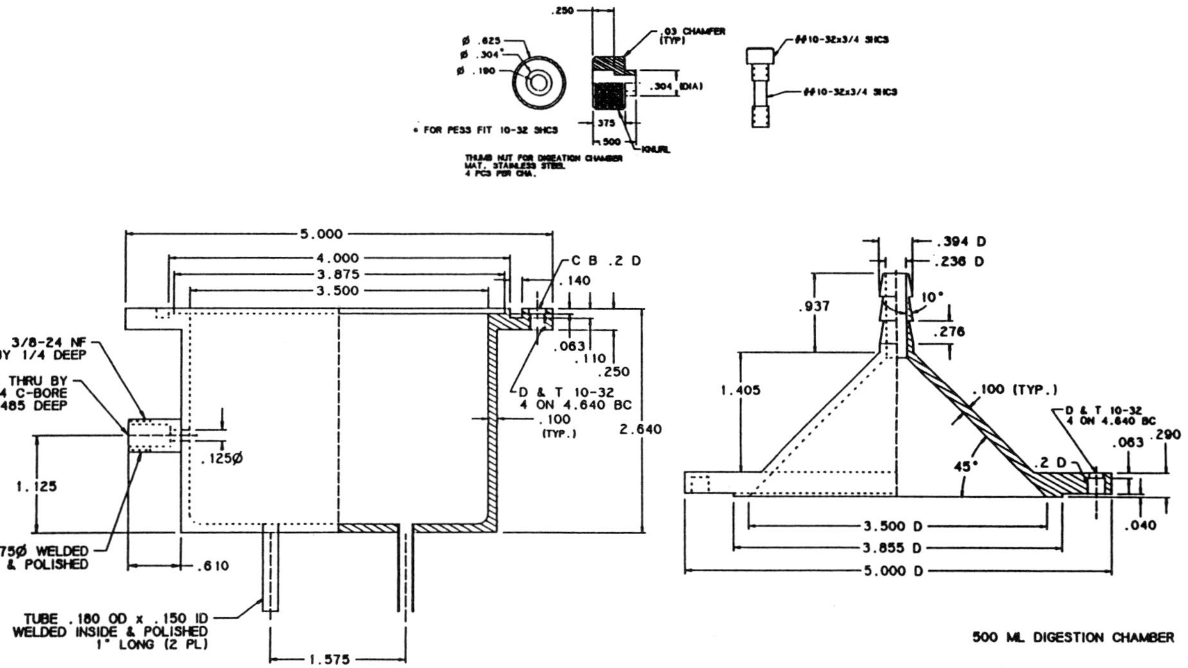

Figure 1 Blueprint of the isolation chamber for large animals and human islet separation.

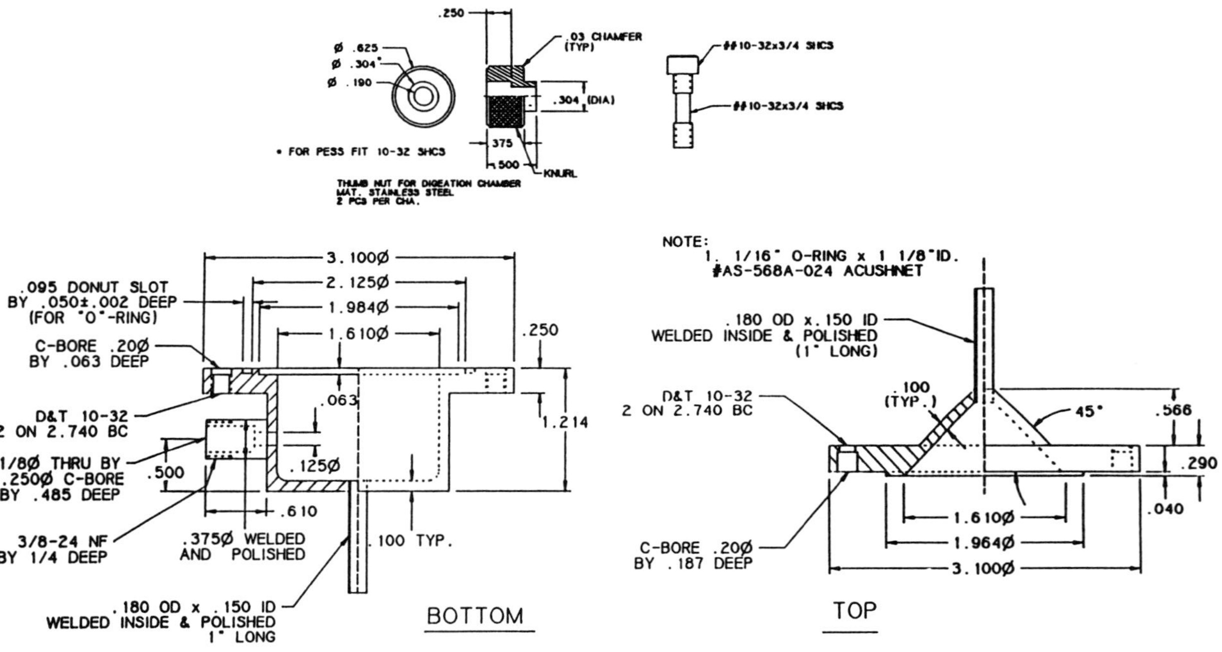

Figure 2 Blueprint of the isolation chamber for rodent islet isolations and for digestion of small pancreatic samples.

windows in the lateral walls of the chamber, directly in opposition allowing, from the anterior window, visualization of the chamber contents when illuminated by a light source near the posterior window. In this description of the isolation procedure we will refer to the stainless steel digestion chambers used at our institution for animals and human isolations. The selection of materials used to construct the chamber are important. For instance, varying thermal conductivity across the chamber walls can result in differing heat exchanges between the intra-chamber recirculating solution and the air. For example, a plastic chamber will exchange less heat with the room air during shaking, therefore, will require a lower temperature setting in the water bath to raise intrachamber temperature at a desired rate. Chamber materials can also alter the movement of the glass beads. In addition, for any given oscillation rate and amplitude of the digestion chamber, the rebound of the glass marbles will be always inferior on a plastic wall compared to a stainless steel wall. Any investigator, who is confident with large animal isolation procedures, can directly use a stainless steel chamber. Detailed blueprints of two stainless steel chamber models are shown in Figures 1 and 2. Currently, 500-ml chambers (Figure 1) are used for both large animal and human islet isolation procedures.

The use of a stainless steel screen is essential in order to retain the pancreas in the lower portion of the digestion chamber during the isolation process. It can be of differing pore sizes. After testing a variety of pore sizes we found it preferable to use screens of approximately 400 to 500 μ for large animal isolation procedures and smaller size screens (230 to 280 μ) for rodent islet isolations. Solid glass beads (seven for the large chamber; two for the small chamber) of approximately 1 cm in diameter, are used as enhancers during the digestion process. Without the use of marbles, the pancreas would move in unison with the chamber.

The main chemicals required for the islet isolation purification procedure are: Collagenase (e.g., Boehringer Mannheim Biochemical, Type P, Indianapolis, IN), Ficoll powder 400 DL (Sigma Chemical CO., St. Louis, MO), Hanks balance salt solution (HBSS) endotoxin tested, dimethylsulfoxide, diphenylthiocarbazone, Eurocollins kidney perfusion solution (Transplant Technology, Inc., Dallas, TX), Fetal calf serum (FCS), and CMRL 1066 culture medium. Useful equipment, highly recommended for the isolation lab, include a densitometer (DMA 35D, Aston PAAR, U.S.) to accurately determine the gradient densities required for islet purification, a digital temperature meter, and disposable temperature sensors (Mon-a-therm, Vital Hospital, Ballwin, MO). The remaining reagents and equipment are usually available in most cell culture labs.

Numerous, though expensive upgrades can be made in the aforementioned equipment. If clinical trials are planned, the laboratory and euqipment used for human islet preparation must be completely separated from the laboratory and equipment used for animal procedures. In this case, laboratory and equipment should be treated in the same manner as a regular operating room and appropriate sterile procedures and gowning should be adopted. If more than a few hours of treatment of the islets before transplantation are planned (e.g., 1-week *in vitro* culture), it is preferable to use a Class 100 laminar flow, positive pressure room. This will decrease the risk of contaminating the islets between isolation and transplant time. Also, the final islet preparation should never leave the room and islet culturing should be performed within the same laminar flow laboratory. Finally, the islet isolation facility should be temperature regulated, to avoid significant changes in room temperature that might introduce new variables in the isolation and purification process.

III. ORGAN PROCUREMENT

Human pancreases can be obtained from multi-organ donors[34-36] after *in situ* perfusion of the abdominal aorta with 1500 to 2000 ml of University of Wisconsin solution (UWS).

An additional 500 to 1000 ml of UWS are infused directly into the liver via the portal vein, which is encircled below the catheter tip to prevent retrograde leakage. Venous hypertension of the pancreas should be carefully avoided by venting the portal and/or splenic vein. The gland should be immersed in UWS and packed on ice until isolation. Some surgeons prefer to excise body and tail of the pancreas at the beginning of the procurement procedure without any perfusion of the gland.

In large and small animals the pancreas is generally excised without any perfusion of the gland. In rodents, it is preferable to immediately inject the collagenase solution before pancreatectomy and a few seconds after division of the inferior caval vein to exanguinate the pancreas (blood inactivates collagenase). In large animals it is preferable to perform a total or segmental (splenic portion) pancreatectomy, leaving the major vascular supply to the pancreas intact until the last minute. At this time, when the gland is dissected and is ready to be removed, the vessels should be ligated on the donor side only and then quickly divided. This allows blood outflow from the transected vessels on the organ side.

The transport solution (600 to 800 ml per pancreas) should be refrigerated to 4° before the pancreas is placed in the sterile transport container. UWS is preferable for long-term storage. However, the best purification results are generally obtained with glands with less than 8 hours of cold ischemia. Eurocollins solution can be used for shorter times. When the ischemia time before islet isolation is less than 1 hour, Hanks solution with 10% FCS could be used. Saline solution should be avoided in any case.

IV. ISLET SEPARATION

To separate pancreatic islets from large animals (dog, pig) and human pancreases, we use a modification[26] of the automated method for human islet isolation.[29] This procedure is relatively simple and reproducible in all species tested (man, dog, rabbit, sheep, swine, hamster, rat, mouse). The essential component of this system is a digestion chamber which retains the pancreas during a continuous digestion process. The chamber (Figure 1) consists of a lower cylinder and an upper conical portion and contains seven solid glass marbles (1 cm diameter) and a removable stainless steel screen separating the two portions. The lower portion has two inlet ports and an opening for a temperature sensor. The upper portion of the chamber has an outlet at the apex. Therefore, particles progressively released from the digesting pancreas are continuously removed and saved from further mechanical and enzymatic action. The 50-ml volume mini-chamber is built with the same design as the 500-ml chamber, only proportionally reduced (Figure 2). The main difference from the larger model is the single inlet, a smaller outlet port and only two marbles.

After distension with the collagenase solution, human and other large mammal pancreases of less than 120 g weight, can easily be loaded into the 500-ml digestion chamber. It is preferable to cut the gland at the neck and distend, in a retrograde fashion, the body and tail (splenic portion of dog and pig glands) and, in an anterograde fashion, the head (duodenal portion of dog and pig glands) previously occluded by clamping the duodenal outlet of the duct(s). The only exception is the adult swine pancreas, which can be significantly distended by the collagenase solution injection. We therefore prefer processing only the splenic portion of adult swine pancreases of more than 90 g, after division of the gland at the level of the annular ring around the portal vein. If the whole gland needs to be processed (e.g., autotransplantation), it is preferable to digest separately the splenic and duodenal portions.

Rodent pancreases are cannulated at the hepatic hilum, previously occluded with a clamp at the duodenal outlet, using a 24-gauge angiocath (rats) or a 27-gauge needle (mice).

After cannulation of the pancreatic duct, 4 to 10 ml (rodents) or 350 ml (large animals and man) of Hanks solution containing collagenase (Boeringher-Mannheim, Type P) are

Table 1 **Flow Diagram for Isolation of Pancreatic Islets**

Heat the chamber to 42°C: recirculation mode with Hanks solution with 2% FCS.

Dissect and discard peripancreatic fat, lymph nodes, vessels.

Divide the pancreas in two portions and cannulate the ducts with angiocatheters (18G for human; 20G for pig and dog, 24G for rodents [for rodents cannulation of the duct at the hepatic hilum]), clamp duodenal outlets.

Distend the pancreas with 350 ml of Hanks (large animals) or 4 to 10 ml (rodents) at 28°C containing 1 to 2 mg/ml (dog, pig, rodents) or 2 to 3 mg/ml (human) collagenase (Boehringer Mannheim, Type P).

Load distended pancreas and the solid glass beads (n = 7, 1 cm diameter) into the digestion chamber (n = 2 in the small chamber).

Position the stainless steel screen (pore size of approximately 400 μ for human, pig, and dog, and 280 μ for rodents).

Assemble chamber and start recirculation phase (flow rate 60 to 80 ml/min; 15 to 20 ml/min with mini-chamber for rodents) with gentle shaking to maintain a uniform intrachamber temperature.

Intrachamber temperature should raise 1 to 2°C/min until 37 to 38°C are reached.

Shaker at 10 cm excursion, 300 oscillation/min for 5 s/min.

Take samples every 2 min (1 ml in small Petri dishes) to monitor digestion.

When most of the islets appear free, switch to dilution mode: Hanks with 10% FCS is pumped into the system (flow rate = 250 to 300 ml/min for large animals; 50 to 60 ml/min for rodents) and the dispersed pancreas is collected in a 2-l flask preloaded with 1-l of Hanks with 10% FCS (1-l flask with 100 ml Hanks for rodents).

Shaker at 10 cm excursion, 100 to 320 oscillation/min.

Stop collection when no more islet tissue appears at a sample (after 15 to 30 min of dilution).

injected through the duct. In large animals, the temperature of the collagenase solution should be 28 to 32°C before intraductal injection. In rodents, due to the small dimension of the gland a cold solution (4°C) should be infused in the duct if several animals are processed at the same time. A small pancreas allows rapid equilibriation to the temperature of the intra chamber solution, thus resulting in an even digestion process. In contrast, in large animals, injection of a cold collagenase solution will result in a temperature gradient between the inner portion of the gland and its surface. This differential temperature produces uneven collagenase activities on different portions of the gland. This results in a faster digestion of the superficial parenchyma, while the central portion of the gland is underdigested. If samples are taken from the chamber to monitor digestion, a premature decision to switch to the dilution mode may be made, thereby leaving a significant portion of the gland undigested.

In rodents, dogs, and pigs we use a 1 to 1.5 mg/ml concentration of collagenase. Collagenase, at 2 to 3 mg/ml is used for human islet isolations. The wide variability in enzymatic activity between different lots of collagenase, even within the same collagenase type, makes it difficult to indicate a precise enzyme concentration to be used.

The pancreas is loaded into the previously described stainless steel digestion chamber (Figure 1 and 2) and the islets are isolated during a continuous digestion process that lasts 30 to 45 min (Table 1 and Figure 3).

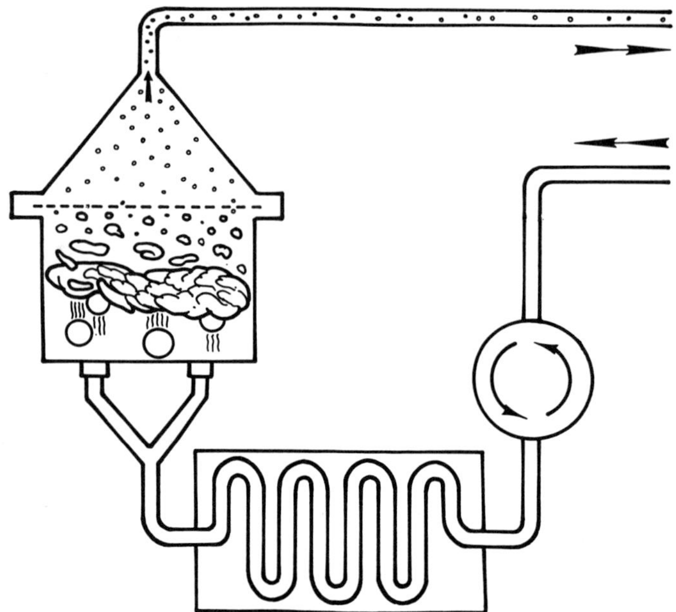

Figure 3 Schematic representation of the automated method for pancreatic islet isolation. A peristaltic pump propels the fluid through the heating circuit to the digestion chamber, in which the pancreas is digested, progressively releasing the islets.

The main modifications of this isolation procedure, compared to the automated method previously described,[29] are that the isolation chamber now has a 500-ml volume and an outlet port diameter of 6 mm. In addition, the pore size of the screen is increased from 280 to 400 to 500μ for human islet isolations, and the cooling system is eliminated as well as the heating circuit bypass, resulting in a simpler isolation apparatus (Figure 3).

During the recirculation phase (flow rate 85 ml/min), intrachamber temperature is increased at a rate of 2°C/min by passing the solution through a stainless steel coil immersed in a 45°C water bath until reaching 37 to 38°C. Partial withdraw of the coil from the water bath, once 37°C is reached, will prevent overheating the intrachamber contents. It is preferable to avoid any bypassing of the heating circuit to avoid overheating the solution which remains in the heated circuit during bypass. This would result in a flow of hot solution into the chamber upon restoration of the flow in the heating circuit. A slow increase in intrachamber temperature is necessary to allow equilibration of superficial and inner glandular temperatures during digestion thus avoiding uneven enzymatic action.

The contents of the chamber are gently mixed by slow shaking. Samples are first removed at 6 min and every 2 min thereafter to monitor digestion. After approximately 15 to 30 min of recirculation, the digestion is stopped by both dilution (4°C Hanks with 10% FCS, 400 ml/min flow rate) and cooling. The main factors influencing the decision on when to dilute are: (1) amount of tissue in the sample, (2) percentage of free islets, and (3) appearance of the islets and/or pH of the solution. As a general rule, it is appropriate to switch to the dilution mode whenever most of the islets are free from the surrounding exocrine tissue and also when significant amounts of tissue are present in a sample (1 ml can easily contain 50 to 100 islets and 1000 to 2000 exocrine aggregates). Whenever signs of overdigestion appear (fragmented islets, several ducts and/or ganglia, <pH), dilution is started even if a significant number of islets are still embedded. Upon initiation of the dilution phase, the chamber is connected to a shaker with an oscillation amplitude of 10

cm and a variable rate of 100 to 320 oscillation/min. The precise rate will be determined by the species and age of the pancreas used. As a general rule, porcine glands and young age pancreases require gentler shaking than human or older pancreases from the same species.

During the dilution mode, the digested tissue is rapidly collected in a 2-l sterile flask (containing 1 l of 4°C Hanks with 10% FCS) with subsequent volumes collected directly into 250-ml conical plastic bottles. The total dilution phase lasts 15 to 30 min. The digested tissue containing the islet cells is then pelleted by centrifugation (400 × g for 3 min at 4°C) using 250-ml conical plastic bottles. Pellets are resuspended in Hanks with 10% FCS into one or two bottles. After an additional centrifugation and washing, the final pellets are processed for purification.

V. ISLET PURIFICATION

It is important that the whole purification procedure be performed with keeping the tissue and the gradients on ice, since characteristics of the solutions constituting the gradients could damage metabolically active islets. It is therefore preferable to avoid purification procedures at room temperature when ficoll gradients are used.

Eurocollins solution is used as the vehicle for the Ficoll powder[17] (Ficoll DL-400, Sigma, St. Louis MO). Eurocollins-Ficoll at densities of 1.108, 1.096, 1.037 should be used in a three-layer discontinuous gradient[18] for human and dog islet purification. In rodent and pig islet purification, an additional layer at a density of 1.069 should be used. The digested pancreatic tissue is bottom-loaded with the 1.108 layer.

A manual (tubes) or a semi-automated (COBE) method can be used for discontinuous gradient purification of the islet tissue. A cell separator (COBE 2991, Lakewood, CO) can be used for centrifugation of the gradients and detailed descriptions on the use of the COBE are available.[37,38]

Autoclavable glass tubes or sterile transparent plastic tubes can be used instead of the COBE whenever small volumes of digested tissue need to be purified. This is the case in rodent islet isolations, or in any experimental situation in which the pellet of tissue to process is less than 10 ml. Generallly speaking, the maximum amount of tissue that can be processed in a tube for gradient purification should be 0.5, 1, and 3 ml for tubes of 15, 50, and 250 ml, respectively. The volume of gradient to be loaded in the bottom layer is 4, 15, and 80 ml for tubes with volume of 15, 50, and 250 ml. After even dispersion of the tissue within the bottom layer, the gradients will be layered on top of the first gradient containing the islets. The volume for each of the remaining gradients will be 3, 10, and 50 ml for tubes of 15, 50, and 250 ml volume, respectively. The choice of tube size will be determined by the species of the islets and by the volume of digested tissue to be purified. For rodent islets, 15-ml round-bottom siliconized glass tubes are preferable. For large animal purification procedures, both 50- and 250-ml round-bottom tubes are generally used. If the size of the pancreas will produce a prepurification pellet of less than 16 ml, 50-ml tubes are desirable, while for larger preparations the 250-ml tube will be more convenient. It is important to evenly resuspend the pelleted tissue within the densest layer. It is loaded after complete removal of the supernatant from the tubes. An uneven dispersion of the digested tissue within the gradient will reduce the postpurification recovery of islets and/or a decrease the purity of islets in the final preparation. When rodent islets are processed, a vortex device can be used to enhance the dispersion of endocrine and nonendocrine particles within the gradient. We prefer not to use any vortex device in large animal islet purification in order to avoid potential fragmentation of the islet tissue. This problem is exceptionally important in porcine islet isolation. Pig islets are particularly predisposed to fragmentation, especially when young animals (3 to 12 months) are used. In these cases, the tissue should be gently resuspended by slow hand-shaking or by careful aspiration through a 10-ml pipette with a wide pipette tip.

After the gradients are properly loaded, the islets are purified by centrifugation (4°C) at 800 × g for 10 (15- and 50-ml tubes) or 13 min (250-ml tubes).

The islets can be retrieved from the first and second gradient interfaces by aspiration with a Pasteur pipette (15- and 50-ml tubes) or with a 10 to 25-ml pipette (250-ml tubes). It is preferable to collect the layers separately since they could include islets with different degrees of purity.

The purity in the first layer (1.037 to 1.096 interface) is generally higher (70 to 95%) than the second layer (1.096 to 1.108 interface) which can also include embedded islets (islets enclosed within a thin rim of exocrine cells). The purity of the second layer is variable. It is generally between 40 and 60% endocrine particles, but in some cases almost all of the islets can be retrieved from the first layer and the second layer can be discarded without any significant loss in islets. The final islet preparation should then be washed three times with Hanks solution with 10% FCS by centrifugation initially at 800 × g for 3 min (first wash with significant ficoll still present) and then at 400 × g for 3 min (second and third wash). The last wash can be performed with culture medium (e.g., CMRL-1066) if the final preparation is to be cultured before transplantation.

VI. ASSESSMENT OF THE ISOLATED ISLET

It is necessary to have accurate methods for assessment of islet preparations, for otherwise comparisons of the results obtained by different investigators will be impossible.

The problem is less for techniques described for the isolation of islets from the pancreas of small animals, such as the rodent, where it is possible to actually count the numbers of islets retrieved. In larger mammals and man, where the yields obtained have been too large to count directly, development of precise and reproducible techniques for assessing islet yield in use in different laboratories is necessary. With the above difficulties in mind a workshop was organized at the 2nd Congress on Pancreas and Islet Transplantation in Minneapolis, to develop a consensus on criteria for assessment of islets isolated from the pancreas of large mammals, including humans.[39]

A. ISLET NUMBER, VOLUME, AND PURITY

The sampling technique represents a critical factor that could affect the results of an islet count as well as insulin content from any islet preparation.[39] Since islets settle rapidly in any container, care must be taken to properly resuspend the preparation before sampling to ensure a representative sample. It is preferable also to collect multiple samples to minimize counting errors (e.g., five aliquots of 50 μl after suspension of the islet preparation in a 200-ml flask). If the preparation is not needed for transplantation it is also wise to collect a larger sample that may be more representative (e.g., 1 ml from 200-ml islet suspension).

Even though experience facilitates the ability to distinguish an islet from a clump of exocrine cells, ganglia, lymph nodes, or so-called "membrane balls" at the microscopic level, it has been uniformly recommended that a specific stain for immediate detection of islet tissue should be used in any sample. For this purpose dithizone (DTZ) seems to be the most appropriate. An easy and rapid way to stain islets is to add a few drops of a DTZ solution (prepared as follows: add 50 mg DTZ to 5 ml dimethylsulfoxide [DMSO], stock solution, and dilute 1 ml of this solution with 20 ml of Hanks with 2% fetal calf serum [FCS]) to a sample contained in a petri dish. The solution must be freshly prepared and filtered before use. After staining the sample it is possible to count the islets that will appear red. Since islet volume is approximately proportional to the cube of the islet radius, it is crucial to divide the islets into diameter classes. For this purpose it is possible

Table 2 **Determination of Equivalent Islet Number**

Islet diameter (μ)	Conversion into islets of 150 μ diameter
50–100	n / 6
101–150	n / 1.5
151–200	n × 1.7
201–250	n × 3.5
251–300	n × 6.3
301–350	n × 10.4
350–400	n × 15.8
>400	n × 23.0

Note: For example if 60 islets of 50 to 100 μ diameter are counted in a sample, they will score only 10 (60/6) when converted into Equivalent Islet Number (EIN). In contrast, 10 islets of 201 to 250 μ diameter will score 35 (10 × 3.5). The sum of the EIN for the different diameter classes will provide the total EIN for the sample of islet preparation.

to use a calibrated grid in the eyepiece of the phase contrast microscope. As a matter of routine, most workers use 50-μ diameter range increments without considering particles smaller than 50 μ, since their contribution to the total volume of the preparation would not be significant. Table 2 indicates the relative conversion factors for islets in order to convert into standards of 150-μ diameter. These factors make it possible to convert the total islet number from any preparation into Equivalent Islet Number (EIN). In place of EIN, the total islet volume of the final preparation can also be estimated. EIN can be useful for those who not are familiar with conversion of the volume of a preparation into islet number. As a general rule, the more descriptive the results of an isolation are, the better their interpretation.

The dithizone stain described earlier is an easy method to roughly estimate the approximate degree of purity of the islets in any preparation.

B. MORPHOLOGICAL ASSESSMENT

The appropriate morphological assessment of isolated islet preparations should include three different aspects: confirmation of islet identification, assessment of the morphological integrity of the isolated and purified islets, and the assessment of the purity of the islet preparation.[39]

The accurate identification of islets by stereomicroscopy should be confirmed by subsequent light microscopic examination demonstrating positive aldehyde-fuchsin staining or immunoperoxidase insulin-labeling of B cells or, on a more complex level, by electron microscopic evaluation showing characteristic granules within the endocrine cells. Another method that permits rapid, accurate, and simple confirmation of islet identification within 3 to 4 hours is by double antibody indirect immunofluorescence of cryostat sections.

In the analysis of the morphological features, attention should be paid to the signals of traumatic damage such as irregularly shaped or fragmented islets, discontinuous plasma membranes, mitochondrial swelling with inner membrane disruption, cytoplasmic vacuolation due to massive dilatation of cisternal of the endoplasmic reticulum, reduction of cytoplasmic and nuclear matrix density or pyknotic nuclei.

Scanning electron microscopy and confocal microscopy are additional approaches to study surface changes of isolated islets.

C. VIABILITY ASSAYS

The use of fluorometric inclusion and exclusion dyes, together have been used to simultaneously quantitate the proportion of cells that are intact or that are damaged.[39]

The combination of inclusion and exclusion dyes acridine orange (AO) and propidium iodide (PI), have been extensively studied as an assay of islet viability. The dyes have minimal background fluorescence, and when used in optimal concentrations (AO: 0.67 μmol/l; PI: 75 μmol/l), stain living cells green and dead cells red. Utilizing this fluorometric method, viable and nonviable whole islets may be differentiated, as well as viable and nonviable components within an islet. The stability, cytotoxicity, and reproductivity of the assay have been demonstrated on isolated islets derived from mice, rats, dogs, and humans.

D. *IN VITRO* STUDIES

The isolation of large numbers of pure human islets is not the only consideration for successful transplantation studies. The isolation process itself may alter the functionality of the islets. It is thus crucial that the isolated islets be not only shown to be viable but also able to respond appropriately to glucose stimulation.[39]

Perfusion of islets with glucose provides a dynamic profile of the characteristics of glucose-mediated insulin release from prestored and newly synthesized stores, as well as the ability of the cells to down regulate insulin secretion after the glycemic challenge is interrupted.

The minimum values that should be reported are the absolute levels of insulin secretion during: the prechallenge basal period; the first 10 min (first phase) and last 50 min (second phase) of high glucose challenge; and the last 30 min after return to low glucose.

Although perfusion data provides a useful guide to islet viability, the quantity and kinetics of insulin release do not necessarily predict islet performance after implantation. Therefore, the ultimate test of viability is the functional activity after transplantation into a diabetic recipient.

E. *IN VIVO* STUDIES

Islet viability can be assessed *in vivo* by transplantation into diabetic nude (athymic) rodents.[39] Such animals have a deficient immune system due to congenital thymic aplasia and are unable to reject transplanted xenogeneic tissue.

Therefore, nude mice and/or nude rats can be made diabetic with streptozotocin, and demonstrate long-term reversal of the blood glucose following transplantation of xenogeneic islets. Furthermore, the engrafted islets can be tested for their ability to responded appropriately to a glucose challenge.

ACKNOWLEDGMENTS

The Author wishes to thank Mr. Ramoin E. Poo for his invaluable assistance in the design and realization of the isolation apparatus. His generous contribution to islet transplantation will not be forgotten.

REFERENCES

1. **Harris, M.I., Hanaman, R.F., Eds.,** *Diabetes in America,* National Institutes of Health, Bethesda, MD, 1985, 85.
2. **La Porte, R.E., Fishbern, H.A., Drash, A.L., et al.,** The Pittsburgh insulin dependent diabetes mellitus (IDDM) registry. The incidence of insulin dependent diabetes mellitus in Allegheny County, Pennsylvania (1965-1976), *Diabetes,* 30, 279, 1981.

3. **Bennet, P.H.,** Epidemiology of diabetes mellitus, in *Diabetes Mellitus: Theory and Practice,* Rifkin, H. and Porte, D., Jr., Eds., Elsevier, New York, 1990, 357.
4. **Krolewski, A.S., Warram, J.H., Rand L.I., et al.,** Epidemiologic approach to the etiology of type 1 diabetes mellitus and its complications, *N. Engl. J. Med.,* 317, 1390, 1987.
5. **Goetz, F.C., Elick, B., Fryd, D., and Sutherland, D.E.R.,** Renal transplantation in diabetes, *Clin. Endocrinol. Metab.,* 15, 807, 1986.
6. **Eisenbarth, G.S.,** Type 1 diabetes mellitus: a chronic autoimmune disease, *N. Engl. J. Med.,* 314, 1360, 1986.
7. **Dubernard, J.M. and Sutherland, D.E.R.,** Introduction and history of pancreatic transplantation, in *International Handbook of Pancreas Transplantation,* Dubernard, J.M. and Sutherland, D.E.R., Eds., Kluwer Academic Publishers, Dordrecht, 1989.
8. **Mullen, Y., Clare-Salzler, M., Stein, E., and Clark, W.,** Islet transplantation for the cure of diabetes, *Pancreas,* 4, 123, 1989.
9. **Hering, B.J., Bretzel, R.G., and Federlin, K.,** Current status of clinical islet transplantation, *Horm. Metab. Res.,* 20, 537, 1988.
10. **Lacy, P.E.,** Islet transplantation, in *The Diabetes Annual,* Alberti, K.G.M.M. and Krall, L.P., Eds., Elsevier, Amsterdam, 1990, 245.
11. **Gray, D.W.R. and Morris, P.J.,** Developments in isolated pancreatic islet transplantation, *Transplantation,* 43, 321, 1987.
12. **Ricordi, C. and Starzl, T.E.,** Cellular transplants, *Transplant. Proc.,* 23, 73, 1991.
13. **Minkowski, O.,** Weitere Mittheilungen uber den Diabetes mellitus nach Exstirpation des Pankreas, *Berl. Klin. Wochenschr.,* 29, 90, 1892.
14. **Williams, P.W.,** Notes on diabetes treated with extract and by grafts of sheep's pancreas, *Br. Med. J.,* 2, 1303, 1894.
15. **Moskalewski, S.,** Isolation and culture of the islets of Langerhans of the guinea pig, *Gen. Comp. Endocrinol.,* 5, 342, 1965.
16. **Lacy, P.E. and Kostianovsky, M.,** Method for the isolation of intact islets of Langerhans from the rat pancreas, *Diabetes,* 16, 35, 1967.
17. **Scharp, D.W., Lacy, P.E., Santiago, J.V., et al.,** Insulin independence after islet transplantation into type I diabetic patient, *Diabetes,* 39, 515, 1990.
18. **Tzakis, A., Ricordi, C., Alejandro, R., et al.,** Pancreatic islet transplantation after upper abdominal exenteration and liver replacement, *Lancet,* 336, 402, 1990.
19. **Scharp, D.W., Lacy, P.E., Ricordi, C., et al.,** Human islet transplantation in patients with type I diabetes, *Transplant. Proc.,* 21, 2744, 1989.
20. **Warnock, G.L., Kneteman, N.M., Ryan, E., Seelis, R.E.A., Rabinovitch, A., and Rajotte, R.V.,** Normoglycemia after transplantation of freshly isolated and cryopreserved pancreatic islets in type I (insulin-dependent) diabetes mellitus, *Diabetologia,* 34, 55, 1991.
21. **Altman, J.J., Cugnenc, P.H., Tessier, C., et al.,** Epiploic flap: a new site for islet implantation in man, *Horm. Metab. Res.,* 25(Suppl.), 136, 1990.
22. **Ricordi, C., Tzakis, A., Carroll, P., et al.,** Human islet isolation and allotransplantation in 22 consecutive cases, *Transplantation,* 53, 407, 1992.
23. **Ricordi, C., et al.,** Human islet allotransplantation in 18 diabetic patients, *Transplant. Proc.,* 24, 961, 1992.
24. **Gray, D.W.R., Warnock, G., Sutton, R., Peters, M., McShane, P., and Morris, P.J.,** Successful autotransplantation of isolated islets of Langerhans in the cynomolgus monkey, *Br. J. Surg.,* 73, 850, 1986.
25. **Warnock, G.L. and Rajotte, R.V.,** Critical mass of purified islets that induce normoglycemia after implantation into dogs, *Diabetes,* 37, 467, 1988.
26. **Ricordi, C., Socci, C., Davalli, A.M., et al.,** Isolation of the elusive pig islet, *Surgery,* 107, 688, 1990.

27. **Alejandro, R., Curfield, R.G., Scheinvold, F.L., et al.,** Natural history of intrahepatic canine islet cell autografts, *J. Clin. Invest.,* 78, 1339, 1986.
28. **Gray, D.W.R., McShane, P., Grant, A., and Morris, P.J.,** A method for isolation of islets of Langerhans from the human pancreas, *Diabetes,* 33, 1055, 1984.
29. **Ricordi, C., Lacy, P.E., Finke, E.H., Olack, B., and Scharp, D.W.,** An automated method for the isolation of human pancreatic islets, *Diabetes,* 37, 413, 1988.
30. **Scharp, D.W., Lacy, P.E., Finke, E., and Olack, B.J.,** Low-temperature culture of human islets isolated by the distension method and purified with Ficoll or Percoll gradients, *Surgery,* 102, 869, 1987.
31. **Rajotte, R.V., Warnock, G.L., Evans, M., and Dawidson, I.,** Isolation of viable islets of Langerhans from collagenase-perfused canine and human pancreata, *Transplant. Proc.,* 19, 916, 1987.
32. **Alejandro, R., Noel, J., Latif, Z., et al.,** Islet cell transplantation in type I diabetes mellitus, *Transplant. Proc.,* 19, 2359, 1987.
33. **Sutherland, D.E.R.,** Pancreas and islet transplantation II. clinical trials, *Diabetologia,* 20, 435, 1981.
34. **Starzl, T.E., Todo, S., Tzakis, A., et al.,** Abdominal organ cluster transplantation for the treatment of upper abdominal malignancies, *Ann. Surg.,* 210, 374, 1989.
35. **Tzakis, A., Todo, S., and Starzl, T.E.,** Upper abdominal exenteration with liver replacement: a modification of the cluster procedure, *Transplant. Proc.,* 22, 273, 1990.
36. **Starzl, T.E., Miller, C., Broznick, B., and Makowka, L.,** An improved technique for multiple organ harvesting, *Surg. Gynecol. Obstet.,* 165, 343, 1987.
37. **Lake, S.P., Basset, P.D., Larkins, A., et al.,** Large-scale purification of human islets utilizing discontinuous albumin gradient on IBM 2991 cell separator, *Diabetes,* 38(Suppl.), 143, 1989.
38. **Alejandro, R., Strasser, S., Zucker, P.F., and Mintz, D.H.,** Isolation of pancreatic islets from dogs. Semiautomated purification on albumin gradients, *Transplantation,* 50, 207, 1990.
39. **Ricordi, C., Gray, D.W.R., Hering, B.J., et al.,** Islet isolation assessment in man and large animals, *Acta Diabetol. Lat.,* 27, 185, 1990.

Section IV

Heart Transplantation

Chapter 12

Heterotopic Heart Transplantation in Rodents

Frances A. Chapman, Donald V. Cramer, and Leonard Makowka

CONTENTS

I. INTRODUCTION

The use of rodents, particularly rats and mice, for microvascular surgical experiments in organ transplantation has many advantages. The experiments can be performed in a cost-effective manner, a limited number of personnel are required, and the availability of genetically defined inbred lines has allowed for the development of a variety of specialized surgical models. The major histocompatibility complex (MHC) of both the rat and mouse, for example, has been isolated in congenic inbred lines. The data derived from transplantation experiments performed in these congenic lines form the fundamental understanding of the role of this important gene complex in the stimulation of rejection. Experiments conducted in rodents have also provided important data on graft ischemia and preservation, the pathogenesis of rejection, immunosuppression, and the physiologic changes seen following transplantation of vascularized organs. The experimental data accumulated from rodent models has often provided the only available data for evaluating the application of many of these techniques to the clinical exchange of organ grafts in humans.

The technique of heterotopic vascularized cardiac transplantation (HHTx) in the rat was first described by Abbot et al.[1] in 1964. The coronary circulation in this model was established by using an end-to-end anastomosis of the donor thoracic aorta to recipient abdominal aorta, and donor pulmonary artery to recipient inferior vena cava. When this surgical technique was used, the recipient abdominal aorta and inferior vena cava were divided below the renal vessels. The lack of blood supply to the lower extremities resulted in a high incidence of recipient complications, including paraparesis and paraplegia. Careful selection of the donor and recipient body size to ensure matching of vessel diameters was an important consideration for the technical success of this operation. Despite these limitations, however, the use of this procedure established: (1) heart

0-8493-3629-5/94/$0.00+$.50

transplantation without mechanical methods for resuscitation was possible in the rat using two vascular anastomoses; (2) monitoring of graft function could be made by simple external palpation or electrocardiographic (ECG) recording; (3) rejection could occur without death of the recipient; and (4) syngeneic transplants could survive indefinitely. Ono and Lindsey[2] refined this model by employing end-to-side anastomoses of the donor vessels to recipient aorta and inferior vena cava, thereby eliminating the need for a complete transection of the recipient vessels. Graft and recipient survival increased dramatically to approximately 90% following this modification. A nonsuture technique for heterotopic heart transplantation in a cervical location in rats was subsequently reported by Hereon and colleagues in 1971.[3] In this model, the recipient external jugular and common carotid are anastomosed to the donor aorta and pulmonary artery using an end-to-end nonsuture cuff technique. The location of the heart graft in the subcutis of the neck allows for visual inspection of graft viability and ECG measurements can be made directly over the graft without interference of the bowel or abdominal adhesions. Thrombosis of the carotid artery, however, occurs in approximately 30% of the graft recipients.

A technique for heterotopic heart transplantation in the mouse was developed by Corry et al.[4] in 1973. The surgical approach described is the same procedure as that used in the rat, albeit technically more demanding. The primary advantage of this surgical model is the use of a larger number of genetically well-defined inbred strains of mice. The use of the mouse also allows for the utilization of highly specific immunological reagents for detailed examination of the immunopathology of the allograft reaction. Graft viability post-transplantation is evaluated using either ECG or direct abdominal palpation.

II. TECHNIQUES

A. ANATOMY

While several anatomical differences exist between the rat heart and those of higher mammals, they do not influence studies of transplantation rejection and are of more importance for experiments designed to investigate the hemodynamics of cardiac blood flow. The primary differences in the heart of rats and higher mammals include: (1) the distribution of coronary artery blood flow, (2) evidence for a dual blood supply in the rat to the right and left atrium of the heart, and (3) heart to body weight ratios that constantly decrease due to the continued increase in body size of the rat throughout its lifetime.[5]

1. Systemic Circulation

The rat heart has a dual blood supply that consists of coronary and extracoronary (cardiacomediastinal) arteries. The right and left coronary arteries originate at the root of the ascending aorta and cross the lateral wall of the right ventricle, and the anterior lateral wall of the left ventricle, rather than follow the atrioventricular grooves or anterior interventricular sulcus. The extracoronary blood supply arises from the right and left internal mammary arteries where it flows into the right and left cardiacomediastinal arteries to supply the atria. Unlike other species, the extracoronary perfusion is a major source of blood supply for the right and left atria.

2. Pulmonary Circulation

Blood from the systemic circulation enters the right atrium via the right and left superior vena cava (SVC) and the inferior vena cava (IVC). The azygous vein drains into the left SVC and blood enters the coronary sinus region with the IVC after crossing the posterior surface of the heart from the left. Blood flow then passes into the right ventricle, is pumped into the lungs for oxygenation via the pulmonary artery, and returns to the left

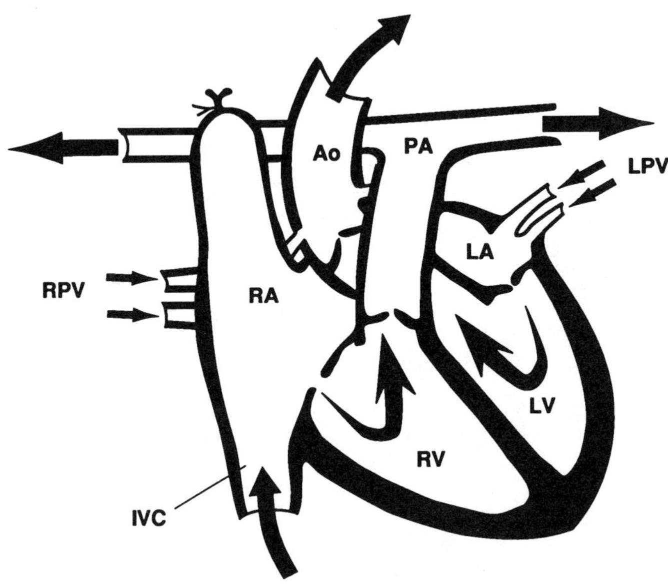

Figure 1 Schematic representation of the normal pulmonary circulation. Ao: aorta; IVC: inferior vena cava; RPV: right pulmonary veins; RA: right atrium; PA: pulmonary artery; LA: left atrium; LV: left ventricle; RV: right ventricle; LPV: left pulmonary veins. (From Silver, S.J., *The Rat in Microsurgery*, Williams & Wilkins, Baltimore, MD, 1980, 161. With permission.)

atrium through the pulmonary veins. From the left atrium, blood enters the left ventricle and is pumped throughout the body via the thoracic aorta (Figure 1).

B. DONOR OPERATION

All procedures are performed using clean but not aseptic techniques and standard microsurgical instruments. A Leitz operating microscope with a magnification of 3.2 to 8.0 or a 2.5 × magnification loupe is used for both the donor and recipient operation. The donor is anesthetized with 100 mg/kg Ketamine, 10 mg/kg Xylazine, and 0.05 mg/kg Atropine administered by intraperitoneal injection. Additional anesthesia may be provided using inhalation anesthesia (i.e., methoxyflurane). The thoracic area is clipped to remove hair and the surgical site is prepared with alternating 1% betadine and 70% ethanol scrubs. The donor thoracic cavity is entered via a butterfly incision of the rib cage near the mid-axillary line. The diaphragm is cut free and the rib cage is reflected anteriorly and secured to allow for adequate exposure. Gauze is placed over the exposed ribs to prevent damage to the heart during the donor operative procedure. The heart is rapidly cooled and flushed using 3 ml of lactated Ringers (LR) solution (4°C) containing 200 U/ml heparin and infused via the IVC. The IVC is then occluded with a bulldog clamp and the pericardial sac removed. Gauze cooled with LR is used to provide caudal retraction of the heart and the SVC is clamped to prevent the inflow of blood. Angled Castroveijo scissors are placed close to the base of the heart and the ascending thoracic aorta and pulmonary artery are exposed and simultaneously transected approximately 4 mm distal to their origin (Figure 2). The residual blood is gently massaged out of the ventricles. The IVC and SVC are then ligated with 5-0 silk ties and divided. A single ligature is used to secure the pulmonary veins and the heart is carefully freed from mediastinal connective tissue and lungs. The heart is then placed into cooled LR while the recipient is prepared for transplantation. The total ischemic time for the donor operation should not exceed 5 to 7 min.

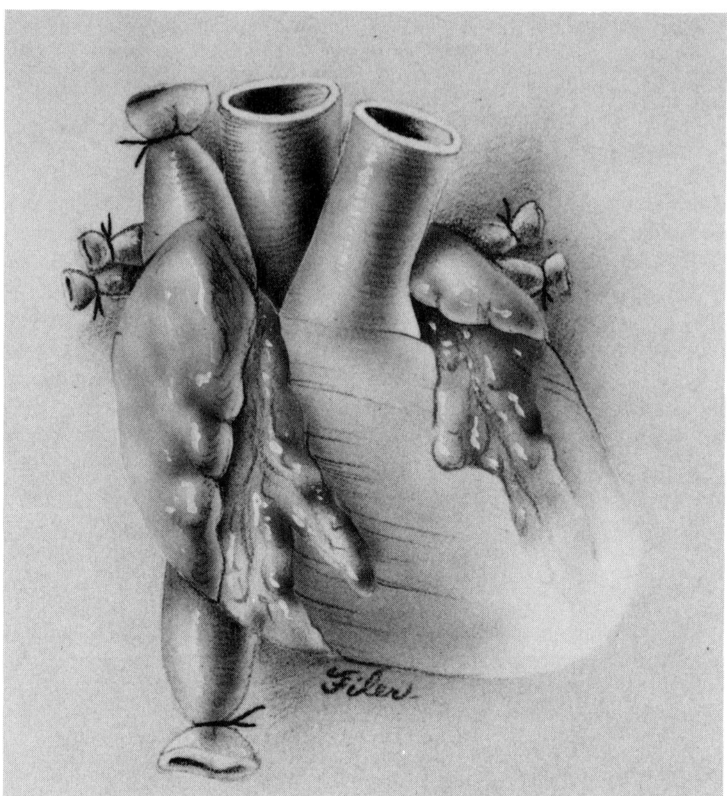

Figure 2 Appearance of the donor heart following the harvest procedure.

C. RECIPIENT OPERATION

The recipient is anesthetized and prepared for surgery as described above. A vertical midline incision is made from the xiphoid process to the pubis, retraction of the abdominal wall is performed, and the intestines are wrapped in warm moist gauze and reflected to the right. The connective tissue is removed from the aorta and IVC posterior to the renal vessels over a length of approximately 6 to 8 mm. The aorta is gently rotated to expose the minor lumbar vessels which are cauterized. A distal and proximal 3-0 cotton tie is placed below the renal arteries in order to occlude the aorta and IVC or, if preferred, a curved bulldog may be used to simultaneously occlude both vessels. The adventitia is dissected away from the aorta and a vertical incision of approximately 2 to 4 mm in length is made in the aorta 4 mm below the renal vessels. The vessel lumen is flushed with LR. The donor heart is placed in the recipient's left abdominal cavity and covered with gauze soaked with cooled LR. Two stay sutures of 10-0 Novafil on a TE-70 needle are placed at opposite corners of the aortic incisions to approximate the donor and recipient vessels. An end-to-side anastomosis of the anterior wall of the donor thoracic aorta to the recipient infrarenal aorta is performed using a continuous suture technique. The surgical table is rotated 180°, the heart is reflected to the animal's right, and the posterior wall of the aortic anastomosis is completed in the same fashion. A vertical incision is then made in the IVC and an end-to-side anastomosis of the posterior wall of the donor pulmonary artery to recipient IVC is performed through the anterior opening using a continuous suture technique of the donor pulmonary artery to recipient IVC. The anterior wall is then completed in the same fashion (Figure 3). The warm ischemic time for the vascular anastomoses should not exceed 15 to 20 min. In

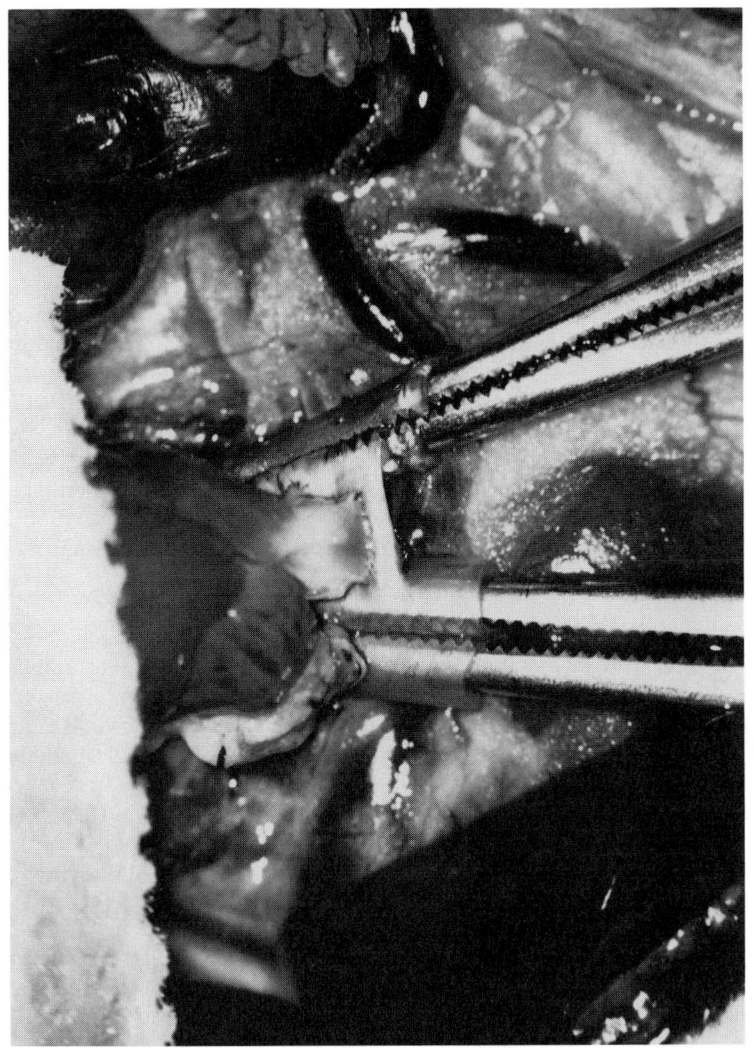

Figure 3 Completion of the venous anastomosis.

experienced hands, the entire surgical procedure for the donor and recipient can be completed in 40 to 50 min.

1. Establishment of Blood Flow

Once the transplantation procedure has been completed, the re-establishment of blood flow to the graft is an important component of the operation. Blood flow is established by first removing the distal aortic-IVC tie to allow preclotting in the suture holes and to minimize blood loss during this initial period. The proximal tie is then released (Figure 4) and hemostasis for any persistent bleeding is accomplished by gentle compression with a cotton-tipped applicator stick at the site of the hemorrhage. Patency of the vessel anastomoses can be established by direct observation. Additional sutures and reclamping should be avoided since these manipulations may result in increased number of technical failures due to vessel thrombosis. If bleeding should persist, gentle pressure with a cotton-tipped applicator or the use of a hemostasis accelerating agent, such as Surgicel (Johnson & Johnson, New Brunswick, New Jersey), is recommended. Recipient oxygenated blood immediately perfuses the coronary vessels and an effective ventricular contraction should occur within 30 sec. Figure 5 diagramatically illustrates the placement of the donor graft in the recipient abdominal cavity.

While the HHTx is a nonphysiologic, nonload heart model because of the altered cardiac blood flow, it is an important model for studying a wide variety of transplantation-related issues. The model represents a left ventricular bypass with a deletion of normal pulmonary circulation for reoxygenation of the blood through the lungs. Blood enters the transplanted graft via the aortic circulation, is perfused through the coronary arteries, veins, and enters the right atrium. Blood is then directed to the right ventricle and enters into the IVC through the pulmonary artery (Figure 6). Blood flow through the heterotopic graft is approximately 5% of the total blood volume per minute.[6] One common complication of the altered blood flow in the graft is the occurrence of thrombi in the lumen of the left and right ventricles.

2. Determination of Graft Function

The function of the HHTx can be monitored by abdominal palpation or ECG recordings. Either procedure is effective and the decision regarding the method of choice depends upon the individual preference of the investigator and the experimental application of the model. The determination of the cessation of graft function for both methods should be confirmed with a laparotomy. More recently, the use of P-31 nuclear magnetic resonance (NMR) spectroscopy has been proposed for the diagnosis of cardiac allograft rejection (see below).

Palpation for establishing the graft function can be easily accomplished through the abdominal wall of the recipient without anesthesia. The functional activity of the graft is recorded using a scale of 0 to 4+. This method is precise and allows for accurate determinations of rejection and survival to be made without recipient sedation or anesthesia. One of the difficulties of using this method are faint ventricular pulsations that may occur after the donor heart has stopped due to recipient aortic circulation in the transplanted heart.

ECG recordings are performed in anesthetized animals with the ECG leads connected to 25 gauge needles inserted subcutaneously in the lower abdomen and lower limbs of the rat[7] or mouse.[8] The results of the ECG recordings can be simultaneously compared to abdominal palpation. One difficulty in the use of the ECG for interpreting alterations in rat cardiac function is the rapid heart rate (350 to 450 beats per minute in the host) and the poorly defined S-T segment of the ECG.[5] This method does, however, allow a more accurate determination of the exact time of rejection. Abnormalities, such as bradycardia and arrhythmias, have been reported with ECG measurements conducted within several

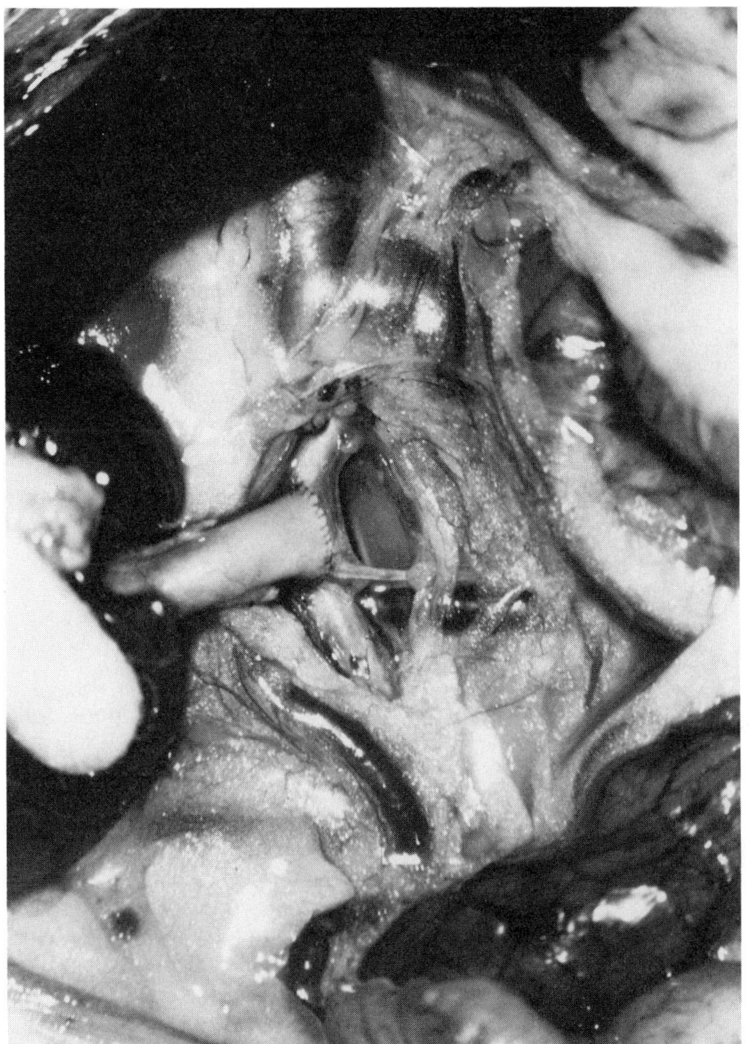

Figure 4 Reestablishment of cardiac blood flow in the transplanted graft.

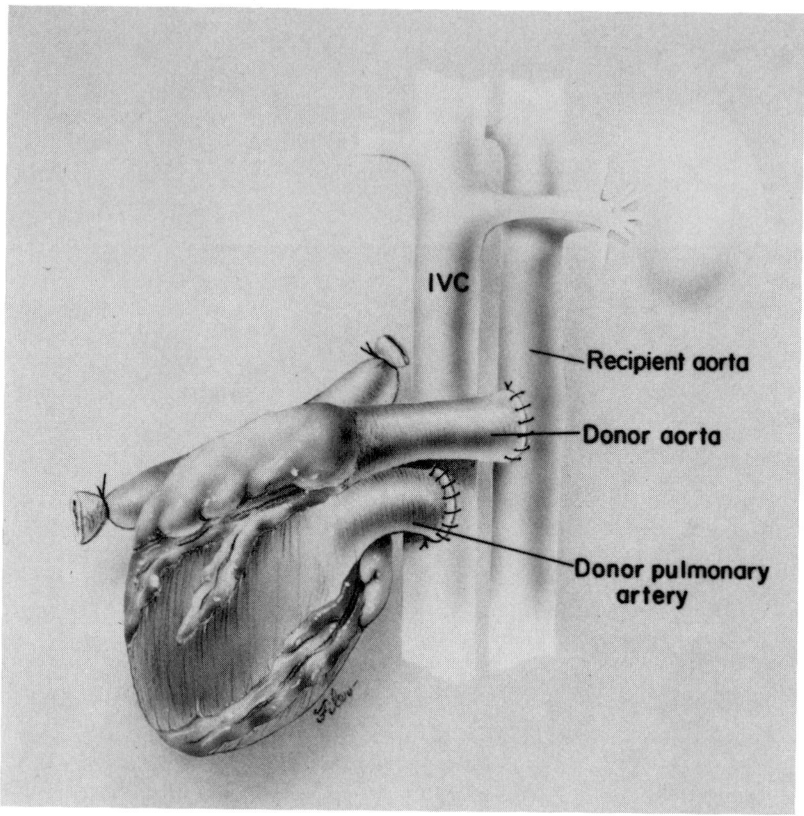

Figure 5 Diagramatic representation of cardiac graft placement.

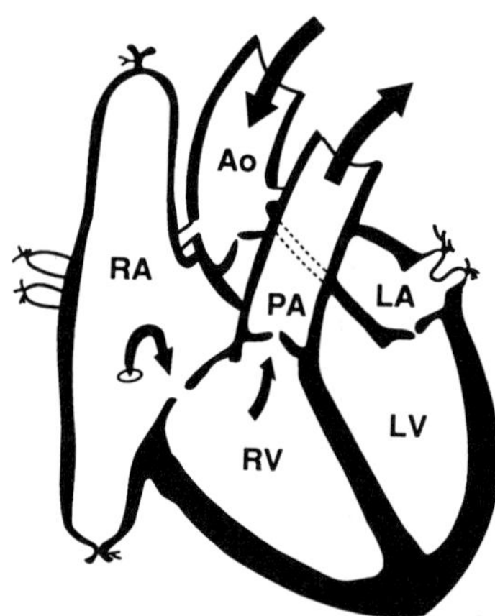

Figure 6 Schematic representation of the altered cardiac blood flow following transplantation. (From Silber, S.J., *The Rat in Microsurgery,* Williams & Wilkins, Baltimore, MD, 1980, 161. With permission.)

hours following transplantation. A normal wave form will generally resume during the next few days and remain normal in syngeneic recipients for as long as 1 year posttransplantation.[9] ECGs obtained from allotransplanted recipients demonstrate a gradual decrease in the voltage of ventricular complexes (R-wave voltage). The alteration in electrical activity of the transplanted graft correlates well with histopathologic evidence of rejection.[9]

NMR spectroscopy technology has been used in isolated experimental reports to assess cardiac organ function in the rat and generally relies on the study of myocardial metabolism in the transplanted graft as a method to more sensitively define allograft rejection.[10] NMR spectra are obtained using a 1.89 Tesla 30-cm horizontal bore magnet with a computer-interfaced spectrophotometer. Changes in phosphocreatinine, adenosine 5-triphosphate, inorganic phosphate, and intracellular pH are evaluated and are compared to histologic evidence of rejection. On this basis, rejecting allografts displayed relatively low levels of phosphocreatinine and high levels of phosphate as compared to syngeneic controls. It has been proposed that NMR technology would provide an effective, non-invasive measurement of clinical graft rejection.

III. EXPERIMENTAL APPLICATIONS

The heterotopic cardiac transplant procedure in rodents has been an important model for studying a wide variety of issues in solid organ transplantation. These studies have provided the basic experimental data for subsequent investigations of the same experimental problems in higher species, including man. The most important of these problems include studies of organ preservation/perfusion, the immune mechanisms that mediate allograft rejection, and donor/recipient issues for xenotransplantation.

A. ORGAN PRESERVATION

The use of the rodent heterotopic heart model for organ preservation has included studies to evaluate the efficacy of new preservation[11] and cardioplegia solutions,[12] and to evaluate the degree of myocardial injury present following hyperkalemic cardioplegia and prolonged hypothermic ischemia.[13] Graft viability following transplantation, enzymatic markers of tissue injury, and histologic evaluation of the heart are most commonly used as indicators of graft function and preliminary conclusions may be reached on the effectiveness of the procedure. McGregor and colleagues, for example, reported improved myocardial protection using the St. Thomas' cardioplegia solution as measured by serum creatinine kinase activity and long-term histologic changes observed in the transplanted hearts.[12] The University of Wisconsin solution (Viaspan, DuPont Merck Pharmaceutical Company) has also been evaluated in the HHTx model and these studies have demonstrated a significant difference in the degree of cardiac injury after 12 and 18 hours of static cold preservation when compared to the traditional Stanford solution.[11] These comparisons were based upon cardiac graft survival, adenosine 5-triphosphate levels, and histologic evaluations of the transplanted hearts. Although Viaspan proved to be a superior vehicle, the degree of morphologic myocardial injury present at 3 days post-transplantation raised considerable doubt as to whether the preserved hearts could support the circulation in an orthotopic position. Despite the "nonphysiological" nature of the HHTx model, investigations of this type have often provided important background information for future testing in larger animal models.

B. GRAFT SURVIVAL/REJECTION

One of the widest applications of the HHTx model has been in the evaluation of the immunological effector mechanisms that are responsible for mediating cardiac allograft rejection.[14] These include the production of antibody against graft antigens, cellular immune reactions, such as cell-mediated lymphocytotoxicity responses, DTH reactions, and nonspecific tissue damage due to inflammation. The acute rejection of cardiac allografts in both the clinical and experimental setting is associated with a heavy infiltration of the graft with inflammatory cells including T lymphocytes (T_h and T_c subpopulations), macrophages, and cells that phenotypically appear to be natural killer (NK) cells. Our current understanding of the specific effector mechanisms that these infiltrating cells represent in rejecting vascularized allografts is primarily based on the use of experimental models of adoptive transfer in rats followed by the heterotopic heart transplant.[15,16] Included within this general category of studies of allograft reactivity are a number of detailed studies of the influence of MHC and non-MHC histocompatibility differences on graft survival in rats[17] and mice.[18] The use of the HHTx model has demonstrated that vascularized grafts display reduced sensitivity to non-MHC histocompatibility differences when compared to skin or cardiac fragment grafts and have established the importance of Class II MHC differences in graft rejection.

The HHTx model in rats has also been used to evaluate the influence of pharmacologic agents on the modification of T- and B-cell mediated immune responses to allografts. Included among these studies are those designed to investigate the role of antibody,[19] cytokine elaboration,[20] the immunological basis of tolerance induction,[21] and the inflammatory mediators responsible for the immune reaction seen following heart transplantation.[22] The results of these and similar experiments have illustrated the role of activated $CD4^+$ and $CD8^+$ T lymphocytes in the pathogenesis of allograft rejection.

A number of the pharmacological studies conducted in rodents have included evaluations of new immunosuppressive drugs after cardiac allotransplantation.[23] Many of the pioneering studies of new immunosuppressive agents, including Cyclosporin A(CsA), 15-deoxyspergualin, FK-506, and Brequinar sodium, were initially tested for their ability to prevent graft rejection using the HHTx model in rats.[24-27] These studies provided initial baseline information for further testing in larger species, and for many, the basic efficacy of these drugs has proven to be consistent when tested in the clinics.

The HHTx model in rats has also proven to be useful for studying selected complications of the transplantation procedure, particularly cardiac graft arteriosclerosis. Cramer et al.[28] have demonstrated that the arteriosclerosis associated with cardiac transplantation is the result of histocompatibility differences between donor and recipient animals. The proliferative arterial lesions are very similar to those seen in patients with heart grafts that have survived for prolonged periods of time. Moreover, recipient modification with either immunosuppressive agents or sensitization to donor tissues, alters the severity and incidence of these lesions. The experiments suggest that the pathogenesis of the arteriosclerotic lesions is the result of a low-grade chronic rejection, rather than factors associated with immunosuppressive protocols.

C. XENOTRANSPLANTATION

The HHTx model in rodents has been used extensively to investigate the mechanism of concordant and discordant xenograft rejection and to examine the use of a variety of methods that would allow for graft acceptance. The exchange of organs between different small laboratory species may exhibit patterns of rejection that result in accelerated (mouse-to-rat; hamster-to-rat) or hyperacute (guinea pig-to-rat; hamster-to-guinea pig) rejection of the transplanted graft. The role of immunosuppressive agents,[29] total lymphoid irradiation,[30] antibody,[31] inflammatory mediators,[32] and recipient modifications have been widely studied in several of these xenograft combinations.

IV. CONCLUSIONS

The heterotopic heart transplant model in rodents has been widely used for experimental studies of organ transplantation. These experiments have provided important data on the (1) pathogenesis of graft rejection, (2) examination of organ ischemia/reperfusion injury and graft preservation, (3) evaluation of new immunosuppressive therapies, and (4) development of new technologies for application in the transplantation of vascularized organs. These investigations have provided important baseline information for additional studies performed in larger animal models and for the clinical exchange of organ grafts in humans. The surgical model requires limited personnel, it is cost-effective, highly reproducible, and may be perfected in a relatively short period of time. The model consists of two surgical end-to-side vascular anastomoses of the cardiac vessels and allows for the rejection of the graft without compromising the status of the recipient. Graft function can be easily assessed by abdominal palpation or ECG recordings and does not depend upon biochemical analysis of bodily fluids. Moreover, the recipient surgical procedure can either be performed using abdominal or cervical vessels allowing for the transplantation of more than one graft, including those from different species.

REFERENCES

1. **Abbot, C.P., Lindsey, E.S., Creech O., and DeWitt, C.W.,** A technique for heart transplantation in the rat, *Arch. Surg.,* 89, 645, 1964.
2. **Ono, K. and Lindsey, E.S.,** Improved technique of heart transplantation in rats, *J. Thorac. Cardiovasc. Surg.,* 57, 225, 1969.
3. **Hernon, I.,** A technique for accessory cervical heart transplantation in rabbits and rats, *Acta Pathol. Microbiol. Scand.,* 79, 366, 1971.
4. **Corry, R.J., Winn, H.J., and Russel, P.S.,** Primarily vascularized allografts of hearts in mice, *Transplantation,* 16, 343, 1973.
5. **Bishop, S.,** Cardiovascular research, in *The Laboratory Rat,* Academic Press, New York, 1980, 161.
6. **Silber, S.,** Microvascular surgery, in *The Rat in Microsurgery,* Williams & Wilkins, Baltimore, MD, 1979, 62.
7. **Abbott, C.P., DeWitt, C.W., and Creech, O.,** The transplanted rat heart: histologic and electrocardiographic changes, *Transplantation,* 3, 432, 1965.
8. **Superina, R.A., Peugh, W.N., Wood, K.J., and Morris P.J.,** Assessment of primarily vascularized cardiac allografts in mice, *Transplantation,* 42(2), 226, 1986.
9. **Hernon, I.,** The iso- and allotransplated rat heart, histological, electrocardiographic and serological Observations, *Acta. Pathol. Microbiol. Scand.,* 80(Sect. A), 9, 1972.
10. **Haug, C.E., Shapiro, J.I., Chan, L., and Weil, R.,** P-31 nuclear magnetic resonance spectroscopic evaluation of heterotopic cardiac allograft rejection in the rat, *Transplantation,* 44, 175, 1987.
11. **Makowka, L., Zerbe, T.R., Chapman, F., Qian, S., Sun, H., Murase, N., Kormos, R., Snyder, J., and Starzl, T.E.,** Prolonged rat cardiac preservation with UW lactobionate solution, *Transplant. Proc.,* 21, 1350, 1989.
12. **McGregor, C.G., McCullum, H.M., Hannon, J., Smith, A.F., and Muir, A.L.,** Long-term effects of cold cardioplegia myocardial protection in the rat, *J. Thorac. Cardiovasc. Surg.,* 87, 913, 1984.
13. **McGregor, C.G., Hannan, J., Smith, A.F., Muir, A.L., and Whearby, D.J.,** A study of cold cardioplegia myocardial protection in rats: an experimental model using uptake of technetium 99m pyrophosphate and enzyme activity as parameters of injury, *Cardiovasc. Res.,* 17, 70, 1983.
14. **Cramer, D.V.,** Cardiac transplantation: immune mechanisms and alloantigens involved in graft rejection, *Crit. Rev. Immunol.,* 7, 1, 1987.
15. **Hall, B.M., Pearce, N.W., Gurley, K.E., and Dorsh, S.E.,** Specific unresponsiveness in rats with prolonged cardiac allograft survival after treatment with cyclosporin, *J. Exp. Med.,* 171, 141, 1990.
16. **Wakely, E., Amram, C., Chapman, F., and Corry, R.J.,** Prolonged mouse heart allograft survival following adoptive transfer of transfused-induced suppressors, *Transplant. Proc.,* 19, 1445, 1987.
17. **Katz, S.M.,Leibert, M., Gill, T.J., Kunz, H. W., Cramer, D.V., and Guttmann, R.D.,** The relative roles of MHC and non-MHC genes in heart and skin allograft survival, *Transplantation,* 36, 96, 1983.

18. **Stepowski, S.M., Peugh, W.N., Wood, K.J., and Morris, P.J.,** The role of class I and class II MHC antigens in the rejection of vascularized heart allografts in mice, *Transplantation,* 44, 753, 1987.
19. **Diamantstein, T., Volk, H.D., Tilney, N.L., and Kupiec-Weglinski, J.,** Specific immunosuppressive therapy by monoclonal anti-IL2 receptor antibody and its synergistic action with cylcosporin, *Immunobiology,* 172, 391, 1986.
20. **Lowry, R.P., Wang, K., Vernooe, B., and Harcus, D.,** Lymphokine transcription in vascularized mouse heart grafts: effect of tolerance induction, *Transplant. Proc.,* 21, 721, 1989.
21. **Hall, B.M., Jelbart, M.E., and Dorsch, S.E.,** Suppressor T cells in rats with prolonged cardiac allograft survival after treatment with cyclosporin, *Transplantation,* 37, 595, 1984.
22. **Makowka, L., Chapman, F., Cramer, D., Sher, L., Podesta, L., Howard, T., and Starzl, T.E.,** The role of inflammatory reactions in xenotransplantation, in *Xenograft 25,* Hardy, M.A., Ed., Elsevier, Amsterdam, 1989, 159.
23. **Bordes-Aznar, J., Lear, P.A., Strom, T.B., Tilney, N.L., and Kupiec-Weglinski, J.W.,** Kinetics of cyclosporin-A-induced unresponsiveness to cardiac allografts in rats, *Transplant. Proc.,* 15, 500, 1983.
24. **Kostakis, A.J., White, D.J.G., and Calne, R.Y.,** Toxic effects in the use of Cyclosporin A in alcoholic solution as an immunosuppressant of rat heart allografts, *IRCS Med. Sci.,* 5, 243, 1977a.
25. **Suzuki, S., Kanashiro, M., and Ameniya, H.,** Effect of a new immunosuppressant, 15-deoxyspergualin, on heterotopic rat heart transplantation, in comparison with cyclosporin A, *Transplantation,* 44, 483, 1987.
26. **Murase, N., Todo, S., Lee, P.H., Chapman, F., Nalesnik, M.A., Makowka, L., and Starzl, T.E.,** Heterotopic heart transplantation in the rat receiving FK-506 alone or with cyclosporin, *Transplant. Proc.,* 19(Suppl. 6), 71, 1987.
27. **Cramer, D.V., Chapman, F.A., Jaffee, B.D., Jones, E.A., Knoop, M., Hreha-Erias, G., and Makowka, L.,** The effect of a new immunosuppressive drug, Brequinar sodium, on heart, liver, and kidney allograft rejection in the rat, *Transplantation,* 53, 303, 1992.
28. **Cramer, D.V., Chapman, F.A., Harnaha, J.B., Qian, S., and Makowka, L.,** Cardiac transplantation in the rat. II. Alteration in the severity of donor graft arteriosclerosis by modulation of the host immune response, *Transplantation,* 50, 554, 1990.
29. **Cramer, D.V., Chapman, F.A., Jaffee, B.D., Hreha-Erias, G., Yasunaga, C., Wu, G.D., and Makowka, L.,** The effect of a new immunosuppressive drug, Brequinar sodium, on concordant hamster-to-rat cardiac xenografts, *Transplant. Proc.,* 24, 720, 1992.
30. **Knechtle, S.J., Halperin, E.C., and Bollinger, R.R.,** Combined total lymphoid irradiation and cyclosporin promote xenograft survival, *Transplant. Proc.,* 19, 1137, 1987.
31. **Marquet, R.L., Heystek, G.A., and Van Leersum, R.H.,** Xenogeneic heart transplantation for the hamster to the rat, The effect of nonspecific and specific immunosuppressive treatment, *Int. J. Microbiol.,* 2, 113, 1980.
32. **Filipponi, F., Michel, A., and Houssin, D.,** Prolongation of guinea pig-to-rat xenograft survival with BN 52063, a specific antagonist of platelet-activating factor, *Ital. J. Surg. Sci.,* 19, 325, 1989.

Chapter 13

Heart, Lung, and Heart-Lung Transplantation in the Dog and the Pig

Ryo Saito and Paul F. Waters

CONTENTS

I. INTRODUCTION

Heart-lung transplantation has now been successfully performed in humans with end-stage cardiorespiratory failure. Many problems must still be solved but the development of clinical heart-lung transplantation has largely been based on the development of the surgical procedures.

0-8493-3629-5/94/$0.00+$.50

The first successful experimental heart-lung transplantation was Demikhov's in 1947. He transplanted the heart and lungs in dogs without using either hypothermia or cardiopulmonary bypass by making appropriate vascular connections.[1] Only 2 of 67 animals survived for 5 days. In 1953, heart-lung transplantation using the methods of hypothermia with circulatory arrest were performed in dogs by Neptune, with survival up to 6 hours.[2] In 1957, heart-lung transplantation using cardiopulmonary bypass in dogs was reported by Webb and Howard, with survival ranging from 75 min to 22 hours.[3]

In the initial experimental orthotopic heart transplantation (Demikhov, 1951;[1] Webb, 1957[3]), all vessels associated with the heart were anastomosed in dogs. At least eight vascular (two venae cavae, four pulmonary veins, the pulmonary trunk, and aorta) anastomoses were performed in these cases. The first successful orthotopic heart transplantation in dogs achieved by Lower and Shumway in 1959,[4,5] was simpler and more reliable. In these cases, four anastomoses (the right atrium, the left atrium, the aorta, and the pulmonary trunk) were performed, and the heart conducting system remained undamaged.

The initial experimental single lung transplantation was performed in dogs by Demikhov in 1947,[1] Juvenelle in 1951,[6] and Hardin in 1954.[7] Experimental *en bloc* double lung transplantation in dogs with cardiopulmonary bypass was reported in 1972 by Vanderhoeft[8] and without cardiopulmonary bypass in 1976 by Grosjean.[9] In these cases the operations were performed via right thoracotomy. In 1986, *en bloc* double lung transplantation in dogs using cardiopulmonary bypass via median sternotomy was performed by Dark.[10] Sequential double lung allotransplantation in dogs was reported in 1971 be Veith[11] and in 1972 by Kondo.[12] Subsequently, few experimental sequential double lung allotransplants have been reported, although clinical sequential double lung allotransplantation has been performed since 1985.[13] Since 1964,[14-17] many sequential double lung autotransplants have been performed in the dog.

Total cardiopulmonary denervation, however, seems to preclude prolonged survival in the dog when compared to primates,[18,19] and orthotopic heart-lung transplantation has not resulted in long-term survival in the dog.

Thus initial experimental heart-lung transplant investigation utilized a canine model. However, public sentiment against their use in medical research have forced investigators to seek alternate large animals. Recently pigs (including piglets), have been used in experimental heart-lung transplantation. The pig, however, is not an ideal animal to study double lung and heart-lung transplantation (see Section II.A.2). Heart transplantation[20] and single (left) lung transplantation[21] have been performed in the pig model.

II. TECHNIQUES

A. ANATOMY

1. Anatomic Characteristics of the Dog

In the dog the anatomy of the heart, great vessels, and lungs are similar to the human (Figure 1). The right lung has four lobes: cranial lobe, middle lobe, caudal lobe, and accessory lobe. The left lung has two lobes: cranial lobe and caudal lobe. The dog has both left and right main bronchi as in the human and is a suitable model for all types of heart-lung transplantation. In the dog it is possible to perform both single left lung transplants and single right lung transplants. A single left transplant is easier than the right, because the left main bronchus is longer. The dog is a suitable model for all kinds of heart-lung transplantation. Usually two size-matched animals between 10 and 40 kg are used for heart-lung transplantation. In dogs, heartworm negative animals should be used.

2. Anatomic Characteristics of the Pig

In the pig the anatomy of the heart and great vessels are similar to the human, except the hemiazygos vein flows to the right atrium directly. The tracheobronchial tree and lungs,

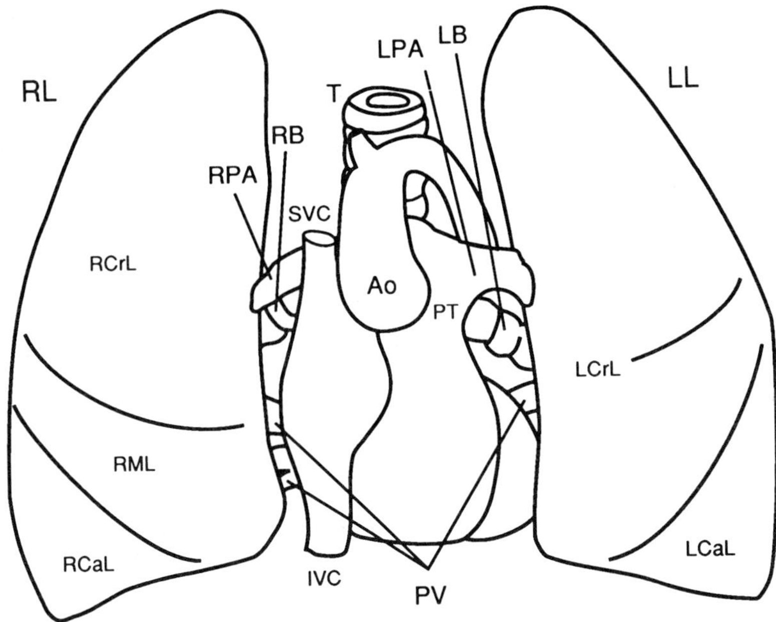

Figure 1 Anterior view of the canine heart-lung. (Ao) Aorta, (SVC) superior vena cava, (IVC) inferior vena cava, (PT) pulmonary trunk, (RPA) right main pulmonary artery, (LPA) left main pulmonary artery, (PV) pulmonary veins, (T) trachea, (RB) right main bronchus, (LB) left main bronchus, (RL) right lung, (LL) left lung, (RCrL) cranial lobe of right lung, (RML) middle lobe of right lung, (RCaL) caudal lobe of right lung, (LCrL) cranial lobe of left lung, and (LCaL) caudal lobe of left lung.

however, differ significantly from the human and the pig. The right lung has three lobes: cranial lobe, caudal lobe, and accessory lobe, and the left lung has two lobes: cranial lobe and caudal lobe. More importantly, the right cranial lobe bronchus branches from the trachea directly, about five cartilage rings above the carina (Figure 2). Therefore, it is difficult to perform right lung transplantation, double lung transplantation, or heart-lung transplantation making this animal a poor model for such investigation. Single (left) lung transplantation and heart transplantation may be performed in the pig.

B. ANESTHESIA

After an overnight fast, the animals are premedicated with intramuscular ketamine hydrochloride (20 to 30 mg/kg) and atropine sulfate (0.05 to 0.06 mg/kg). Following adequate sedation, veins (superficial veins of limbs in the dog, ear veins in the pig) are cannulated and anesthesia induced with intravenous sodium pentobarbital (25 to 30 mg/kg in the dog, 8 mg/kg in the pig) and paralyzed with intravenous pancuronium bromide (0.1 mg/kg). The animals are then intubated and mechanically ventilated: tidal volume 12 ml/kg, rate 8 to 12 breaths/min in the dog, tidal volume 15 ml/kg, rate 18 to 20 breaths/min in the pig. Anesthesia is maintained with additional doses of sodium pentobarbital (total dose should not exceed 35 mg/kg) and pancuronium bromide, or inhalation anesthetic (halothane, enflurane or isoflurane). If required, high oxygen concentrations (40 to 100%) and positive end-expiratory pressure (3 to 10 cmH_20) may be used.

The animal's temperature, arterial pressure, central venous pressure, pulmonary arterial pressure, and electorcardiogram should be monitored continuously. Arterial blood gases, blood electrolytes, cardiac output, and urine volume should be regularly observed. The right carotid or femoral artery is canulated to monitor arterial pressure and to sample

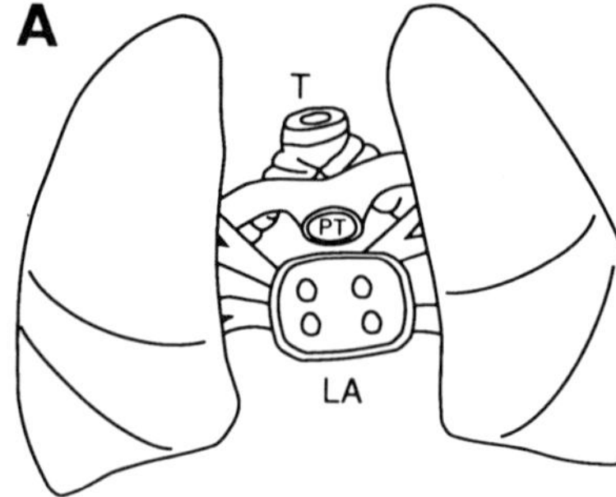

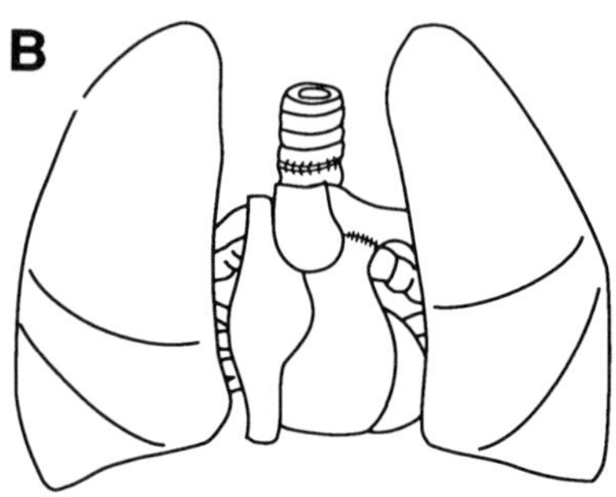

Figure 2 Anterior view of the swine tracheobronchial tree and lungs. (T) Trachea, (RL) right lung, (LL) left lung, (RCB) right cranial lobe bronchus, (RIB) right intermediate bronchus, (LB) left main bronchus, (RCrL) cranial lobe of right lung, (RCaL) caudal lobe of right lung, (LCrL) cranial lobe of left lung, and (LCaL) caudal lobe of left lung.

the arterial blood, and a Swan-Ganz catheter is placed from the right jugular vein to monitor central venous pressure, pulmonary arterial pressure, and cardiac output.

C. TECHNIQUES OF ANASTOMOSES

Usually, anastomoses of aorta, left atrium, right atrium, and pulmonary artery are performed with running 4-0 to 6-0 atraumatic nonabsorbable monofilament sutures (polypropylene, nylon, etc.). Occasionally, absorbable sutures are used to anastomose the vessels for examination of the growth potential of heart-lung allografts after pediatric heart-lung transplantation.

In tracheal and bronchial anastomoses, both 5-0 to 3-0 atraumatic absorbable and nonabsorbable sutures can be used usually with the running suture technique. If the animals will live long after transplantation, absorbable sutures should be used.

D. CARDIOPULMONARY BYPASS

Cardiopulmonary bypass and an oxygenator are required for experimental orthotopic heart transplantation or *en bloc* double lung transplantation.

Before the procedure, intravenous bretylium (3 mg/kg) is administered to prevent ventricular arrhythmias. After exposure of the heart, the animal is heparinized (3 mg/kg). Two cannulae (20 French) are inserted through separate purse-string sutures (4-0 braided suture) in the right artrial appendage and directed into the superior and inferior venae cavae. A canula (12-13 French) is placed in the distal ascending aorta or the femoral artery. The superior and inferior venae cavae are isolated and caval snares passed around each. The flow drains blood from the venae cavae to an external oxygenator, and back to the aorta or the femoral artery (Figure 3). Flow on bypass is kept at 50 to 80 ml/kg and the animal is cooled to 28 to 30°C.

E. ORTHOTOPIC HEART TRANSPLANTATION

1. Donor Operation

The donor animal is given intravenous heparin (3 mg/kg) after median sternotomy and exposure of the heart. The superior and inferior venae cavae are divided and the aorta and pulmonary arteries are separated. The inferior vena cava is ligated and its anterior wall is opened. The heart is exsanguinated and the aorta cross-clamped. Cardioplegic solution (4°C) is infused into the aortic root. The aorta is divided at the origin of the innominate artery and the pulmonary artery at its bifurcation. The pulmonary veins are transsected individually at the level of the pericardial reflection. The donor heart is immersed in 4°C saline and transferred to the recipient operating table. Before implantation, the donor heart is prepared by opening the posterior wall of the left atrium.

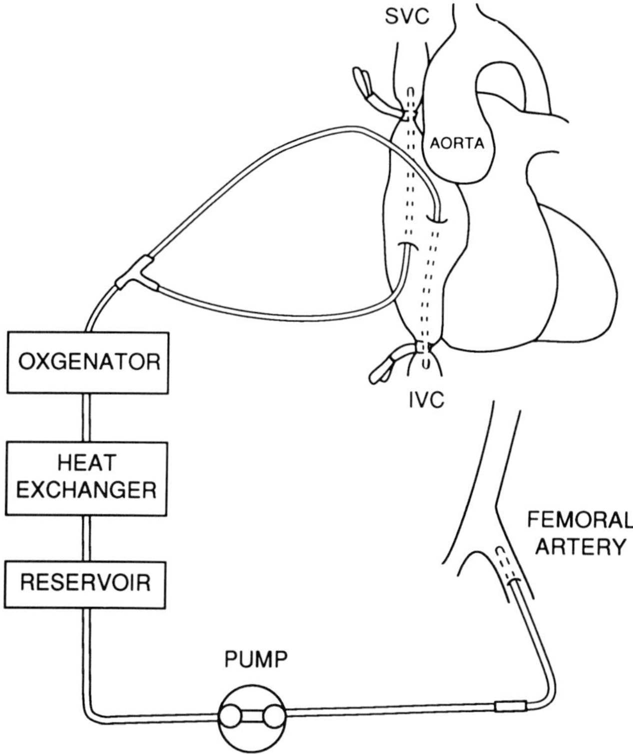

Figure 3 Cardiopulmonary bypass circuit. The flow drains blood from the vena cava to an external oxygenator, and back to the femoral artery.

2. Recipient Operation

The recipient heart is exposed through a midsternal incision avoiding phrenic nerve injury. After instituting cardiopulmonary bypass, the aorta is cross-clamped and transsected. The atrial incision is started at the right atrial appendage and division of the heart is performed along the atrio-ventricular groove just below the coronary sinus. The left atrial, right atrial, aortic, and pulmonary arterial anastomoses are performed with 5-0 to 6-0 running suture. The caval snares are released after completion of the pulmonary anastomosis, and air is aspirated from the ascending aorta before removal of the aortic cross-clamp (Figure 4).

Cardiac rhythm is restored spontaneously or by internal defibrillation. The animal is given isoproterenol (and dopamine) infusions to maintain heart rate and blood pressure, allowing the animal to be weaned from bypass. Protamine (1 mg/100 U heparin) is administered after decannulation. The incision is closed after placing a chest tube in each pleural cavity.

F. ORTHOTOPIC HEART-LUNG TRANSPLANTATION

1. Donor Operation

The donor preparation is the same manner as for the orthotopic heart transplant donor. After infusion of cardioplegic solution into the aorta, the superior vena cava and inferior vena cava are divided. The aorta is divided at the origin of the innominate artery, and the end of the trachea (or both the left and right main bronchi) is divided. The *en bloc* donor

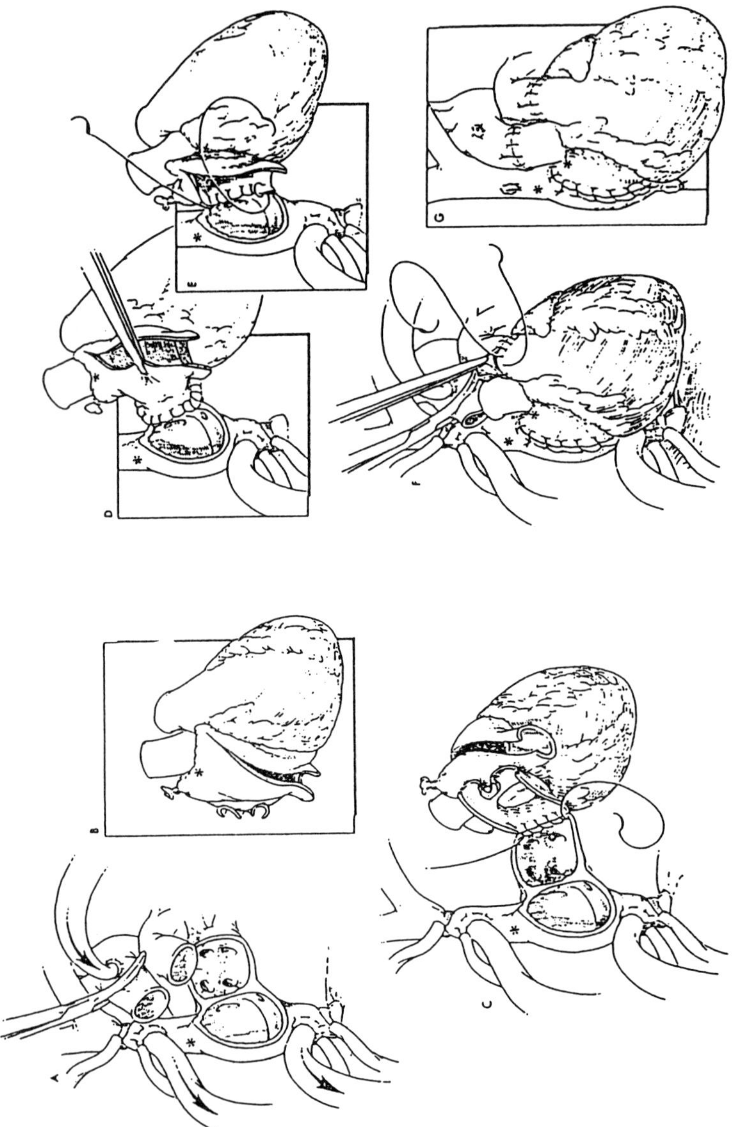

Figure 4 (A to G) Technical steps in orthotopic heart transplantation. (From Cooley, D.A., *Techniques in Cardiac Surgery*, 2nd ed., W.B. Saunders, Philadelphia, PA, 1986, With permission.)

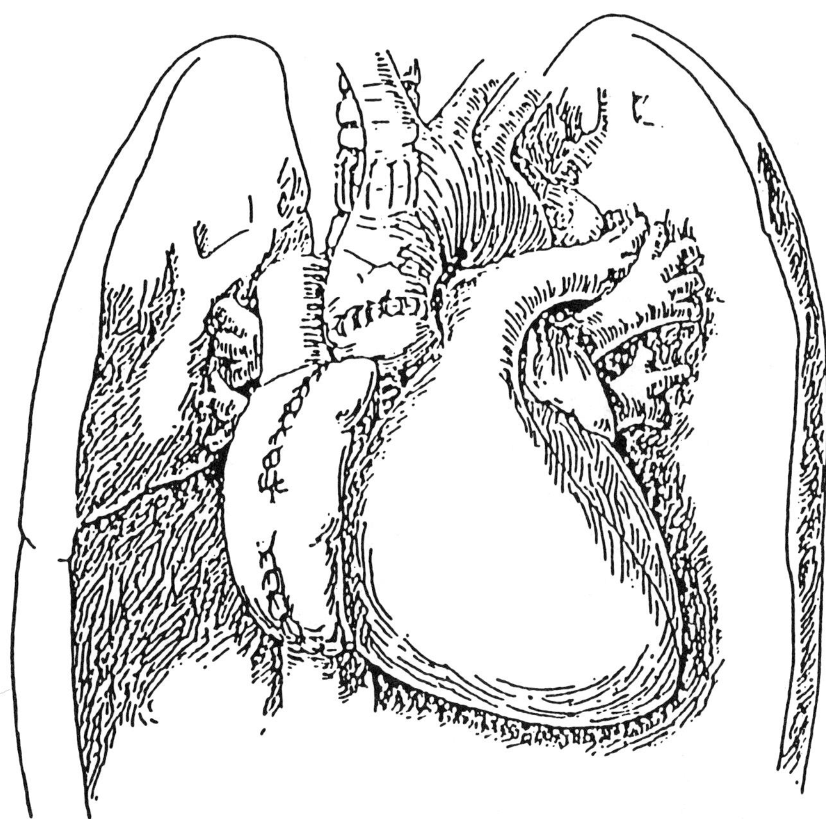

Figure 5 A completed orthotopic heart-lung transplantation. (From Patterson, G. A., *The Surgical Clinics of North America,* Vol. 68, No. 3, W.B. Saunders, Philadelphia, PA, 1988. With permission.)

heart-lung is prepared in this way. Before implantation of the heart-lung, the posterior wall of the left atrium is incised.

2. Recipient Operation

The recipient heart-lung is exposed through a midsternal incision. After employing cardiopulmonary bypass, the aorta is cross-clamped and transsected. The heart and lungs are removed, and the donor heart-lung bloc is implanted. Anastomoses of the trachea (or the left and right main bronchi), the aorta, and the right atrium are performed with a running suture (Figure 5). Weaning from cardiopulmonary bypass and closure of the incision follow the same technique as for orthotopic heart transplantation.

G. SINGLE (LEFT) LUNG TRANSPLANTATION

1. Donor Operation

The donor chest cavity is opened in the left fifth intercostal space or via median sternotomy. The left main bronchus, pulmonary artery, and pulmonary vein (left atrial cuff) are divided to ensure adequate length, after separation from the hilar tissues and administration of intravenous heparin (3 mg/kg). For lung preservation studies, however, the donor left lung is harvested after preservation of the entire *en bloc* heart-lung preparation.

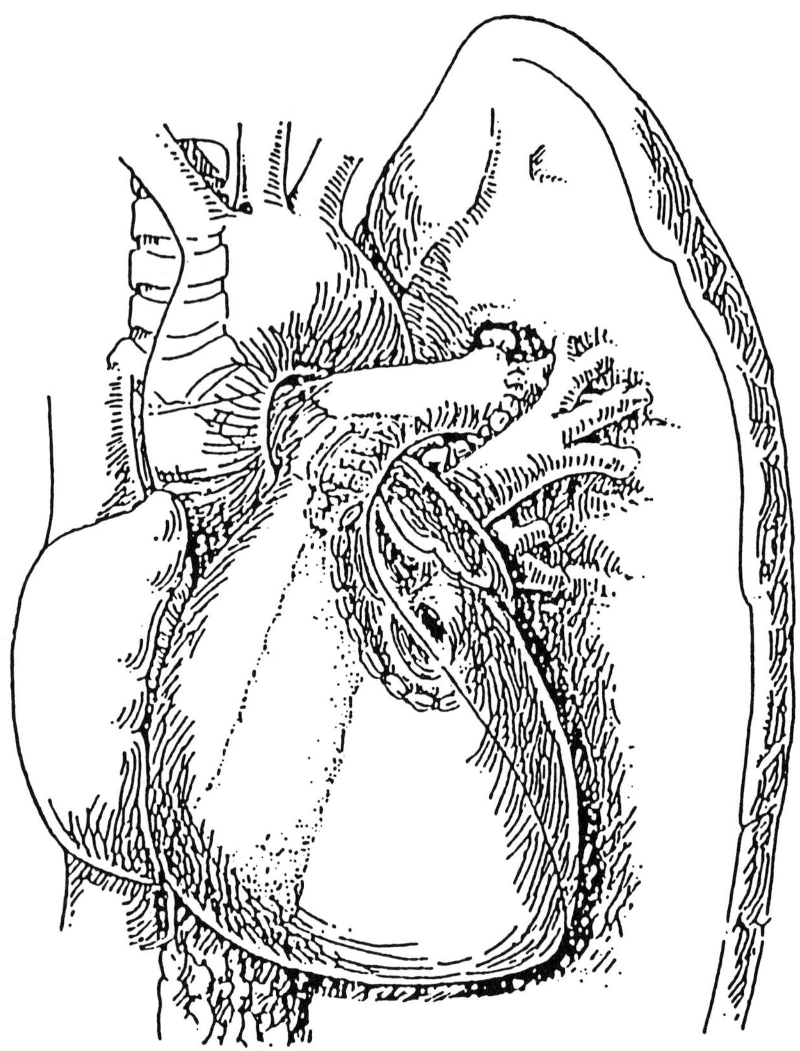

Figure 6 A completed left lung transplantation. (From Patterson, G.A., *The Surgical Clinics of North America,* Vol 68, No. 3, W.B. Saunders, Philadelphia, PA, 1988. With permission.)

2. Recipient Operation

The recipient left main bronchus, pulmonary artery, and pulmonary vein (left atrial cuff) are isolated from the hilar tissues. The left main bronchus and pulmonary artery are clamped at their origins. The pericardium around the left pulmonary vein is incised and the vein clamped in the left atrial area. All of them are divided 5 to 10 mm distal to the clamps. Anastomoses of the left pulmonary vein (left atrial cuff), pulmonary artery, and bronchus are performed with 4-0 to 6-0 running suture (Figure 6). The incision is closed after placing a chest tube in the left pleural cavity.

H. *EN BLOC* DOUBLE LUNG TRANSPLANTATION

1. Donor Operation

The donor animal is given intravenous heparin (3 mg/kg) after median sternotomy and exposure of the heart and lungs. The donor lungs with left atrium and pulmonary trunk are harvested *en bloc* after removal of the heart.

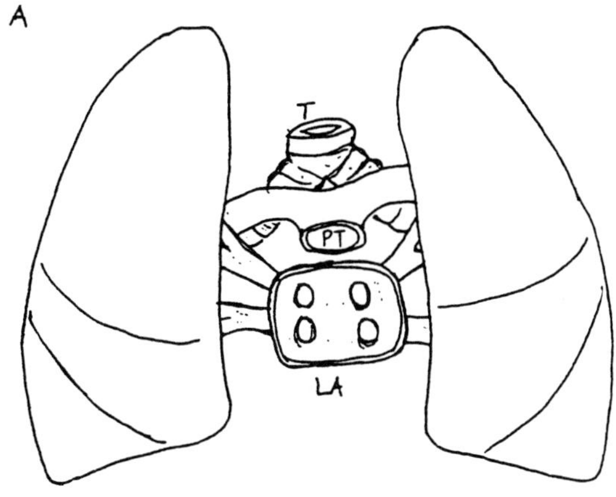

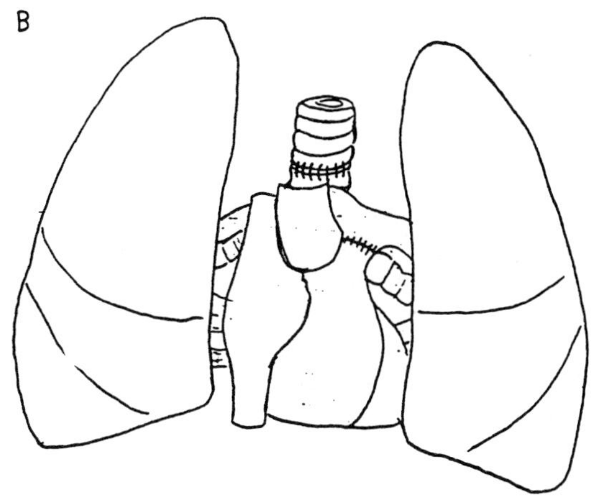

Figure 7 (A) The donor double lung graft. (T) Trachea, (PT) pulmonary trunk, and (LA) left atrium. (B) A completed *en bloc* double lung transplantation.

2. Recipient Operation

The recipient heart and lungs are exposed through a midsternal incision with care taken to avoid phrenic nerve injury. After employing cardiopulmonary bypass, the end of the trachea (or the left and righ main bronchi), the pulmonary trunk and the left atrium are divided to remove the both lungs. The donor double-lung-bloc is transferred to the recipient. The left lung is inserted through the mediastinum. Anastomoses of the pulmonary trunk , the left atrium, and the trachea (or the left and right main bronchi) are performed with a running suture (Figure 7). Weaning from cardiopulmonary bypass and closure of the incision are made in the same manner as orthotopic heart transplantation.

I. SEQUENTIAL DOUBLE LUNG TRANSPLANTATION

Left lung transplantation is performed first, and then immediately after left lung transplantation or a few weeks later, right lung transplantation is performed in the same manner.

III. APPLICATIONS

A. REJECTION

Heart-lung allografts are usually acutely rejected 1 to 3 weeks after transplantation in untreated control animals. The histologic features of mild rejection in the heart allografts are characterized by a perivascular and mild interstitial infiltration of mononuclear cells in the absence of myocyte necrosis. Those of moderate rejection are focal areas of myocyte necrosis with an increasing mononuclear infiltrate extending into the interstitium. Severe acute rejection of the heart allografts exhibit increased myocyte necrosis, hemorrhage, and a more prenounced infiltrate including neutrophils.

The histologic features of mild rejection in the lung allografts include a focal and mild lymphocyte infiltration around vessels and bronchioles without interstitial changes. Those of moderate rejection are multifocal to diffuse and severe lymphocytic infiltration around blood vessels and airways with extension to the adjacent interstitium.

Severe acute rejection of lung allografts is characterized by diffuse and severe lymphocytic infiltration in the graft with destruction of the pulmonary parenchyma and hemorrhage. A number of studies, both clinical and experimental,[22,23] indicate that lung allograft rejection tends to occur before heart rejection in heart-lung procedure. Also the rate of heart rejection in heart-lung transplant recipients is significantly less than that seen in patients who have received a heart allograft alone.

Occlusive arteriosclerosis occurs in 30 to 40% of patients by 5 years following heart transplantation. Up to 50% of patients who have survived the first year after heart-lung transplantation develop bronchiolitis obliterans (OB). Occlusive arteriosclerosis and OB are the most common and severe long-term complications of heart-lung transplantation. Although their exact etiology still remains unknown, many investigators believe that they may represent forms of chronic rejection.

B. IMMUNOSUPPRESSION

In initial experimental heart-lung transplantation, azathioprine, methotrexate, adrenocorticosteroids, antilymphocyte globulin were used for immunosuppressive therapy. These therapies, however, have not resulted in long-term survival in the dog. More recently, cyclosporine (10 to 20 mg/kg/day), prednisone (0.2 mg/kg/day), and small doses of azathioprine (2 mg/kg/day) are the most common immunosuppression regimen. Serum levels of cyclosporine are maintained at 200 to 250 ng/ml. Pulse therapy of methylprednisolone (10 to 20 mg/kg/day) is instituted when acute rejection occurs.

Experimentally, newer immunosuppressive agents (FK 506, prostaglandin analogues, anti-CD 4 antibodies, anti-TNF antibodies) are being studied in the canine and swine heart-lung transplant models.

C. USE OF EACH SURGICAL PROCEDURE IN RESEARCH

Experimental heart, lung, and heart-lung transplantation models are being used to study a variety of important questions.

1. Preservation

Although an organ with short ischemic time is acceptable, in clinical transplantation a shortage of donors frequently requires distant procurement. It is therefore important to establish improved methods of heart-lung preservation. What type of perfusates and temperature are suitable for organ preservation? How many hours is a donor organ able to be preserved? How is the viability of the organ determined? These are some areas of current investigation in experimental heart-lung preservation. The dog and pig are appropriate for organ preservation studies in this area.

2. Reperfusion Injury (Heart), Reimplantation Response (Lung)

The reperfusion injury after preservation may induce myocardial dysfunction following heart transplantation. The reimplantation response is one of the main causes of early morbidity in lung and heart-lung transplantation. It is believed that reperfusion injury, denervation, interruption of lymph flow, and surgical trauma lead to these changes in the lung. The mechanism of the reperfusion injury in the heart and the reimplantation response in the lung are important areas of investigation.

3. Bronchial and Tracheal Wound Healing

Poor healing of the tracheal or bronchial anastomosis is a serious complication in the early stages of lung/heart-lung transplantation. To prevent tracheobronchial complications the anastomosis may be protected by omental wrapping (omentopexy). In *en bloc* double lung transplantation, bronchial anastomoses in both right and left main bronchi instead of trachea may improve results. Inadequate healing of tracheal or bronchial anastomosis, however, remains a serious complication.

4. Rejection Monitoring

The history of organ transplantation is the history of the struggle with allograft rejection. The results of solid organ transplantation have rapidly improved with the introduction of cyclosporine. The control of rejection, however, remains a problem. Methods to detect the early stages of rejection include; endomyocardial biopsy, electrocardiograph monitor, ultrasonic cardiography, myocardial scintigraphy, magnetic resonance imaging, immunological-immunohistological methods for heart transplantation, chest x-ray film, perfusion scan, ventilation scan, transbronchial lung biopsy (TBLB), bronchoalveolar lavage (BAL), open lung biopsy, and immunological-immunohistological methods for lung transplantation. Biochemical and immunological methods using peripheral blood are also being studied. Not only the diagnosis of rejection, but its differentiation from infection requires further study in human heart and lung transplantation.

5. Organ Function after Transplantation

As prolonged survival after heart-lung transplant is achieved, studying organ function over time in the experimental laboratory becomes more important. Questions remain concerning long-term aspects of the heart-lung procedures. The influence of heart-lung denervation and the growth potential of the allografts after pediatric transplantation are areas of current interest.

As described in Section III.A, OB and occlusive arteriosclerosis are major, severe complications in patients who are long-term survivors of heart and lung transplantation. No effective methods have been established to prevent their development. The dog and pig provide suitable animals to study the development of OB and occlusive arteriosclerosis and to test the effectiveness of new therapeutic strategies of prevention.

IV. CONCLUSION

The use of dogs has been essential in development of heart-lung transplantation for clinical application. The dog is a suitable animal for experimental heart, lung, and heart-lung transplantation. Total cardiopulmonary denervation, however, seems to preclude prolonged survival in the dog. Pigs have been used in selective types of transplantation of the lung and heart. Single left lung transplantation and heart transplantation may be performed in the pig and may be advantageous in situations where limitations on availability of dogs may be an issue. The dog and pig will remain an important experimental model for heart-lung transplantation because of their size and anatomical and functional relevance to ethical transplantation.

REFERENCES

1. **Demikhov, V.P.,** Some essential points of the techniques of transplantation of the heart, lung and other organs, in *Experimental Transplantation of Vital Organs,* Medgiz State Press for Medical Literature in Moscow, Moscow, 1960, chap. 2, translated from Russian by Basil Haigh, Consultants Bureau, New York, 1962.
2. **Neptune, W.P., Cookson, B.A., Baily, C.P., et al.,** Complete homologous heart transplantation, *Arch. Surg.,* 66, 174, 1953.
3. **Webb, W.B. and Howard, H.S.,** Cardiopulmonary transplantation, *Surg. Forum,* 8, 313, 1957.

4. **Lower, R.R., Stofer, R.C., and Shumway, N.E.,** Homovital transplantation of the heart, *J. Thorac. Cardiovasc. Surg.,* 41, 196, 1961.
5. **Shumway, N.E., Lower, R.R., and Stofer, R.C.,** Transplantation of the heart, *Adv. Surg.,* 2, 265, 1966.
6. **Juvenelle, A.A., et al.,** Pneumonectomy with reimplantation of lung in dog for physiologic study, *J. Thorac. Cardiovasc. Surg.,* 21, 111, 1951.
7. **Hardin, C.A. and Kittle, C.F.,** Experience with transplantation of lung, *Science,* 119, 97, 1954.
8. **Vanderhoef, P., Dubois, A., Lauvau, N., et al.,** Bloc allotransplantation of both lungs with pulmonary trunk and left atrium in dogs, *Thorax,* 27, 415, 1971.
9. **Grosjean, O., Leroux, G., Schepense-Rocoux, G., and Leruth, P.,** Double lung transplantation through a right thoracotomy and without extracorporeal circulation, *Acta Chir. Belg.,* 4, 427, 1976.
10. **Dark, J.H., Patterson, G.A., Cooper, J.D., et al.,** Experimental *en bloc* double-lung transplantation, *Ann. Thorac. Surg.,* 42, 394, 1986.
11. **Veith, F.J., Sinha, S.B.P., Torres, M., and Richards, K.,** Bilateral simultaneous canine lung allotransplantation, *Ann. Surg.,* 174, 48, 1971.
12. **Kondo, Y., Isin, E., Cockrell, J.V., and Hardy, J.D.,** One-stage bilateral allotransplantation of canine lungs, further studies, *J. Thorac. Cardiovasc. Surg.,* 64, 897, 1972.
13. **The Toronto Lung Transplantation Group,** Sequential bilateral lung transplantation for paraquat poisoning, *J. Thorac. Cardiovasc. Surg.,* 64, 897, 1972.
14. **Lempert, N. and Blumenstock, D.A.,** Survival of dogs after bilateral reimplantation of the lungs, *Surg. Forum,* 15, 179, 1964.
15. **Slim, M.S., Yacoubian, H.D., Wilson, J.L., Rubeitz, J.L., and Gandhur-Manymheh, L.,** Successful bilateral reimplantation of canine lungs, *Surgery,* 55, 676, 1964.
16. **Alian, F. and Cayirili, M.,** One-stage reimplantation of both lungs in the dog, *JAMA,* 212, 863, 1970.
17. **Fujimura, S., Parmley, W.W., Tomoda, H., Norman, J.R., and Matloff, J.M.,** Hemodynamic alterations after staged and simultaneous bilateral lung autotransplantation in dogs, *J. Thorac. Cardiovasc. Surg.,* 63, 527, 1972.
18. **Halgin, J., Telander, R.L., Muzzall, R.E., et al.,** Comparison of lung autotransplantation in the primate and dog, *Surg. Forum,* 14, 196, 1963.
19. **Nakae, S., Webb, W.R., Theodorides, T., and Gregg, W.L.,** Respiratory function following cardiopulmonary denervation in dog, cat and monkey, *Surg. Gynecol. Obstet.,* 125, 1285, 1967.
20. **Calne, R.Y., English, T.A.H., Dunn, D.C., McMaster, P., Wilkins, D.C., and Herbertson, B.M.,** Orthotopic heart transplantation in the pig, the pattern of rejection, *Transplant. Proc.,* 8, 27, 1976.
21. **Harjula, A. and Baldwin, J.C.,** Lung transplantation in the pig with successful preservation using prostaglandin E-1, *Appl. Cardiol.,* 2, 397, 1987.
22. **McGregor, C.G.A., Baldwin, J.C., Jamieson, S.W., Billingham, M.E., Yousem, S.A., Bruke, C.M., Oyer, P.E., Stinson, E.B., and Shumway, N.E.,** Isolated pulmonary rejection after combined heart-lung transplantation, *J. Thorac. Cardiovasc. Surg.,* 90, 623, 1985.
23. **Prop, J., Kuijpers, K., Petersen, A.H., Bartels, H.L., Nieuwenhius, P., and Wildvuur, Ch.R.H.,** Why are lung allografts more vigorously rejected than hearts?, *Heart Transplant.,* 4, 414, 1985.

Chapter 14

Heart Transplantation in Primates

David K.C. Cooper, Yong Ye, and Marek Niekrasz

CONTENTS

0-8493-3629-5/94/$0.00+$.50

I. INTRODUCTION

Several non-human primate species are suitable for experimental transplantation studies.[1] Our own experience has been largely with the Chacma baboon (*Papio ursinus)* and the African green (vervet) monkey (*Cercopithecus aethiops),* and, to a certain extent, with the olive baboon (*Papio anubis).* Other species commonly used include the cynomolgus monkey (*Macaca fascicularis)* and other macaque monkeys, such as the rhesus (*Macaca mulatta).*

The surgical techniques used in the primate are largely those that have been developed in other experimental animals, particularly the dog. These techniques, and the experimental work that has been carried out utilizing them, have been extensively reviewed elsewhere.[2,3]

In brief, experimental work on cardiac transplantation has evolved through several phases, though there has been considerable overlap between them. Firstly, animals were given a second, often parasitic, heart which enabled certain physiological, pharmacological, and pathological studies to be made. The locus chosen was usually the neck, though the abdomen and inguinal regions have also been used. Suggestion was then made that an additional heart might act as an auxiliary pump in certain circumstances, and this led to the evolution of techniques of inserting the donor heart into the chest in circuit with the recipient organ. With the advent of hypothermia and the pump oxygenator, total excision and replacement of the heart became feasible, and, finally, when technical and physiological problems had been studied and minimized, efforts were made to combat the immune response with immunosuppressive drugs.

A. THE NON-HUMAN PRIMATE IN EXPERIMENTAL CARDIAC TRANSPLANTATION

Amongst the earliest to use primates as experimental animals for work in this general field were Willman and his colleagues in 1965, who investigated the autotransplanted heart.[4] Haglin et al. had explored lung autotransplantation in 1963,[5] and, in 1967, Nakae and his colleagues[6] investigated the problem of pulmonary function following transplantation of the heart and both lungs, carrying out a small number of these procedures in monkeys. Castaneda's group were also interested in cardiopulmonary autotransplantation in primates in the early seventies.[7] Later in the seventies, Losman and Barnard, in Cape Town, developed their technique of heterotopic placement of the donor heart in the chest, and carried out physiological studies in the baboon[8] before embarking on a clinical program of heterotopic heart transplantation.

In recent years, partly due to restrictions on the use of the dog in laboratory studies, but mainly due to the advantage of the non-human primate as an experimental animal in immunological work, the baboon and other non-human primates have become increasingly popular in surgical laboratories.

When cyclosporine first became available at the end of the seventies, Reitz and his colleagues carried out important studies on heart and heart-lung transplantation at Stanford using rhesus and cynomolgus monkeys.[9,10] This work was a preliminary to the first use of cyclosporine in clinical heart transplantation, and, perhaps even more significantly, provided a valuable experience before embarking on a clinical program of heart-lung transplantation.

Testing of methods of storage of the heart by orthotopic heart allo- and auto-transplantation was carried out in the baboon by the Cape Town group in the early 1980s.[11-15] This group also initiated a series of studies on aspects of immunosuppressive therapy and immune modulation involving cardiac allografting in the baboon using heterotopic heart transplantation in the neck.[16-24] Both the Cape Town group and others have recently investigated newer immunosuppressive agents, such as FK 506 and 15-deoxyspergualin in non-human primate models.[25-27]

With regard to xenotransplantation between closely related primate species, again the Cape Town group, with their ready supply of locally procured baboons and vervet monkeys, would appear to be amongst the earliest in this field,[22,24,28-33] along with the Columbia-Presbyterian group in New York, who used the cynomolgus monkey-to-baboon model.[34-36] The Cape Town group has also used the baboon as the recipient of discordant cardiac xenografts with the pig as donor.[24,32,37-39] Recently, the Minneapolis group has carried out a similar study in the rhesus monkey.[40]

B. THE NON-HUMAN PRIMATE AS DONOR FOR MAN

The non-human primate has also been used as the donor in a small number of cardiac transplants in man.[41-44] Both Hardy et al and Marion used chimpanzees as donors of orthotopic heart transplants in 1964 and 1968, respectively.[41,42] Barnard and his colleagues used both chimpanzee and baboon hearts for heterotopic heart transplantation in emergency situations in 1977,[43] and, most recently, Bailey and his colleagues used a baboon heart in the well-known Baby Fae case in 1984.[44] Details of these experiences, all of which were unsuccessful, have been discussed elsewhere.[45]

The chimpanzee and baboon were also donors of kidneys and livers transplanted into man in the early sixties,[46-54] and baboons have also been used for the treatment of hepatic failure in man by cross-circulation techniques.[55,56]

II. THE BABOON AS AN EXPERIMENTAL ANIMAL IN TRANSPLANT RESEARCH

As our own experience is largely with baboons, the role of this species in cardiac transplant research will be emphasized and discussed in some detail. Baboons most frequently used in such studies include *Papio ursinus, P. cynocephalus* and *P. anubis.* Many of the points raised, however, apply equally to other species of non-human primates that are commonly used for experimental surgical purposes.

As an experimental animal, the baboon has several major advantages and some disadvantages (Table 1).

A. ADVANTAGES

The baboon's size and anatomy make it an easy animal for operative procedures. A male Chacma baboon may grow to over 30 kg in weight, though most adults are in the range of 12 to 25 kg. The baboon's anatomy is very similar to human anatomy, and its tissues are generally strong and ideally suited to surgical procedures. With regard to thoracic anatomy, the only significant difference from a surgical point of view is that the baboon

Table 1 **Advantages and Disadvantages of the Baboon in Transplant Research**

Advantages		Disadvantages
Size	Phylogenetic closeness to man	Lack of availability
Anatomy		High cost of purchase
Physiology		High cost of maintenance
Immune system		Potential carrier of serious viral infection
Tolerates anesthesia and surgical procedures		Suffers from "stress"
Good "surgical" tissues		Requires sedation to be handled
Tolerates drug therapy		Slow reproduction
Tolerates total lymphoid irradiation		If kept in colonies, may injure each other
ABO blood groups		Susceptible to certain diseases (e.g., tuberculosis) transferred from human handlers or other baboons
Lymphocytes can be typed		
Usually outbred populations		

has only two, rather than three major vessels arising from the aortic arch. These are the brachiocephalic, which divides into three branches (right subclavian, right and left carotids), and the left subclavian.

Its physiology, including the immune system, is again similar to that of the human, and therapeutic measures in relation to organ transplantation that are successful in the baboon have a high likelihood of being similarly successful in human subjects. In general, it tolerates drug therapy well and, in experienced hands, has been shown to tolerate total lymphoid irradiation successfully.[57-59]

The baboon's saliva and tissues (though not its red blood cells) can be typed according to the human ABO blood group system[60-62] and therefore this animal is valuable in any study involving blood group compatibility or incompatibility.[20-22,29,30] There are in addition, however, simian blood types (monkeys' own blood groups defined by specific antibodies raised in primate animals) that can be identified and which should be considered in experiments of an immunologic nature.[60-63]

To our knowledge there have not been recent major efforts to type baboons or other non-human primates according to lymphocyte groups, though much work in this area was performed in the 1960s and 70s.[64,66] The major histocompatibility complexes (MHC) of non-human primates are similar to those of the HLA system.[64,65] There has been some recent work attempting to tissue type baboons using human antisera.[44,67,68] Lymphocytotoxic and hemagglutinating antibodies, if present, can be readily demonstrated in baboon sera by standard techniques.

In general, the baboon does not interfere with surgical wounds, in contradistinction to some other experimental animals, such as the dog. The baboon also has the advantage of living for approximately 30 years in captivity, enabling long-term experimental work to be carried out if required.

B. DISADVANTAGES

A major disadvantage of the baboon, and indeed of almost all non-human primates, is their lack of ready availability worldwide. Routine use of these animals in Europe and North America may be made difficult by lack of availability and by the considerable expense of purchase. At the current time an adult baboon of approximately 12 kg costs in the region of $1500 in the U.S., which can be compared with a cost of less than $300 in Southern Africa. Cost of maintenance is also higher than for most other comparable experimental animals, such as the dog or pig.

There are periods when the baboon cannot be imported into Europe and/or North America due to fears of serious viral infection (of which it can be a carrier),[69-73] and this can seriously delay and complicate experimental series. There is some evidence, however, that colony-bred baboons are less likely to be serologically positive for viral infection than wild-caught baboons.[71] Such viral infection in human contacts is exceedingly rare, and in general has not proven to be a major problem.

Indeed, the baboon is more likely to be infected by man than vice versa; for example, the non-human primate is highly susceptible to the transfer of tuberculosis from its handlers. Because of the risk of carrying tuberculosis and infecting other baboons, it is customary for the baboon to be kept in quarantine for a period of 4 to 6 weeks after arriving at a research center before it is actually used for experimental studies.[1]

Baboons kept in colonies are prone to fight with each other and, therefore, unless caged separately, trauma of this nature can inhibit experimental studies from time to time.

After transportation or other periods of "stress" the baboon can be susceptible to gastrointestinal upsets (e.g., diarrhea), which are frequently associated with bacterial infections.[1] Although these can usually be easily treated, they can lead to dehydration and, if severe, to ulcerative colitis. Parasitic infection of the gut is a relatively uncommon cause of gastrointestinal disorder.

Finally, unlike the dog, the baboon is not an easy animal to handle under experimental conditions, and, in general, requires sedation before even minor procedures, such as blood sampling, can be performed. Techniques are available, however, to allow intravenous drug infusion and blood sampling without the need for sedation (Figure 1).[74]

C. MANAGEMENT

Maintenance of baboons and other non-human primates for surgical research purposes is a major undertaking requiring specialized facilities and skilled staff. A review of this subject is beyond the scope of this chapter, but details of the care required have been summarized by Richter et al.,[1] who also provide numerous helpful baseline values for various physiological and metabolic parameters in these animals. Another valuable source of information is provided by Benirschke.[75]

D. CHOICE OF DONOR AND RECIPIENT PAIRS

When orthotopic cardiac transplantation is being considered, baboons of similar size are chosen; technical aspects of the procedure are facilitated by using baboons in excess of 10 kg body weight, though the surgery can be successfully performed on much smaller animals. For heterotopic heart transplantation performed in the neck, using a technique based on the work of Carrel and Guthrie[76] and Mann et al.,[77] a recipient baboon that is significantly larger than the donor is chosen, but a recipient of greater than 12 kg can be paired with a donor of less than 10 kg in most cases.

For xenografting, a number of closely related primate species are available for studies of a concordant nature. These include the African green (vervet) monkey, the cynomolgus monkey, and other macaque monkeys (e.g., rhesus), all of which are of a size comparable to that of a small baboon (approximately 2 to 8 kg). They could therefore be used in either orthotopic or heterotopic transplantation. When heterotopic heart transplantation has been performed, it has been usual to use the (larger) baboon as the recipient and the (smaller) monkey as the donor. These monkey species can also be blood typed according to the ABO system, and therefore ABO compatibility between recipient and donor can be assured.[60,61,63]

When xenotransplantation between widely disparate (discordant) species is planned, almost any other commonly available experimental animal such as the pig, sheep, or dog would prove suitable as donors as long as there is an adequate size ratio between donor and recipient. Blood group systems in these animals differ greatly from that of the

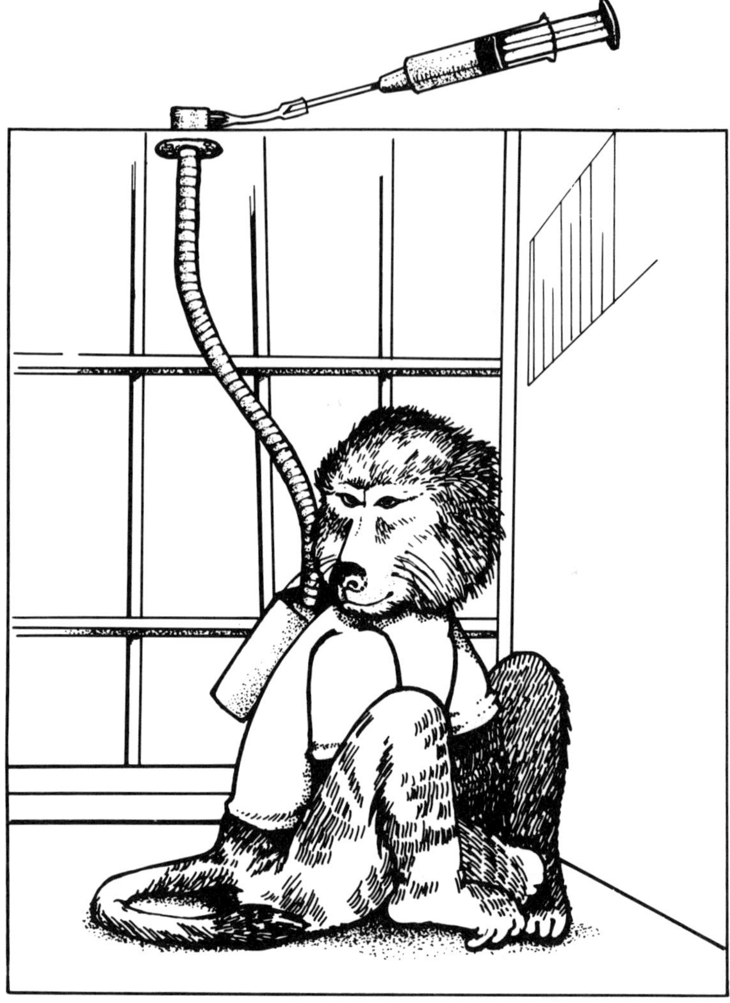

Figure 1 Drawing showing system allowing continuous or intermittent hemodynamic monitoring, drug and fluid infusion, and blood sampling in the baboon without the need for sedation. Femoral venous and/or arterial cannulae are inserted, tunnelled subcutaneously to the middle of the back, and brought out through a protective flexible metal tube (held in place by a jacket).

baboon, but fortunately blood group matching does not appear to play an important role in discordant xenografting.[37,39,78]

E. ANESTHESIA AND MONITORING

Premedication is generally with ketamine hydrochloride (5 to 20 mg/kg i.m.). The addition of pancuronium (0.1 mg/kg i.v.), atropine (0.05 mg/kg i.v.), and morphine may facilitate intubation, but is not essential. Endotracheal intubation is carried out and positive-pressure ventilation continued with a mixture of oxygen (2 to 4 l/min) and nitrous oxide (2 to 6 l/min) in a ratio of 1:1 to 1:2. The addition of halothane (0.5% to 2%) is usually required to maintain and deepen anesthesia, but can result in hypotension (which may require correction with intravenous inotropic therapy).

Infection prophylaxis is usually carried out with a suitable antibiotic (e.g., a cephalosporin) given intravenously immediately before the operative procedure and intramuscularly at intervals for the subsequent 24 to 48 hours. The operation should be performed under surgically sterile conditions, though the baboon has proved relatively resistant to infectious complications.

During orthotopic transplantation it is essential to monitor the baboon's vital signs by continuous electrocardiography, arterial and venous pressure monitoring, and the recording of urine output. During heterotopic transplantation in the neck, monitoring of arterial pressure is advisable; this can be by indwelling arterial cannula or sphygmomanometer cuff.

Posttransplant monitoring should be continued until the baboon is breathing spontaneously and is clearly waking up. Subsequent monitoring is difficult, though not impossible (see below), and fortunately is frequently unnecessary as the baboon usually recovers rapidly from even major surgical procedures. It is not unusual for a baboon to be sitting in its cage eating and drinking within a few hours of a successful orthotopic heart transplant.

Prolonged hemodynamic monitoring can be obtained by the insertion of long arterial and/or venous catheters in the femoral region. The distal ends of the catheters are inserted into small branches of the artery and/or vein and passed up into the descending aorta and/or inferior vena cava or right atrium. The proximal ends of the cannulae are tunnelled subcutaneously to the scapula region where they are brought out through the skin and passed through a protective, flexible metal tube or tether (maintained in position by a relatively tight-fitting jacket) which is fixed to the roof of the cage (Figure 1).[74] This system enables continuous or intermittent drug infusion and fluid administration, monitoring of arterial and venous pressures, and blood sampling. Such a system can be maintained for several weeks if so desired, and obviates the need for sedation for such procedures.

The surgical technique selected will depend on the aim of the study (Table 2).

III. SURGICAL TECHNIQUES

A. HETEROTOPIC HEART TRANSPLANTATION IN THE NECK

This has been the most common technique used in non-human primates. It requires the least in the way of surgical facilities and equipment, and is a relatively simple and quick procedure. As such, it is the ideal transplant to screen new methods of immunosuppressive therapy or immune modulation.

The operative technique is a modification of that described by Mann et al.[77] The neck has been the site most commonly chosen, though the abdominal vessels can be utilized.[2,27,40] As it is the most frequently used surgical technique in experimental cardiac transplantation in non-human primates, a detailed description will be given here.

Table 2 **Indications for the Common Experimental Surgical Techniques of Cardiac Transplantation**

Heterotopic (HHT)	- Screening of immunosuppressive drug therapy and manipulation of the immune system
Orthotopic (OHT)	- Confirmation of results obtained following HHT - Assessment of methods of storage of the heart
Auto	- Physiological studies - Long-term assessment of methods of heart storage
Heart-lung	- Investigation of immunosuppressive drug therapy and immunomodulation - Assessment of methods of storage of the lungs

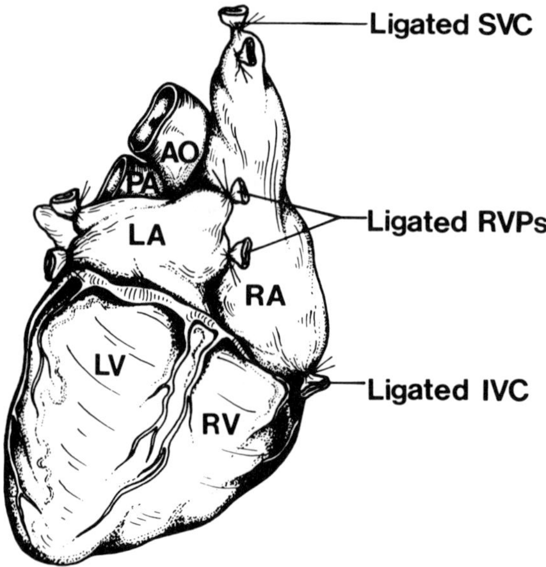

Figure 2 Heterotopic heart transplantation — preparation of the excised donor heart. A posterior view of the donor heart is shown. The superior (SVC) and inferior (IVC) venae cavae and all pulmonary veins (VPs) have been ligated. (RA) Right atrium, (RV) right ventricle, (PA) pulmonary artery, (LA) left atrium, (LV) left ventricle, and (AO) aorta.

1. Donor Operation

With the animal supine, and with limbs extended, the donor thorax is opened by a midline sternotomy. The anatomic differences between baboon and man with regard to the branches of the aortic arch have been outlined above. The azygos vein is generally small and enters the superior vena cava (SVC) low, often almost at the origin of the right atrium. This vein is doubly ligated and divided. The superior and inferior (IVC) venae cavae and ascending aorta are mobilized.

Heparinization of the donor is followed by (1) double ligation and division of the SVC, (2) division of the IVC (after clamping at the diaphragm), thus decompressing the right side of the heart, and (3) division of two or more pulmonary veins, decompressing the left side of the heart. The proximal aortic arch is immediately clamped, and cold (4°C) cardioplegic solution (15 ml/kg) is infused into the aortic root. Cold saline is applied over the heart to fill the pericardial cavity and cool the myocardium externally. Ventilation is discontinued.

Cardiac contractions usually cease within 30 to 60 sec. The heart should be gently massaged to ensure that it is decompressed through the divided IVC and pulmonary veins. If the heart distends, further pulmonary veins should be divided to ensure decompression. The cold saline within the pericardial cavity should be sucked out and replaced continuously during infusion of the cardioplegic solution. (The myocardial septal temperature at the end of infusion of cardioplegia and topical cooling has been measured at approximately 17 to 18°C.).

The heart is excised as follows. The remaining pulmonary veins are divided. The cardioplegic catheter is withdrawn and the ascending aorta is divided at the origin of the brachiocephalic artery. The main pulmonary artery is divided at the origin of the right and left branches. The heart is lifted up and the remaining mediastinal tissue posterior to the atria is divided carefully, ensuring that cardiac structures are not damaged. The heart is removed from the pericardial cavity and placed in a bowl of cold saline. (After 1 to 2 min immersion in cold saline, the myocardial temperature has been documented to fall to approximately 12 to 13°C.)

It is beneficial, though not essential, to create a small atrial septal defect which, after transplantation, will allow the left atrium to drain into the right atrium, and therefore provide an outlet for left-sided blood if the transplanted left ventricle cannot eject against carotid artery pressure. An atrial septal defect can be fashioned by passing a hemostat through the inferior vena cava orifice and atrial septum, emerging through a pulmonary vein orifice.

The pulmonary venous and IVC orifices are individually ligated (Figure 2), and the heart stored under hypothermic conditions until the recipient has been prepared. (Viability of the heart cannot be assured if the total ischemic time at 4 to 10°C is longer than a maximum of 4 hours).

2. Recipient Operation

a. Position

The baboon is placed on its back, with its left shoulder pulled down (by strapping the left arm over the right leg). The head is turned to the right and the chin is slightly elevated. The left sternomastoid muscle should then be obvious.

b. Incision

An oblique incision is made (through the skin and underlying platysma muscle) from a point posterior to the left ear over the mastoid process, inferiorly and anteriorly across the sternomastoid muscle, to the left sternoclavicular joint. The anterior edge of the sternomastoid muscle is identified, and this muscle is retracted posteriorly; it is sometimes necessary to partially divide it horizontally across its belly from anterior to posterior in order to expose the underlying vessels satisfactorily. A self-retaining retractor is inserted to retract the sternomastoid and the skin, thus exposing the left common carotid artery and the left internal jugular vein.

c. Anatomy and Dissection

The left common carotid artery and the left internal jugular vein are identified; their relative positions are variable, the vein being sometimes anterior and sometimes posterior to the artery. Both vessels are exposed for a length of 5 to 8 cm. The left vagus and left recurrent laryngeal nerves are dissected off the vessels, care being taken not to damage these nervous structures. The vagus is usually lateral to the artery, and the recurrent laryngeal medial to both vessels.

d. Insertion of Donor Heart into Recipient

Using small vascular clamps the common carotid artery is clamped superiorly and inferiorly. An incision is made in its anterolateral wall and extended with fine scissors, the length of the incision approximating to the diameter of the donor ascending aorta. The blood in the isolated section of the carotid artery is flushed out with saline.

The donor heart (preferably covered by cold wet swabs to help maintain a low myocardial temperature) is then placed in the anterior aspect of the neck so that the divided ends of the aorta and pulmonary artery lie at right angles to the recipient carotid artery and jugular vein. The donor vessels may need to be trimmed to a suitable length.

The donor ascending aorta is then anastomosed end-to-side to the recipient carotid artery using a continuous suture of 5-0 polypropylene (Figure 3). The donor ascending aorta is then clamped with a small vascular clamp as close to the aortic valve as possible. Gauze swabs are placed around the site of anastomosis, and the distal carotid clamp is removed, followed immediately by the proximal clamp. (The clamp on the donor ascending aorta prevents both air and blood from entering the donor coronary arteries or left ventricle, and thus prevents myocardial rewarming and contractions at this stage).

Manual pressure is applied to the swabs around the anastomosis and maintained for 4 to 5 min, by which time hemostasis should have been achieved. If there is significant bleeding, a further suture or sutures may be required; it is frequently preferable to temporarily reclamp the carotid artery under these circumstances.

The recipient jugular vein is then occluded superiorly and inferiorly with small vascular clamps. An incision is made in its anterolateral aspect at a site which will allow the donor pulmonary artery to lie undistorted. The length of the incision is related to the diameter of the donor pulmonary artery. The lumen of the jugular vein is flushed with saline. The donor pulmonary artery is anastomosed end-to-side to the jugular vein using a continuous 5-0 polypropylene suture (Figure 3). A small vascular clamp is then placed

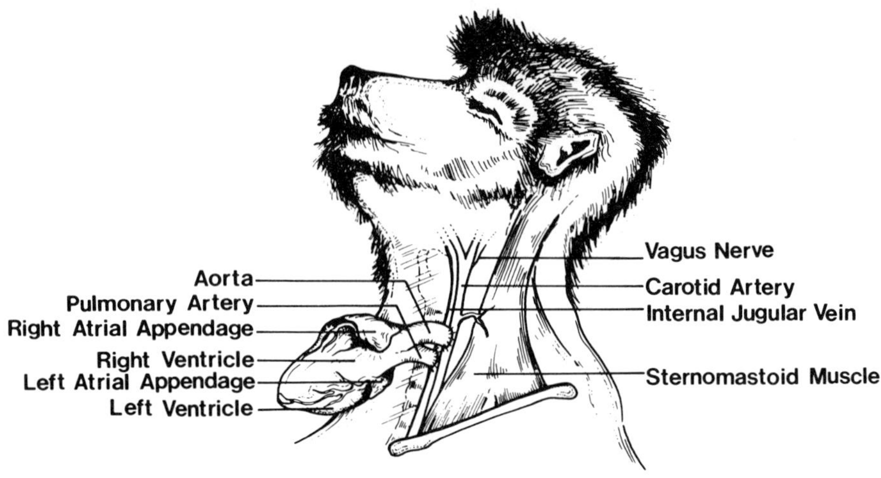

Figure 3 Drawing of baboon with heterotopic heart transplant in the neck. At the end of the procedure, the heart would lie in a subcutaneous pouch covered only by skin and platysma.

across the donor pulmonary artery just distal to the pulmonary valve, at which time the clamps on the jugular vein are released. Again, manual pressure on swabs at the site of anastomosis is usually sufficient to control any bleeding and achieve hemostasis.

The pulmonary artery-jugular vein anastomosis is rather more difficult than the arterial anastomosis, It is possible to incorporate too much of the wall of the jugular vein in the suture line, and thus occlude, or at least distort, the vein at one end or the other. Very small bites of the jugular vein are preferable.

The vascular clamp across the donor pulmonary artery is released first, followed by that across the donor aorta; this allows blood to circulate through the coronary arteries (Figure 4). The heart is supported so that the apex is uppermost, and air needles are inserted into the apices of both right and left ventricles to allow escape of any air. Generally at this stage the heart begins to beat spontaneously and air will be forced through the needles or needle holes. It is rarely necessary to place a suture at these needle hole sites; clot usually seals them. Occasionally, coordinated contractions do not occur, and ventricular fibrillation is seen, requiring electrical defibrillation using a shock of 5 to 10 J.

It is wise to allow the heart to beat for a minute or two before making a final search for any bleeding from (1) the anastomoses and (2) the posterior aspect of the left atrium at the sites of the ligated pulmonary veins.

e. Closure

Once it is clear that the heart is functioning satisfactorily, consideration is given as to where and how it will lie best in the neck. It is important to avoid distortion of the vessels involved, and this is particularly likely to happen with regard to the internal jugular vein or pulmonary artery. It is also important to avoid pressure on these latter vessels by the donor heart or by the donor aorta, or impairment of venous drainage may result.

A pouch, just large enough to accommodate the donor heart, is then fashioned by blunt and sharp dissection deep to the platysma at a suitable site in the anterior aspect of the neck. The heart is seen to lie satisfactorily in the pouch before any attempt is made to close the skin over it. It may be necessary to undermine the skin and platysma in various directions in order to provide enough loose skin to cover the heart satisfactorily. It is essential that the skin should not compress the heart, or tamponade may occur.

The skin is loosely closed over the heart using suitable interrupted sutures. At each end of the incision a small drainage hole is left, allowing serous fluid to drain from the neck

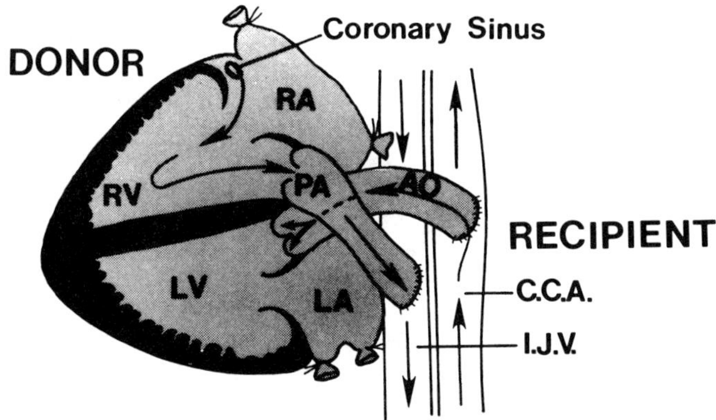

Figure 4 Diagram of circulation through donor heart following heterotopic heart transplantation in the neck in the baboon. Blood from the recipient common carotid artery (CCA) passes in a retrograde direction down the donor aorta (AO) and perfuses the myocardium via the coronary arteries. Donor coronary venous return from the coronary sinus passes through the right atrium (RA) and right ventricle (RV) into the pulmonary artery (PA) and thence to the recipient internal jugular vein (IJV). The donor myocardium is therefore perfused passively by the recipient circulation and does not support any significant workload. Ejection of blood from the donor left ventricle (LV) takes place when the donor LV pressure rises above the recipient CCA pressure (systemic or diastolic). The surgical formation of an atrial septal defect in the donor heart facilitates drainage of the left side of the heart into the right atrium and prevents over-distention of the left atrium (LA) or LV in the early reperfusion period (when the donor myocardium may not have fully recovered from the ischemic interval). (Modified from Mann, F.C., Priestley, J.T., Markowitz, J., and Yater, W.M., *Arch. Surg.*, 26, 219, 1933.)

during the next 24 hours. On occasions, it may be preferable to fashion a separate drainage hole in the skin at a dependent point. (It is occasionally necessary to insert a suture at these drainage sites after 24 to 48 hours, but the skin usually seals the gap spontaneously.)

3. Immediate Postoperative Care

When the animal is clearly awake and breathing spontaneously, positive pressure ventilation is discontinued and the baboon is allowed to breath a mixture of oxygen and nitrous oxide spontaneously. The endotracheal tube is removed, and the baboon returned to its cage, being placed so that it lies on its right side (i.e., opposite to the side of operation); this will ensure that the donor heart is not compressed. The baboon is allowed water, but no food, until the following morning.

4. Subsequent Postoperative Management

It is sometimes necessary to express some serous fluid from around the heart, thus facilitating its drainage through the openings which were left at each end of the wound. It is not uncommon for the skin to become mildly edematous for a few days. Rarely a baboon will scratch at the suture line and remove sutures, and it may be necessary to insert new sutures under sedation. Similarly, if the heart is rejecting and is enlarging in size, the suture line may occasionally break open in places and require repair.

5. Monitoring for Rejection

The heterotopic heart beat in the neck is monitored visually twice daily. If there is any doubt that the heart is beating satisfactorily, the baboon is sedated with ketamine hydrochloride and the heart palpated. An electrocardiogram can be taken if necessary. Cessation of graft function can readily be detected. Myocardial biopsies can be taken under ketamine sedation either by a percutaneous biopsy technique (e.g., using a Trucut needle), or by a direct minor surgical procedure. (If the tip of the donor right atrial appendage is brought out and sutured to the skin, the insertion of a bioptome to obtain a right ventricular endomyocardial biopsy is also possible).

6. Excision of the Transplanted Heart

Excision of the rejected organ can be performed under ketamine sedation and does not necessitate death of the recipient animal. Unilateral ligation of the carotid artery and jugular vein is well tolerated in the healthy baboon.

7. Abdominal Placement of the Donor Heart

When the donor heart is large in comparison with the size of the recipient, the abdomen may be the preferred site of placement. End-to-side anastomoses are performed between donor aorta–recipient infra-renal abdominal aorta and donor pulmonary artery–recipient IVC.[2,27,40]

8. Selected Results of Heterotopic Heart Transplantation

The results of selected allograft procedures in the baboon are briefly documented in Table 3, which gives some indication of the studies that have been performed in this model.

Table 3 **Selected Results of Cardiac Allotransplantation in the Baboon**

Group	n	Graft survival (mean in days ±SD)	Ref.
1. No IS (control)	10	11.0 (±5.6)	16
2. AZA, MP	7	19.9 (±10.6)	17
3. AZA, MP (low dose)	8	18.3 (±7.6)	17
4. AZA (high dose), MP	8	25.6 (±24.9)	18
5. AZA, MP (low dose), bolus IV MP (days 5, 6 and weekly)	6	15.2 (±6.9)	18
6. Pretransplant splenectomy, AZA, MP	6	10.7 (±2.5)	18
7. AZA, MP, Propranolol	6	17.8 (±13.8)	18
8. AZA, MP, Cyclophosphamide	5	15.0 (±5.1)	18
9. Niridazole	7	13.9 (±5.1)	16
10. Niridazole, AZA, MP	7	23.0 (±13.7)	16
11. AZA, MP, Niridazole metabolite	3	22.0 (±9.2)	18
12. Pretransplant blood transfusion (protocol A)	8	13.3 (±7.6)	19
13. PTBT(A)+AZA, MP+post-transplant AZA, MP	5	23.4 (±2.9)	19
14. PTBT(A)+CSA(A)+post-transplant AZA, MP	6	23.7 (±6.6)	19
15. PTBT (protocol B)+post-transplant AZA, MP	6	24.0 (±9.2)	19
16. DST+post-transplant AZA, MP	7	22.3 (±16.7)	19
17. DS Liver cell extract, AZA, MP	8	5.0 (±3.8)	23
18. NS Liver cell extract, AZA, MP	4	7.8 (±5.7)	23
19. CSA, MP	10	>30.0	21

Note: (SD) standard deviation; (IS) immunosuppressive therapy; (AZA) azathioprine; (MP) methylprednisolone; (PTBT) pretransplant blood transfusion; (CSA) cyclosporine; (DST) donor specific pretransplant blood transfusion; (DS) pretransplant donor specific; and (NS) pretransplant nonspecific (third party).

Table 4 **Selected Results of Cardiac Xenotransplantation in the Baboon**

Group		n	Graft survival (mean in days ±SD)	Ref.
Concordant xenografts				
(vervet monkey-to-baboon)				
1. No IS (control)		9	10.3 (±5.2)	30
2. CSA,AZA,MP		6	13.0 (±8.2)	22
3. CSA,AZA,MP	IV MP therapy for rejection episodes	5	19.0 (±21.8)	31
4. RATG,CSA,AZA,MP	IV MP therapy for rejection episodes	6	43.3 (±18.5)	31
5. 15-DS,CSA,AZA,MP	IV MP therapy for rejection episodes	7	20.1 (±11.5)	31
6. 15-DS,CSA,MP	IV MP therapy for rejection episodes	5	35.6 (±14.2)	31
Discordant xenografts				
(pig-to-baboon)			(Individual experiments)	
7. No IS (control)		4	40,60,180,480 min.	39
8. Pretransplant splenectomy		3	30,360,480 min.	39
9. CSA,AZA,MP		5	15,15,40,75 min. + 5 days	39
10. Antibody adsorption[a]		7	360,480 min. + 0.5,4,4,4, 5 days	39
11. Antibody adsorption[a],CSA , AZA,MP		4	480 min. + 0.5,1,4 days	39

Note: (RATG) rabbit anti-human thymocyte globulin; (15-DS) 15-deoxyspergualin; other abbreviations as for Table 3.

[a] Antibody adsorption was achieved by prior perfusion of recipient baboon blood through donor pig kidneys.

Heterotopic heart transplantation is also valuable as a screening technique to study survival of heart grafts from closely related (concordant) or widely disparate (discordant) animal species; the results of some of these studies are outlined in Table 4. The vervet monkey-to-baboon model[24,28-33] would seem to closely mimic that of the non-human primate-to-man, and the pig-to-baboon model[24,32,37-39] closely mimics the pig-to-man situation. Successful outcomes in these models are likely, therefore, to have application in clinical organ transplantation.

9. Limitations of Heterotopic Heart Transplantation

The results obtained following heterotopic heart transplantation should be confirmed, however, by subsequent orthotopic heart transplantation, particularly when xenotransplantation is being studied. Work from the Columbia-Presbyterian group has demonstrated that in xenografting between closely related species prolonged survival of the heterotopically placed heart is not necessarily confirmed in the orthotopic model. An immunosuppressive regimen that led to survival of the heterotopic heart for relatively long periods[34,35] could not be reproduced in the orthotopic model.[36] The reasons for this remain uncertain, but it is presumed that a mild to moderate degree of rejection did not induce significant cardiac dysfunction in the heterotopic heart (which does not carry a significant workload), but led to early heart failure in the orthotopic situation (where the heart supports the entire circulation of the animal).

B. AUXILIARY HEART TRANSPLANTATION IN THE THORAX

The pioneering work by Losman and Barnard[8] in developing the technique of thoracic heterotopic (or auxiliary) heart transplantation that is still used in a small number of

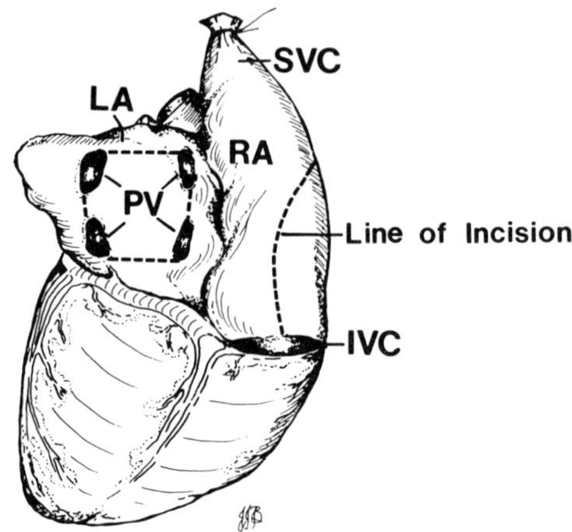

Figure 5 Orthotopic heart transplantation — excised donor heart (posterior view) showing lines of incision. (Abbreviations used in figures in this sequence are as in Figures 2 and 4).

patients undergoing heart transplantation today deserves brief mention. Although developed in the baboon model, this technique is rarely indicated as an experimental procedure.

The surgical methods of donor heart excision and implantation in the recipient have been clearly documented elsewhere[79,80] and will not be elaborated here. The procedure can be varied to support the native left ventricle alone or to provide biventricular support. In the clinical situation, biventricular support is almost uniformly chosen.

C. ORTHOTOPIC HEART TRANSPLANTATION

The technique in the baboon is identical to that in the human subject,[81] and is based on the pioneering work of Lower and Shumway in the dog.[82] Orthotopic heart transplantation is a major surgical undertaking in any experimental animal, though the baboon tolerates cardiopulmonary bypass and major surgery well, and its vascular tissues are excellent in the way they hold sutures.

1. Donor Operation

Excision of the donor heart is identical to that described for subsequent heterotopic heart transplantation into the neck (see above). Subsequent preparation of the donor heart, however, is different.

The right atrial cavity is opened, beginning posteriorly or laterally at the IVC orifice and continuing the incision into the base of the right atrial appendage, thus avoiding the areas of the coronary sinus and the sinoatrial node (Figures 5 and 6). The tissue between the orifices of the 4 pulmonary veins on the posterior aspect of the left atrium is excised, leaving one large opening (Figures 5 and 6).

The heart can then be hypothermically stored until it is required for insertion into the recipient.

2. Recipient Operation

With the baboon lying supine, a median sternotomy is performed, the pericardium opened longitudinally, and its edges retracted.

a. Initiation of Cardiopulmonary Bypass

After heparinization, cardiopulmonary bypass is initiated via cannulae inserted into the ascending aorta, just proximal to the origin of the brachiocephalic artery, and into the SVC and IVC (Figure 7). Snares (snuggers) or clamps are placed around the SVC and IVC to bring about total cardiopulmonary bypass. The animal is cooled to 28°C to 32°C. The aorta is then crossclamped immediately proximal to the aortic cannula.

b. Excision of Recipient Heart

The heart is excised by dividing the right and left atrial free walls and atrial septum close to the atrioventricular groove, leaving a cuff of atrial wall to allow easy suture of the donor heart (Figure 7). Both atrial appendages should be excised to prevent thrombus

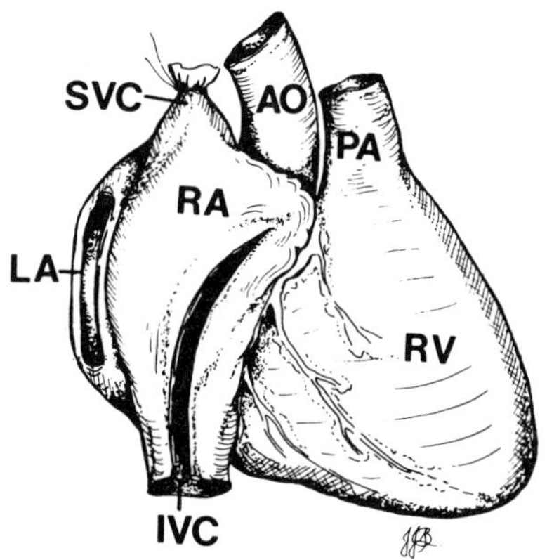

Figure 6 Orthotopic heart transplantation — donor heart (right anterolateral view) prepared for implantation.

formation occurring in these cavities subsequently. The aorta and main pulmonary artery are divided as close to their respective valves as possible (Figure 7). Subsequently, these vessels may be trimmed before being sutured to their counterparts of the donor heart.

The chronological order of division of these structures is unimportant, but a simple and accepted sequence is (1) free wall of right atrium, (2) free wall of left atrium, excluding the superior wall posterior to the origins of the aorta and pulmonary artery, (3) pulmonary artery, (4) aorta, (5) (after retracting the proximal aorta and pulmonary artery anteriorly) the remaining superior wall of the left atrium, and (6) the atrial septum.

c. Anastomosis of Left Atria

The donor heart is then placed (or held by an assistant) over the left side of the divided sternum, parallel to the remnants of the excised recipient heart. The donor heart is rotated 90 to 180° to the left so that its posterior surface faces anteromedially (towards the surgeon if he is standing to the right side of the table); the free walls of both recipient and donor left atria will then lie adjacent to each other (Figure 8). Using a double-ended 4-0 polypropylene suture, the left atrial walls are anastomosed by a continuous suture,

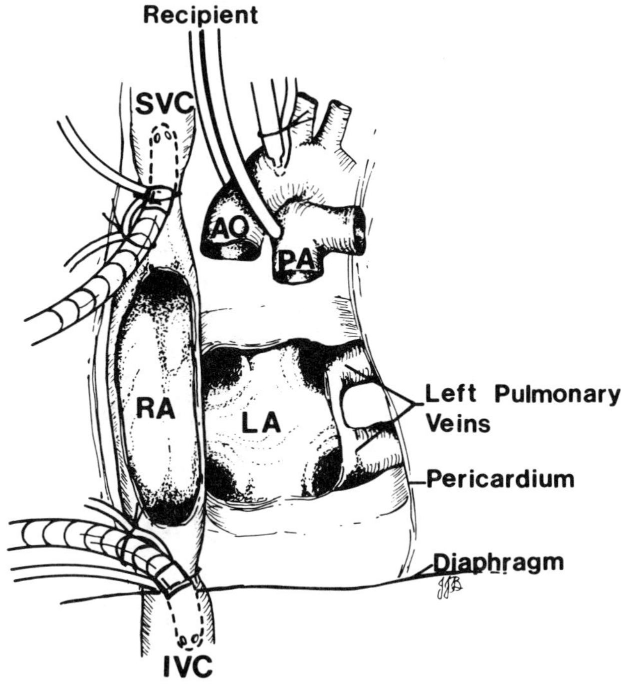

Figure 7 Orthotopic heart transplantation — view of recipient pericardial cavity after excision of the recipient heart.

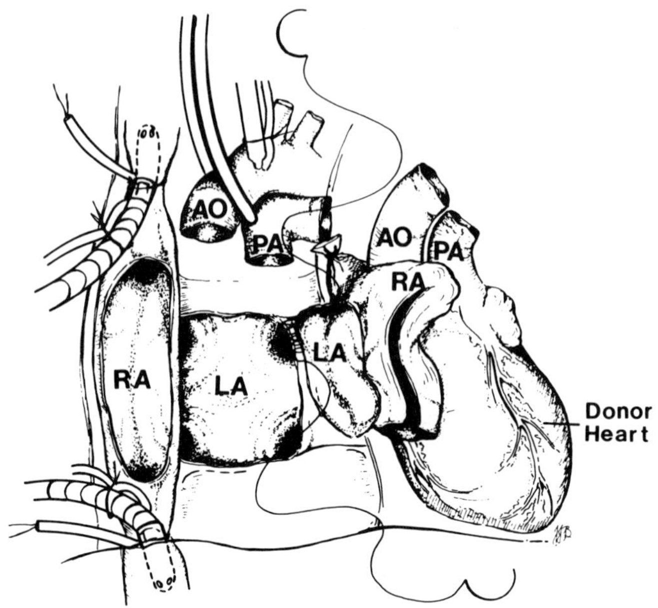

Figure 8 Orthotopic heart transplantation — donor and recipient hearts, showing the beginning of the anastomosis between the two left atria.

beginning at the base of the donor left atrial appendage and at a point close to the caudal end of the recipient left superior pulmonary vein (Figure 9). The suture is continued around the superior and inferior borders of the left atrium onto the atrial septum, and tied in the middle of the septum (Figure 9).

Between the performance of each suture line the pericardium is irrigated with cold saline to maintain a low myocardial temperature.

d. Anastomosis of Right Atria

The two right atria are anastomosed using a double-ended suture of 5-0 or 4-0 polypropylene. The suture is initially placed at the midpoint of the donor septum and at a convenient point on the posterior lip of the incision in the recipient right atrium. The anastomosis is continued first inferiorly and subsequently superiorly, and the two ends of the suture tied at the midpoint of the right atrial free wall (Figure 10). The atrial septum has therefore been sutured twice, once on the left atrial side and once on the right.

e. Anastomosis of Pulmonary Arteries

The two pulmonary arteries are trimmed to their ideal lengths and then anastomosed using a continuous suture of 4-0 polypropylene (Figure 11).

f. Anastomosis of Aortae

The aortae are similarly trimmed and anastomosed by continuous suture using 4-0 polypropylene (Figure 11).

g. Reperfusion and Rewarming

Air needles are placed in the apex of the left ventricle, anterior wall of the right ventricle, ascending aorta, and main pulmonary artery. The SVC and IVC snares or clamps are released and the heart is gently massaged to expel air from the ventricles and major vessels. The lungs are gently ventilated to increase the venous return to the left side of the heart, thus expressing further air, though care must be taken to ensure the heart is not over-distended.

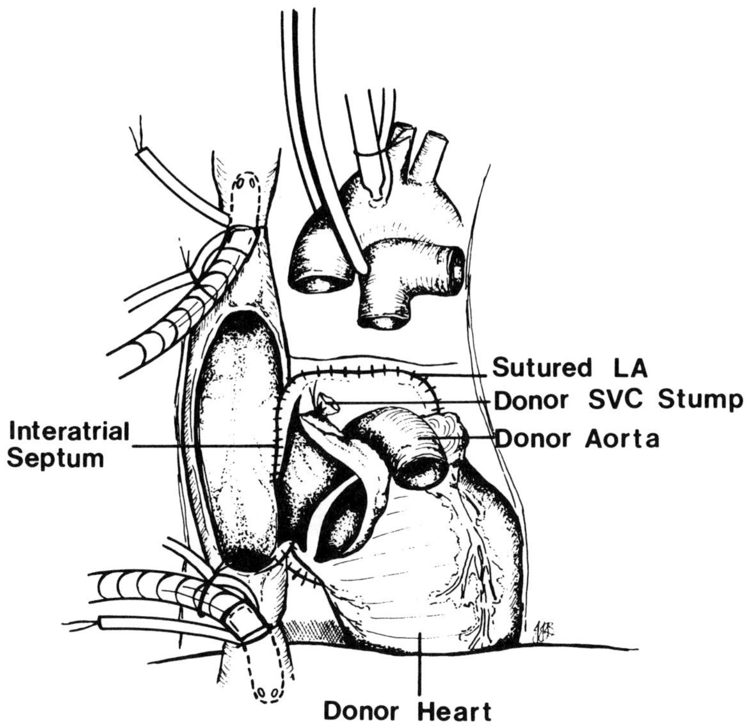

Figure 9 Orthotopic heart transplantation — completed left atrial free wall suture lines; the anastomosis between the two septa is being performed.

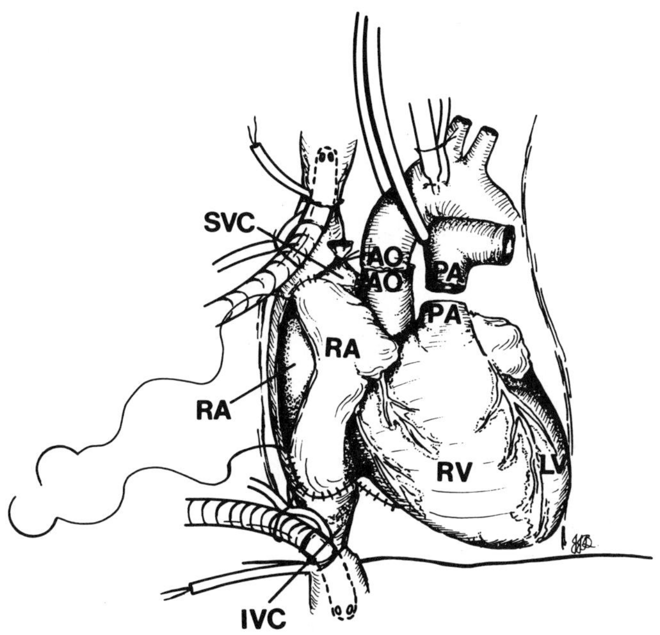

Figure 10 Orthotopic heart transplantation — the septal anastomosis has been completed; the free walls of the two right atria are being anastomosed.

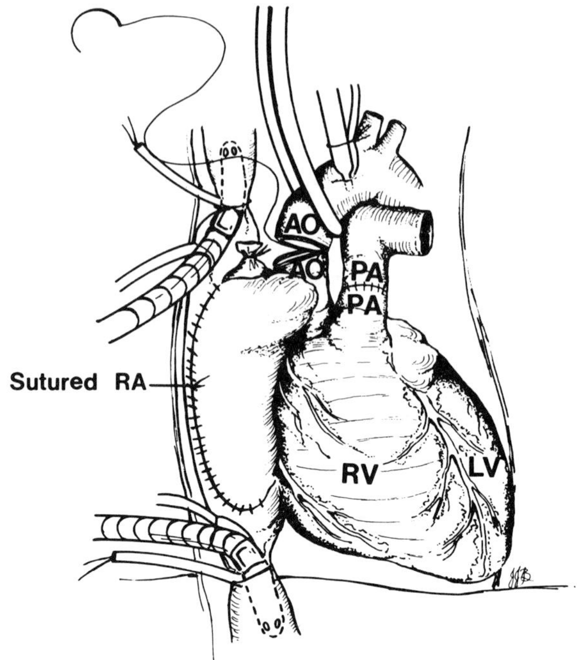

Figure 11 Orthotopic heart transplantation — the right atrial and pulmonary artery suture lines have been completed; beginning of aortic anastomosis.

The aortic crossclamp is released, allowing the coronary arteries of the donor heart to be perfused once again with oxygenated blood. Further efforts are made to expel air from both ventricles and major arteries.

Total body rewarming is begun, and will be associated with the return of either spontaneous coordinated myocardial contractions or vigorous ventricular fibrillation. If ventricular fibrillation is present, electrical defibrillation is necessary. Until coordinated contractions occur, and are sufficient to eject blood and thus decompress the ventricles, the heart should be gently manually decompressed at intervals if distention occurs. If after rewarming the heart remains bradycardic or in temporary heart bloc, ventricular pacing wires are applied and the heart is temporarily paced to stimulate ejection of blood and ventricular decompression.

It is preferable to provide at least 30 to 60 min of pump-oxygenator support after release of the aortic clamp to allow full recovery of the donor heart from its ischemic episode before challenging it with responsibility for support of the circulation. During this period, a careful check of all suture lines is made for bleeding, and further sutures inserted if necessary.

h. Alternative Technique

If it is desired to reduce the ischemic time to a minimum, then the left atrium can be anastomosed first, followed immediately by the aorta (Figure 12), at which time the aortic cross-clamp can be removed to perfuse the coronary arteries. Coronary venous return can be evacuated from the pericardial cavity by suction while the pulmonary artery and right atrial anastomoses are performed.

i. Discontinuation of Pump-Oxygenator Support

When myocardial function is clearly satisfactory, and adequate rewarming has taken place, the ventricular and pulmonary artery air needles are removed and their insertion

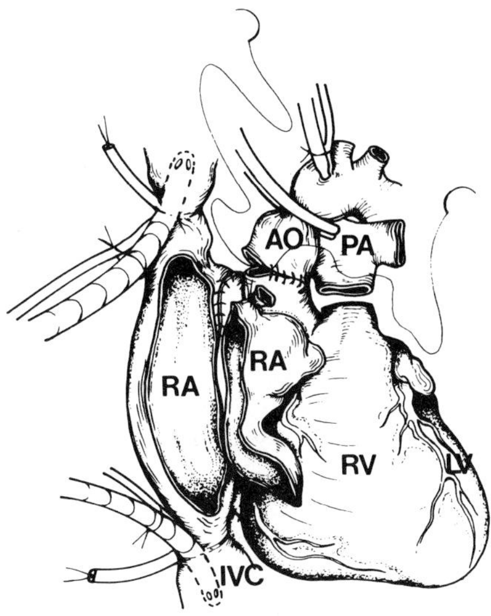

Figure 12 Orthotopic heart transplantation — alternative technique. The left atria have been anastomosed. When the aortic anastomosis is completed, the aortic cross-clamp can be removed, allowing myocardial reperfusion while the pulmonary artery and right atrial anastomoses are carried out.

sites oversewn with 4-0 polypropylene. The SVC cannula is withdrawn into the right atrium, and the IVC cannula removed. Cardiopulmonary bypass is discontinued, and the aortic air vent removed, and its site oversewn. If cardiac performance is satisfactory, then the aortic and SVC cannulae are removed and protamine sulfate administered (Figure 13).

Two drains are inserted, one into the pericardial cavity posterior to the heart and the other into the anterior mediastinum; the pericardium may be left entirely open, or can be closed partially over the aorta. The sternum is reunited with suitable strong sutures and the tissues anterior to the sternum repaired.

j. Potential Complications

The two potential major complications of this operation are bleeding, in view of the extensive areas of anastomosis, and systemic air emboli. In addition, every effort must be made to maintain donor myocardial temperature as low as possible (yet avoid freezing injury) during storage of the heart and until blood reperfusion is commenced.

3. Limitations of Orthotopic Heart Transplantation

Apart from confirming the results of immunosuppressive drug therapy or other immune manipulation obtained following heterotopic transplantation, orthotopic transplantation is also valuable in assessing methods of long-term storage of the heart.[11-15] Indeed, orthotopic transplantation is surely a major, and possibly an essential, test of any such storage system. This evaluation will inevitably be limited, however, by the onset and timing of any acute rejection episode that may develop. Although currently available drug therapy, using various combinations of cyclosporine, azathioprine, and a corticosteroid, can delay acute rejection for relatively long periods in the baboon, there still remain complications, such as infection, that may limit this experimental model.

When information on truly long-term cardiac function is required, therefore, autotransplantation of the heart may be indicated, and has proven to be an ideal model for assessing the efficacy of myocardial preservation systems.

D. AUTOTRANSPLANTATION OF THE HEART

The technique differs from orthotopic allotransplantation in small details.[12]

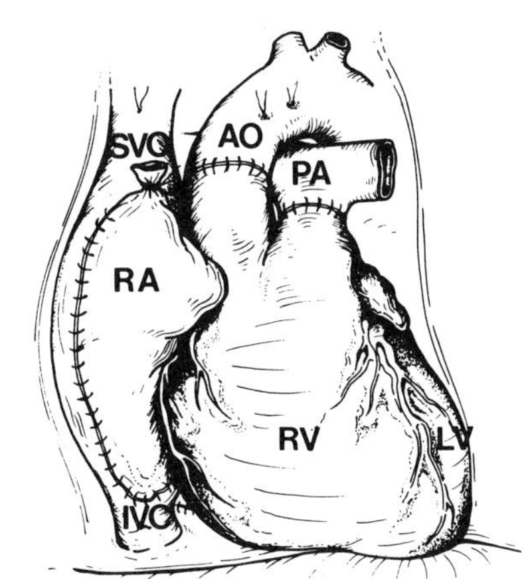

Figure 13 Orthotopic heart transplantation — completed operation. Arterial and venous cannulae have been removed.

1. Excision of the Heart

The recipient's own heart is excised with care being taken to include the sinus node with the heart. The line of excision must therefore be carried into the SVC well above the site of the sinus node. The posterior walls of the right and left atria (containing the orifices of the SVC, IVC, and four pulmonary veins) are left *in situ.* The aorta and pulmonary artery are divided rather more distally than in the usual donor heart (to protect the respective valves and allow a sufficient length of vessel to successfully perform the subsequent anastomoses at reinsertion). In other respects the technique of excision of the heart is identical to that for orthotopic heart allotransplantation.

The excised heart can be stored *ex vivo* for any period of time that is required (for example, 24 to 48 hours[12,14,15]) and can then be replaced in the original baboon. During the period of *ex vivo* heart storage the baboon is maintained alive by an orthotopic heart allotransplant from a donor baboon. This allograft is excised at the time of replacement of the baboon's native heart.

2. Reimplantation of the Heart

The technique of re-implantation is basically identical to that for allografting though there is clearly less atrial wall and arterial tissue available for making the various anastomoses.

3. Indications

This experimental procedure, though it places significant demands on the animal, which is required to undergo orthotopic heart transplantation on two occasions within a relatively short period of time, has been well tolerated in the small series of animals in which it has been carried out. Subsequent survival of baboons has been recorded for periods in excess of 2 years, at which time studies of the heart and circulation were carried out by cardiac catheterization and histological examination of the myocardium.[15]

Autotransplantation allows assessment of both immediate and late function of a heart that has been stored *ex vivo* for a period of time, such function being unmodified by acute or chronic rejection. Such elaborate experimental models would probably prove more difficult and less successful in animals such as the dog or pig which are possibly less tolerant of repeated anesthesia and surgical procedures.

E. TRANSPLANTATION OF THE HEART AND BOTH LUNGS

Early work in heart-lung transplantation was mainly aimed at elucidating the effects of denervation of the lungs.[3] Primates tolerated total denervation of the lungs, and did not depend on the presence of the Hering-Breuer reflex, unlike the dog.[5,6] Heart-lung transplantation in the dog was followed by failure to breath spontaneously, but this was not the case when the procedure was carried out in the primate.[6,7] Long-term survival of 6 to 24 months after heart-lung autotransplantation in baboons was reported by Castaneda and his colleagues in 1972.

1. Orthotopic Heart-Lung Transplantation

Early work in the dog in this field was carried out by Demikhov in the 1940s.[3,83] The technique used was ingenious, as cardiopulmonary bypass had not yet been developed. It is described, along with many other innovative experiments, by Demikhov in a book that has been translated into English, and which is recommended reading for anybody interested in experimental surgery.[83] The technique used clinically today, and currently of value in the experimental primate model, was probably first described (in the dog) by Longmore and his colleagues in 1969,[84] though it was the studies of Reitz et al. (in rhesus and cynomolgus monkeys),[9,10] using a slightly modified technique but with the advantage of cyclosporine therapy, that did much to establish it.

The non-human primate tolerates this major surgical insult surprisingly well.[9,10,85,86] The technique is identical to that used in man, which has been well described,[87-90] and will not be detailed here. In the experimental animal, the procedure has been carried out utilizing cardiopulmonary bypass[9,10,84-86] or deep hypothermia with circulatory arrest.[9]

It was in this model that one of the earliest observations was made of the fact that pulmonary and cardiac rejection rarely occur simultaneously, and that pulmonary rejection frequently develops earlier or in the absence of cardiac rejection.[85,86] Indeed, subsequent observations in both the experimental animal and in man have confirmed that the presence of the allografted lungs is associated with a significant reduction in the incidence of cardiac allograft rejection; the exact reason for this remains uncertain.

Heart-lung transplantation in the baboon has also been utilized in assessing methods of storage of the lung,[91] though experimental single lung transplantation would now appear to be the technique of choice for this purpose.

2. Heterotopic Heart-Lung Transplantation

Demikhov[3,83] and Longmore et al.,[92] using dogs, experimented with techniques of insertion of a heart or heart-lung allograft as an auxiliary intrathoracic pump. In the Longmore studies, the donor heart was grafted along with one or both lungs, the two right atrial appendages being anastomosed, and the donor aorta being joined end-to-side to the recipient ascending aorta. The donor bronchus or trachea was then anastomosed to that of the recipient or, more commonly, brought out as a tracheostomy. To accommodate the donor organs in the chest it was often necessary to remove at least one lobe of the recipient's own lungs.

A similar technique was developed in the baboon by the Cape Town group,[93] and this involved excision of the native left lung with insertion of the donor heart and left lung in the chest. Venous return to the native SVC and IVC was therefore divided, with blood passing through either donor heart/(left) lung or the recipient heart/(right) lung; blood flow was united again in the aorta. Donor cardiac function was found to be good, but it was difficult to assess the donor (left) lung function in view of the presence of the recipient's own (right) lung.

Investigation of this technique took place before either heart-lung or single lung transplantation had become successful techniques, and was originally intended as a method of transplanting a lung in the clinical situation. At that time, transplantation of the heart with the lung was believed to be beneficial as it would allow endomyocardial biopsy to diagnose acute rejection; subsequent experience has demonstrated that such biopsies are not helpful in the diagnosis of lung rejection (see above).

Probably the only indication for experimental heterotopic heart-lung transplantation today is to enable acute rejection of both the heart and lung to be investigated in an animal where the native heart and lung remain *in situ* to be used as control organs.

IV. APPLICATION OF CARDIAC TRANSPLANTATION IN PRIMATES

A. IMMUNOSUPPRESSIVE DRUG THERAPY AND MANIPULATION OF THE IMMUNE SYSTEM

Cyclosporine is not absorbed well when administered to the baboon by the oral route.[94] Bolus intravenous therapy leads to a high initial level, followed by a rapid fall;[94] blood levels can be maintained by continuous intravenous infusion using the system described above.[74] For most purposes, however, the intramuscular route has been found to be the most convenient,[94] though a period of several days may be required for adequate blood levels to be obtained, and relatively large doses in the region of 10 to 25 mg/kg/day may be required to maintain whole blood trough levels that are deemed adequate for immunosuppression.

Both azathioprine and methylprednisolone can be administered intravenously, intramuscularly, or orally with equally satisfactory results. Some preparations of human anti-

Table 5 **Results of Cardiac Allografting and Xenografting Across the ABO Blood Group Barrier**

Group	n	Type of rejection: Hyperacute (antibody-mediated)	Mixed	Acute (cellular)	None	Graft survival (mean in days ±SD)
Allografts						
(baboon-to-baboon)						
1. No IS (control)	10	0	0	10	0	11.0 (±5.6)
2. ABO incompatible;no IS	9	2	2	5	0	12.4 (±11.7)
3. CSA,MP	10	0	0	1	9(1)	>30.0
4. ABO-incompatible;CSA, MP	8	2	1	4(1)	1	22.5 (±12.0)
Concordant xenografts						
(vervet monkey-to-baboon)						
5. No IS (control)	9	5	4	0	0	10.3 (±5.2)
6. ABO-incompatible;no IS	9	3	2	4	0	7.3 (±5.6)
7. CSA,AZA,MP	6	1	0	5(2)	0	13.0 (±8.2)
8. ABO-incompatible;CSA, AZA,MP	5	1	1	3(1)	0	11.4 (±11.2)
9. TLI,CSA,AZA,MP	5	0	1(1)	3(3)	1(1)	16.2 (±9.8)
10. ABO-incompatible;TLI, CSA,AZA,MP	5	2(1)	0	1(1)	2(1)	17.8 (±10.5)

Note: (TLI) pretransplant total lymphoid irradiation; other abbreviations as for Table 3. Figures in parentheses denote number of recipients that died.

thymocyte globulin have been shown to be effective as immunosuppressive agents in the baboon,[31-36] as has anti-Tac antibody.[95] Monoclonal antibodies have recently been demonstrated to prolong cardiac allograft survival in the cynomolgus monkey.[96]

Pretreatment of the recipient baboon by total lymphoid irradiation,[28,57-59] blood transfusion,[19] splenectomy,[18] or other manipulation of the immune system[23,39] can be carried out.

V. CONCLUSIONS

The non-human primate has therefore proved to be an invaluable experimental animal for the study of various problems relating to the transplantation of thoracic organs. The evaluation of immunosuppressive therapies, of storage systems of the heart and lungs, and, in particular, of the role of ABO incompatibility in heart transplantation (Table 5) have all been clarified to a greater or lesser degree in this model. Currently, increasing attention is being paid to the subject of xenotransplantation, and non-human primates are being utilized in both concordant and discordant experimental studies.

In North America and Europe, however, high cost and occasional lack of availability continue to inhibit their maximal use in the surgical laboratory.

VI. ACKNOWLEDGMENTS

The description of the technique of orthotopic heart transplantation, together with Figures 5 through 11, are reproduced from *The Transplantation and Replacement of Thoracic Organs* (edited by D.K.C. Cooper and D. Novitzky) published by Kluwer Academic Publishers, 1990. The authors thank the publishers for permission to reproduce these figures, which were prepared by Jenny Kukielski.

The authors acknowledge the contributions of their many colleagues in Cape Town and Oklahoma City with whom the studies reported in this chapter were performed. The Cape Town studies were supported by the Chris Barnard Fund, the Medical Research Council of South Africa, the Cape Provincial Administration, and the University of Cape Town. The Oklahoma studies have been supported by Chembiomed Ltd. of Edmonton, Canada, and by the Oklahoma HealthCare Corporation.

REFERENCES

1. **Richter, C.B., Lehner, N.D.M., and Henrickson, R.V.,** Primates, in *Laboratory Animal Medicine,* Fox, J.G., Cohen, G.J., and Loew, F.M., Eds., Academic Press, Orlando, FL, 1984.
2. **Cooper, D.K.C.,** Experimental development of cardiac transplantation, *Br. Med. J.,* 4, 174, 1968.
3. **Cooper, D.K.C.,** Transplantation of the heart and both lungs: I. Historical review, *Thorax,* 24, 383, 1969.
4. **Willman, V.L., Cooper, T., Kaiser, G.C., and Hanlon, C.R.,** Cardiovascular response after cardiac autotransplant in primate, *Arch. Surg.,* 91, 805, 1965.
5. **Haglin, J., Telander, R.L., Muzzall, R.E., Kaiser, J.C., and Strobel, C.J.,** Comparison of lung autotransplantation in the primate and dog, *Surg. Forum.,* 14, 196, 1963.
6. **Nakae, S., Webb, W.R., Theodorides, T., and Sugg, W.L.,** Respiratory function following cardiopulmonary denervation in dog, cat, and monkey, *Surg. Gynecol. Obstet.,* 125, 1285, 1967.
7. **Castaneda, A.R., Zamora, R., Schmidt-Habelmann, P., Hornung, J., Murphy, W., Ponto, D., and Moller, J.H.,** Cardiopulmonary autotransplantation in primates (baboons), late functional results, *Surgery,* 72, 1064, 1972.

8. **Losman, J.G. and Barnard, C.N.,** Hemodynamic evaluation of left ventricular bypass with a homologous cardiac graft, *J. Thorac. Cardiovasc. Surg.,* 71, 695, 1977.
9. **Reitz, B.A., Burton, N.A., Jamieson, S.W., Bieber, C.P., Pennock, J.L., Stinson, E.B., and Shumway, N.E.,** Heart and lung transplantation. Autotransplantation and allotransplantation in primates with extended survival, *J. Thorac. Cardiovasc. Surg.,* 80, 360, 1980.
10. **Reitz, B.A., Bieber, C.P., Rainey, A.A., Pennock, J.L., Jamieson, S.W., Oyer, P.E., and Stinson, E.B.,** Orthotopic heart and combined heart and lung transplantation with cyclosporin-A immune suppression, *Transplant. Proc.,* 13, 393, 1981.
11. **Wicomb, W.N., Cooper, D.K.C., Hassoulas, J., Rose, A.G., and Barnard, C.N.,** Orthotopic transplantation of the baboon heart after 20 to 24 hours' preservation by continuous hypothermic perfusion with an oxygenated hyperosmolar solution, *J. Thorac. Cardiovasc. Surg.,* 83, 133, 1982.
12. **Cooper, D.K.C., Wicomb, W.N., Rose, A.G., and Barnard, C.N.,** Orthotopic allotransplantation and autotransplantation of the baboon heart following twenty-four hours storage by a portable hypothermic perfusion system, *Cryobiology,* 20, 385, 1983.
13. **Cooper, D.K.C., Wicomb, W.N., and Barnard, C.N.,** Storage of the donor heart by a portable hypothermic perfusion system: experimental development and clinical experience, *J. Heart Transplant.,* 2, 104, 1983.
14. **Wicomb, W.N., Novitzky, D., Cooper, D.K.C., and Rose, A.G.,** Forty-eight hours hypothermic perfusion storage of pig and baboon hearts, *J. Surg. Res.,* 40, 276, 1986.
15. **Wicomb, W.N., Rose, A.G., Cooper, D.K.C., and Novitzky, D.,** Hemodynamic and myocardial histological and ultrastructural studies in baboons three to twenty-seven months following autotransplantation of hearts stored by hypothermic perfusion for 24 or 48 hours, *J. Heart Transplant.,* 5, 122, 1986.
16. **Boyd, S.T., Cooper, D.K.C., Baigrie, R., Rose, A.G., and Barnard, C.N.,** An investigation of the immunosuppressive effects of niridazole and metronidazole in rat and baboon heterotopic cardiac allograft models, *Transplantation,* 31, 326, 1981.
17. **Cooper, D.K.C., Rose, A.G., and Barnard, C.N.,** Low-dose versus high-dose steroid therapy in the prevention of acute rejection in baboon heterotopic cardiac allografts, *Transplantation,* 34, 107, 1982.
18. **Cooper, D.K.C., Novitzky, D., Lanza, R.P., Rose, A.G., Wicomb, W.N., and Barnard, C.N.,** Immunosuppression in the baboon cardiac allograft model: effects of splenectomy, bolus methylprednisolone, propranolol, high-dose azathioprine, cyclophosphamide, and a niridazole metabolite, *Transplantation,* 38, 299, 1984.
19. **Cooper, D.K.C., Rose, A.G., du Toit, E., Langman, E., and Wicomb, W.N.,** Failure of pretransplant third party and peroperative donor specific blood transfusion to improve heterotopic cardiac allograft survival in baboons, *Transplantation,* 40, 569, 1985.
20. **Cooper, D.K.C., Lexer, G., Rose, A.G., Rees, J., Keraan, M., and du Toit, E.,** Cardiac allograft survival in ABO blood group incompatible baboons, *Transplant. Proc.,* 19, 1036, 1987.
21. **Cooper, D.K.C., Lexer, G., Rose, A.G., Keraan, M., Rees, J., du Toit, E., and Oriol, R.,** Cardiac allotransplantation across major blood group barriers in the baboon, *J. Med. Primatol.,* 17, 333, 1988.
22. **Cooper, D.K.C., Human, P.A., Rose, A.G., Rees, J., Keraan, M., du Toit, E., and Oriol, R.,** Can cardiac allografts and xenografts be transplanted across the ABO blood group barrier?, *Transplant. Proc.,* 21, 549, 1989.
23. **Cooper, D.K.C., Harris, N., Rose, A.G., and Novitzky, D.,** Accelerated cardiac allograft rejection associated with administration of liver cell extract in the baboon, *Transplant. Proc.,* 22, 1966, 1990.

24. **Rose, A.G., Cooper, D.K.C., Human, P.A., Reichenspurner, H., and Reichart, B.,** Histopathology of hyperacute rejection of the heart — experimental and clinical observations in allografts and xenografts, *J. Heart Lung Transplant.,* 10, 223, 1991.
25. **Hildebrandt, A., Meiser, B., Human, P., Reichenspurner, H., Rose, A., Odell, J., and Reichart, B.,** FK 506: short- and long-term treatment after cardiac transplantation in non-human primates, *Transplant. Proc.*, 23, 509, 1991.
26. **Flavin, T., Ivens, K., Wang, J., Gutierrez, J., Hoyt, E.G., Billingham, M., and Morris, R.E.,** Initial experience with FK 506 as an immunosuppressant for non-human primate recipients of cardiac allografts, *Transplant. Proc.,* 23, 531, 1991.
27. **Kapelanski, D.P., Perelman, M.J., Faber, L.A., Paez, D.E., Rose, E.F., and Behrendt, D.M.,** 15-Deoxyspergualin and primate heart transplantation, *J. Heart Transplant.*, 9, 668, 1990.
28. **Cooper, D.K.C., Human, P.A., and Reichart, B.,** Prolongation of cardiac xenograft (vervet monkey to baboon) function by a combination of total lymphoid irradiation and immunosuppressive drug therapy, *Transplant. Proc.,* 19, 4441, 1987.
29. **Cooper, D.K.C., Human, P.A., and Rose, A.G.,** Is ABO compatibility essential in xenografting between closely related species?, *Transplant. Proc.,* 19, 4437, 1987.
30. **Cooper, D.K.C., Human, P.A., Rose, A.G., Rees, J., Keraan, M., Reichart, B., du Toit, E., and Oriol, R.,** The role of ABO blood group compatibility in heart transplantation between closely related animal species; an experimental study using the vervet monkey to baboon cardiac xenograft model, *J. Thorac. Cardiovasc. Surg.,* 97, 447, 1989.
31. **Reichenspurner, H., Human, P.A., Boehm, D.H., Rose, A.G., May, R., Cooper, D.K.C., Zilla, P., and Reichart, B.,** Optimalization of immunosuppression after xenogeneic heart transplantation in primates, *J. Heart Transplant.,* 8, 200, 1989.
32. **Cooper, D.K.C. and Rose, A.G.,** Experience with experimental xenografting in primates, in *Xenograft 25,* Hardy, M.A., Ed., Elsevier, Amsterdam, 1989, p. 95.
33. **Reichenspurner, H., Human, P., Rose, A.G., Reichart, B., and Cooper, D.K.C.,** Effect of pharmacological immunosuppression on donor heart survival in a closely related non-human primate xenograft model, *Transplant. Proc.,* 22, 1086, 1990.
34. **Kurlansky, P.A., Sadeghi, A.M., Michler, R.E., Smith, C.R., Marboe, C.C., Thomas, W.G., Coppy, L.J., Reemtsma, K., and Rose, E.A.,** Comparable survival of intraspecies and cross-species primate cardiac transplants, *Curr. Surg.,* 3, 413, 1986.
35. **Sadeghi, A.M., Robbins, R.C., Smith, C.R., Kurlansky, P.A., Michler, R.E., Reemtsma, K., and Rose, E.A.,** Cardiac xenograft survival in baboons treated with cyclosporine in combination with conventional immunosuppression, *Transplant. Proc.,* 19, 1149, 1987.
36. **Sadeghi, A.M., Robbins, R.C., Smith, C.R., Kurlansky, P.A., Michler, R.E., Reemtsma, K., and Rose, E.A.,** Cardiac xenotransplantation in primates, *J. Thorac. Cardiovasc. Surg.,* 93, 809, 1987.
37. **Lexer, G., Cooper, D.K.C., Rose, A.G., Wicomb, W.N., Rees, J., Keraan, M., and du Toit, E.,** Hyperacute rejection in a discordant (pig to baboon) cardiac xenograft model, *J. Heart Transplant.,* 5, 411, 1986.
38. **Lexer, G., Cooper, D.K.C., Wicomb, W.N., Rose, A.G., Rees, J., Keraan, M., Reichart, B., and du Toit, E.,** Cardiac transplantation using discordant xenografts in a non-human primate model, *Transplant. Proc.,* 19, 1153, 1987.
39. **Cooper, D.K.C., Lexer, G., Rose, A.G., Keraan, M., Rees, J., and du Toit, E.,** Effects of cyclosporine and antibody adsorption on pig cardiac xenograft survival in the baboon, *J. Heart Transplant.,* 7, 238, 1988.
40. **Fischel, R.J., Platt, J.L., Matas, A.J., Perry, E., Dalmasso, A.P., Manivel, C., Najarian, J.S., Bach, F.H., Bolman, R.M., III,** Prolonged survival of a discordant cardiac xenograft in a rhesus monkey, *Transplant. Proc.,* 23, 589, 1991.

41. **Hardy, J.D., Chavez, C.M., Kurrus, F.E., Webb, W.R., Neely, W.A., Eraslan, S., Turner, M.D., Fabian, L.W., and Labecki, J.D.,** Heart transplantation in man; developmental studies and report of a case, *JAMA,* 188, 113, 1964.
42. **Marion, P.,** Les transplantations cardiaques et les transplantations hepatiques, *Lyon Med.,* 222, 585, 1969.
43. **Barnard, C.N., Wolpowitz, A., and Losman, J.G.,** Heterotopic cardiac transplantation with a xenograft for assistance of the left heart in cardiogenic shock after cardiopulmonary bypass, *S. Afr. Med. J.,* 52, 1035, 1977.
44. **Bailey, L.L. Nelsen-Cannarella, S.L., Concepcion, W., and Jolley, W.B.,** Baboon-to-human cardiac xenotransplantation in a neonate, *JAMA,* 254, 3321 1985.
45. **Cooper, D.K.C. and Ye, Y.,** Clinical experience with cardiac xenotransplantation, in *Xenotransplantation,* Cooper, D.K.C., Kemp, E., Reemtsma, K., and White, D.J.G., Eds., Springer-Verlag, Heidelberg, 1991, 541.
46. **Reemtsma, K., McCracken, B.H., Schlegel, J.V., Pearl, M.A., Dewitt, C.W., and Creech, O., Jr.,** Reversal of early graft rejection after renal heterotransplantation in man, *JAMA,* 187, 691, 1964.
47. **Starzl, T.E., Marchioro, T.L., Peters, G.N., Kirkpatrick, C.H., Wilson, W.C., Porter, K.A., Rifkind, D., Ogden, D.A., Hitchcock, C.R., and Waddell, W.R.,** Renal heterotransplantation from baboon to man: experience with six cases, *Transplantation,* 2, 752, 1964.
48. **Hitchcock, C.R., Kiser, J.C., Telander, R.L., and Seljeskob, E.L.,** Baboon renal grafts, *JAMA,* 189, 934, 1964.
49. **Porter, K.A., Marchioro, T.L., and Starzl, T.E.,** Pathological changes in six treated baboon to man renal heterotransplantations, *Br. J. Urol.,* 37, 274, 1965.
50. **Ogden, D.A., Sitprija, V., and Holmes, J.H.,** Function of the baboon renal heterograft in man and comparison with renal homograft function, *J. Lab. Clin. Med.,* 65, 370, 1965.
51. **Reemtsma, K.,** Heterotransplantation, *Transplant. Proc.,* 1, 251, 1969.
52. **Reemtsma, K.,** Renal heterotransplantation fron non-human primates to man, *Ann. N.Y. Acad. Sci.,* 162, 412, 1969.
53. **Starzl, T.E.,** *Experience in Hepatic Transplantation,* W. B. Saunders, Philadelphia, PA, 1969, 408.
54. **Giles, G.R., Boehmig, H.J., Amemiya, H., Halgrimson, C.G., and Starzl, T.E.,** Clinical heterotransplantation of the liver, *Transplant. Proc.,* 2, 506, 1970.
55. **Bosman, S.C.W., Saunders, S.J., Terblanche, J., Harrison, G.G., and Barnard, C.N.,** Cross-circulation between man and baboon, *Lancet,* 2, 583, 1968.
56. **Saunders, S.J., Terblanche, J., Bosman, S.C.W., Harrison, G.G., Walls, R., Hickman, R., Biebuyck, J., Dent, D., Pearce, S., and Barnard, C.N.,** Acute hepatic coma treated by cross-circulation with a baboon and by repeated exchange transfusions, *Lancet,* 2, 585, 1968.
57. **Smit, J.A., Stark, J.H., and Myburgh, J.A.,** Mechanisms of immunologic tolerance following total lymphoid irradiation in the baboon, *Transplant. Proc.,* 19, 490, 1987.
58. **Roslin, M.S., Tranbaugh, R.F., Coons, M.S., Panza, A., Toporoff, B., Alexandopoulous, I., Chen, C-K., Chang, T.H., Norin, A.J., and Cunningham, J.N.,** Long-term survival of heterotopic primate xenografts treated with total lymphoid irradiation and cyclosporine, *Surg. Forum.,* 41, 279, 1990.
59. **Panza, A., Roslin, M.S., Coons, M., Toporoff, B., Strashun, A., Alexandopoulous, I., Chen, C.K., Chang, T., Cunningham, J.N., and Norin, A.J.,** One year survival of heterotopic heart primate xenografts treated with total lymphoid irradiation and cyclosporine, *Transplant. Proc.,* 23, 483, 1991.
60. **Socha, W.W. and Ruffie, J.,** *Blood Groups of Primates: Theory, Practice, Evolutionary Meaning,* A.R. Liss, New York, 1983.

61. **Socha, W.W.,** Blood groups of non-human primates, in *Comparative Primate Biology Vol. I - Systematics, Evolution and Anatomy,* Swindler, D.R. and Erwin, J., Eds., Alan R. Liss, New York, 1986, 299.
62. **Michler, R.E., Marboe, C.C., Socha, W.W., Moor-Jankowski, J., Reemtsma, K., and Rose, E.A.,** Simian-type blood group antigens in non-human primate cardiac xenotransplantation, *Transplant. Proc.,* 19, 4456, 1987.
63. **Socha, W.W. and Moor-Jankowski, J.,** Primate animal model for xenotransplantation: serological criteria of donor-recipient selection, in *Xenograft 25,* Hardy, M.A., Ed., Elsevier, Amsterdam, 1989, 87.
64. **Balner, H.,** Histocompatibility testing in primates, *Vox Sang,* 11, 306, 1966.
65. **Balner, H.,** Choice of animal species for modern transplantation research, *Transplant. Proc.,* 6, 19, 1974.
66. **Borleffs, J.C.C., Marquet, R.L., De By-Aghai, Z., Van Vreeswijk, W., Neuhaus, P., and Balner, H.,** Kidney transplantation in rhesus monkeys: matching for D/DR antigens, pretransplant blood transfusions, and immunological monitoring before transplantation, *Transplantation,* 33, 285, 1982.
67. **Neethling, F.A., Nortman, P.J., and Cooper, D.K.C.,** Histocompatibility matching between humans and baboons, *Transplant. Proc.,* 22, 1067, 1990.
68. **Stark, J.H, Smit, J.A., Neethling, F.A., Nortman, P.J., and Myburgh, J.A.,** Immunological compatibility between the Chacma baboon and man, *Transplantation,* 52, 1072, 1991.
69. **Kalter, S.S.,** Overview of simian viruses and recognized virus diseases and laboratory support for the diagnosis of viral infections, in *Primates: The Road to Self-Sustaining Populations,* Benirschke, K., Ed., Springer-Verlag, New York, 1986, 680.
70. **Van Der Reit, F., Human, P.A., Cooper, D.K.C., Reichart, B., Fincham, J.E., Kalter, S.S., Kanki, P.J., Essex, M., Madden, D.L., Lai-Tung, M.T., Chalton, D., and Sever, J.L.,** Virological implications of the use of primates in xenotransplantation, *Transplant. Proc.,* 19, 4068, 1987.
71. **Human, P.A., Van Der Reit, F., Cooper, D.K.C., Fincham, J.E., Smuts, H.E.M., Reichenspurner, H., Madden, D.L., Sever, J.L., and Reichart, B.,** The virological evaluation of non-human primates for xenotransplantation in *Recent Advances in Cardiovascular Surgery,* Reichart, B., Fasol, R., Odell, J., Von Oppell, U., Reichenspurner, H., and Zilla, P., Eds., R.S. Schulz, Seehang, 1989.
72. Editorial, Ebola-related filovirus infection in non-human primates and interim guidelines for handling non-human primates during transit and quarantine, *MMWR,* 39, 22, 1990.
73. **Kalter, S.S.,** The non-human primate as a potential organ donor for man — virological considerations, in *Xenotransplantation,* Cooper, D.K.C., Kemp, E., Reemtsma, K., and White, D.J.G., Eds., Springer-Verlag, Heidelberg, 1991, 457.
74. **Byrd, L.D.,** A tethering system for direct measurement of cardiovascular function in the caged baboon, *Am. J. Physiol.,* 236, H775, 1979.
75. **Benirschke, K.,** *Primates — The Road to Self-Sustaining Populations,* Springer-Verlag, New York, 1986.
76. **Carrel, A. and Guthrie, C.C.,** The transplantation of veins and organs, *Am. Med. (Philadelphia),* 10, 1101, 1907.
77. **Mann, F.C., Priestley, J.T., Markowitz, J., and Yater, W.M.,** Transplantation of the mammalian heart, *Arch. Surg.,* 26, 219, 1933.
78. **Otte, K.E., Andersen, N., Jorgensen, K.A., Christensen, T., Barfort, P., Starklint, H., Larsen, S., and Kemp, E.,** Xenoperfusion of pig kidney with human AB or O whole blood, *Transplant. Proc.,* 22, 1091, 1990.
79. **Novitzky, D., Cooper, D.K.C., and Barnard, C.N.,** The surgical technique of heterotopic heart transplantation, *Ann. Thorac. Surg.,* 36, 476, 1983.

80. **Novitzky, D. and Cooper, D.K.C.,** Surgical technique of heterotopic heart transplantation, in *The Transplantation and Replacement of Thoracic Organs,* Cooper, D.K.C. and Novitzky, D., Eds., Kluwer, Dordrecht, 1990, 81.
81. **Cooper, D.K.C. and Novitzky, D.,** Surgical technique of orthotopic heart transplantation, in *The Transplantation and Replacement of Thoracic Organs,* Cooper, D.K.C. and Novitzky, D., Eds., Kluwer, Dordrecht, 1990, 75.
82. **Lower, R.R. and Shumway, N.E.,** Studies on orthotopic transplantation of the canine heart, *Surg. Forum.,* 11, 18, 1960.
83. **Demikhov, V.P.,** *Experimental Transplantation of Vital Organs,* Authorized translation from Russian by Haigh, B., Consultant's Bureau, New York, 1962.
84. **Longmore, D.B., Cooper, D.K.C., Hall, R.W., Sekabunga, J., and Welch, W.,** Transplantation of the heart and both lungs, II. Experimental cardiopulmonary transplantation, *Thorax,* 24, 391, 1969.
85. **Cooper, D.K.C., Novitzky, D., Rose, A.G., and Reichart, B.,** Acute pulmonary rejection precedes cardiac rejection following heart-lung transplantation in a primate model, *J. Heart Transplant.,* 5, 29, 1986.
86. **Novitzky, D., Cooper, D.K.C., Rose, A.G., and Reichart, B.,** Acute isolated pulmonary rejection following transplantation of the heart and both lungs: experimental and clinical observations, *Ann. Thorac. Surg.,* 42, 180, 1986.
87. **Reitz, B.A., Wallwork, J., Hunt, S.A., Pennock, J.L., Billingham, M.E., Oyer, P.E., Stinson, E.B., and Shumway, N.E.,** Heart and lung transplantation: successful therapy for patients with pulmonary vascular disease, *N. Engl. J. Med.,* 306, 557, 1982.
88. **Jamieson, S.W., Stinson, E.B., Oyer, P.E., Baldwin, J.C., and Shumway, N.E.,** Operative technique for heart-lung transplantation, *J. Thorac. Cardiovasc. Surg.,* 87, 930, 1984.
89. **Haverich, A., Novitzky, D., and Cooper, D.K.C.,** Selection of the donor; excision and storage of donor organs, in *The Transplantation and Replacement of Thoracic Organs,* Cooper, D.K.C. and Novitzky, D., Eds., Kluwer, Dordrecht, 1990, 273.
90. **Novitzky, D. and Cooper, D.K.C.,** Surgical technique of the recipient operation, in *The Transplantation and Replacement of Thoracic Organs,* Cooper, D.K.C. and Novitzky, D., Eds., Kluwer, Dordrecht, 1990, 289.
91. **Reichart, B.A., Novitzky, D., Cooper, D.K.C, Cunningham, M.S., and Rose, A.G.,** Successful orthotopic heart-lung transplantation in the baboon after 5 hours cold ischemia with cardioplegia and Collins' solution, *J. Heart Transplant.,* 6, 15, 1987.
92. **Longmore, D.B., Cooper, D.K.C., and Sekabunga, J.,** Unpublished, Quoted by Cooper, D.K.C. in Reference 2.
93. **Cooper, D.K.C., Wicomb, W.N., and Barnard, C.N.,** Unpublished data, 1980.
94. **Novitzky, D., Cooper, D.K.C, and Wicomb, W.N.,** A successful method of administering cyclosporin A to the Chacma baboon, *S. Afr. Med. J.,* 68, 737, 1985.
95. **Cooper, M.M., Robbins, R.C., Waldmann, T.A., Gansow, O.A., and Clark, R.E.,** Use of anti-Tac antibody in primate cardiac xenograft transplantation, *Surg. Forum,* 39, 353, 1988.
96. **Flavin, T., Ivens, K., Rothlein, R., Faanes, R., Clayberger, C., Billingham, M., and Starnes, V.A.,** Monoclonal antibodies against intercellular adhesion molecule 1. Prolonged cardiac allograft survival in cynomolgus monkeys, *Transplant. Proc.,* 23, 533, 1991.

Section V

Small Bowel and Multivisceral

Chapter 15

Multivisceral Transplantation in the Rat

N. Murase and T.E. Starzl

CONTENTS

I. INTRODUCTION

The concept of multivisceral transplantation (MVTX), in which multiple organs including the liver, pancreas, stomach, small intestine, and colon are orthotopically transplanted as an *en bloc* graft, was described in dogs by Starzl et al. in 1960.[1,2] During the next 30 years, this challenging procedure was occasionally studied using other experimental animals, such as the pig.[3,4] Clinically, MVTX has been applied to treat patients with the short-bowel syndrome or intestinal hypersecretion and associated hepatic failure caused by total parenteral nutrition (TPN),[4-6] or to treat a patient with an extensive abdominal malignancy.[7] The problems arising in the development of this surgical procedure and postoperative management are largely solved and extended survival has been achieved in numbers. However, many questions remained from the early experience with clinical MVTX. These questions concern the treatment of rejection, questions about graft-versus-host disease (GVHD), and especially the development of associated lymphomas, which were eventually lethal in all but one of the patients who survived the perioperative period.

In order to perform precise experimental work related to clinical transplantation, a MVTX model using inbred rats was developed in our laboratory (Figure 1).[8] The many advantages in using small animals for transplantation studies are well known. The MHC of many rat strains is well defined, the cost of small animals and their maintenance are

0-8493-3629-5/94/$0.00+$.50

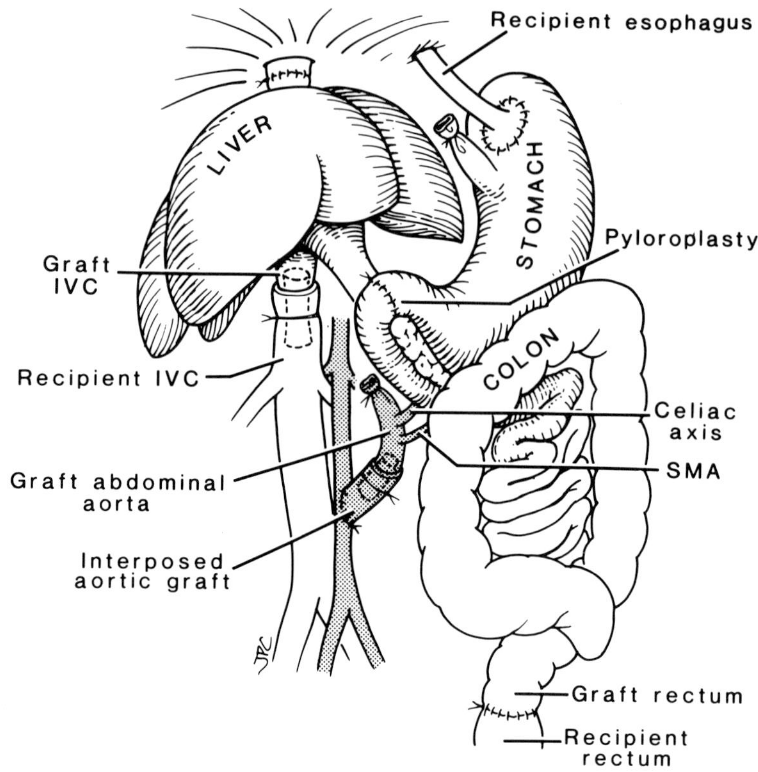

Figure 1 Surgical procedure of multivisceral transplantation. (IVC) Inferior vena cava, (SMA) superior mesenteric artery. (From Murase, N., Demetris, A.J., Kim, D.G., Todo, S., Fung, J.J., and Starzl, T.E., *Surgery,* 108, 880-9, 1990. With permission.)

substantially less than those of large animals, and a large number of observations can be accumulated in a short period of time. Animals with long-term survival can be easily maintained with a minimal space and care investment.

II. TECHNIQUES

A. ANATOMY

Knowledge of visceral anatomy[9,10] is a prerequisite to attempt this kind of procedure.

1. Arterial Circulation

The abdominal aorta begins at the point where it pierces the diaphragm and ends with its division into right and left common iliac arteries (Figure 2). It lies along the midline of the vertebral column, and on many occasions is crossed by the left renal vein. The visceral branches of the abdominal aorta are the celiac (celiac axis), superior mesenteric, inferior mesenteric, superior and inferior adrenal, renal, and internal spermatic (ovarian) arteries. The parietal branches are the inferior phrenic, lumbar, iliolumbar, middle caudal, and common iliac arteries.

The important arteries for this procedure are the celiac and superior mesenteric arteries. The celiac axis originates from the ventral surface of the aorta at the level of the crus of the diaphragm. It has a short trunk and divides into three branches: the left gastric, the splenic, and the hepatic arteries. The left gastric artery turns toward the cardiac opening of the

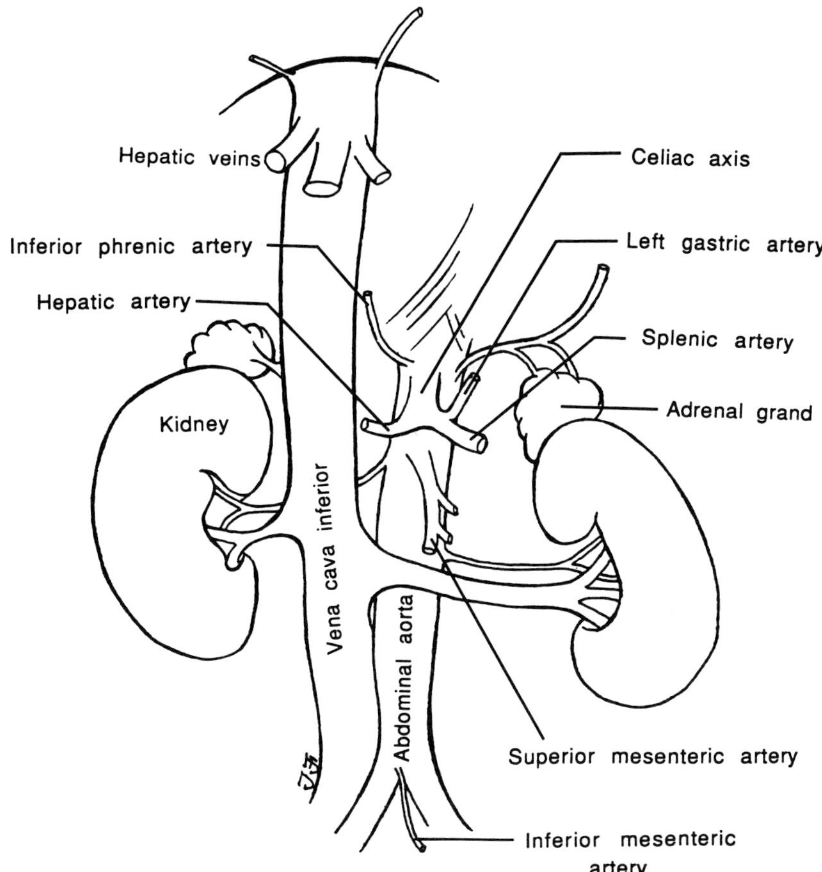

Figure 2 The abdominal aorta and inferior vena cava and their main branches.

stomach and gives branches to both surfaces of the stomach. The splenic artery passes to the left to the spleen, giving branches to the stomach and pancreas as well. The hepatic artery turns right supplying branches via the right gastric (pyloric) artery to the lesser curvature of the stomach and to the greater curvature of the stomach plus the duodenum, and pancreas via the gastroduodenal artery. The hepatic artery continues to the liver. The superior mesenteric artery is the largest of the abdominal aortic branches. It gives an inferior pancreaticoduodenal branch to the pancreas and the first loop of the duodenum, a series of intestinal branches to the jejunum and ileum, and an ileocolic branch to the transverse colon.

The renal arteries arise very close to the superior mesenteric artery, the right one usually higher than the left. Both renal arteries give rise to inferior adrenal branches to the adrenal gland before entering the hilum of each kidney.

The inferior phrenic artery may arise with the inferior adrenal from the renal artery. This is much more common on the right side. The left springs directly from the aorta. The paired internal spermatic arteries of the male leave the aorta a short distance below the renals, or they may arise from the renals. The paired ovarian arteries of the female have the same origin as the spermatics in the male.

The lumbar arteries arise as five pairs from the dorsal surface of the aorta. The first pair arises just below the crura of the diaphragm, the second pair from the level of the renal arteries. The third pair arises just posterior to the spermatics. The fourth pair arises below the right iliolumbar artery, and the fifth pair arises from the middle sacral artery. These arteries turn dorsally and laterally to the psoas and quadratus lumborum muscles. They also send a spinal branch into the vertebral canal.

The inferior mesenteric artery branches from the ventral surface of the aorta close to the origin of the two common iliacs and divides into two main branches, the left colic which runs anteriorly and is distributed along the wall of the descending colon.

2. Venous Circulation

The inferior vena cava (IVC) begins with the junction of the two common iliac veins slightly posterior to the bifurcation of the aorta (Figure 2). Tributaries are lumbar, iliolumbar, spermatic (ovarian), renal, inferior phrenic, and hepatic veins. The lumbar veins, corresponding to the lumbar arteries, join the IVC as a series of four segmental pairs of vessels draining the iliopsoas and quadratus lumborum muscles. The renal veins are more widely separated than arteries, and the left renal opens into IVC well below the right, which is longer than the right and crosses ventral to the aorta in many cases. The inferior phrenic veins in both sides, collecting from the inferior surface of the diaphragm, open directly into the IVC or enter the renal vein. The hepatic vein, collecting branches from the liver, enters the IVC just below the diaphragm.

3. Portal Circulation

The portal system comprises the blood from the digestive tract below the diaphragm and its associated organs, the spleen and pancreas. It passes through the liver before entering the systemic circulation via the hepatic veins and IVC. Those tributaries are splenic, superior mesenteric, and pyloric veins. During the MVTX procedure the portal system is never touched, since it is in the *en bloc* graft.

4. Lymphatic System

The thoracic duct conveys the great mass of lymph and chyle into the blood. It commences in the abdomen by a dilated sac, the cisterna chyli, which is situated upon the front of the second lumbar vertebra, to the right side and behind the abdominal aorta. The cisterna chyli receives lymph from right and left lumbar nodes, renal nodes, and intestinal nodes.

5. The Digestive and Accessory Organs

The esophagus enters the stomach at the lesser curvature through a fold of the limiting ridge, which separates the forestomach from the glandular stomach. The nonglandular forestomach has rumenlike mucosal folds covered with stratified squamous epithelium and serves as a reservoir. A grandular stomach, the corpus, is characterized by gastric pits lined with simple columnar epithelium. The entire small intestine (duodenum, jejunum, and ileum) is about six times as long as the large intestine (colon and rectum), and Peyer's patches can be identified externally and internally along the distal jejunal and ileal portions. The ileum empties into the cecum at the ileocecocolic orifice. The cecum is a large, thin-walled, blind pouch, and is divided into two parts, an apical and a basal portion. The apical portion contains a distinct mass of lymphoid tissue in its lateral wall. The colon runs cranially from the cecum as the ascending colon, crosses the duodenum, and then proceeds laterally as the short transverse colon, which turns caudally as the descending colon to form the rectum in the pelvic region.

Accessory organs associated with the viscera include the spleen, liver, and pancreas. The pancreas is very diffuse and extends from the end of the duodenal loop to the left into the gastrosplenic omentum. The liver is divided into four major lobes, the median (cystic), the right lateral, the left, and the caudate lobes. The rat has no gallbladder.

B. SURGICAL PROCEDURE

1. Animals

Normal healthy animals weighing 200 to 300 g and 250 to 350 g are used as donors and recipients, respectively. Animals are maintained under standard conditions with water and regular rat chow *ad libitum* until surgery.

2. Animal Preparation

Donor animals are treated with oral neomycinphosfate (25 mg/kg/day) for 5 days before surgery to control intestinal bacterial flora. For 2 days before surgery, regular diet is withdrawn, but animals have free access to water and cubed sugar (50 cal/day) to avoid body weight loss. No special preparation is performed for the recipient animals.

3. Operative Procedure

All surgical procedures are clean but not sterile. For the anesthesia, an open system of methoxyflurane is used for both donors and recipients. Oxygen is supplied over the recipient's face during graft implantation.

a. Donor Operation

The abdomen is entered through an entire midline incision. Intestines are gently wrapped with a wet sponge and retracted to the left side of the animal. Throughout the procedure much attention is paid to minimize handling the whole viscera and to keep it moist. The abdominal aorta is isolated from the IVC and left renal vein. A long segment of aorta from bifurcation to the crus of the diaphragm is prepared by individually ligating and dividing all of the aortic branches, excluding celiac and superior mesenteric arteries. This is facilitated by division of the right and left renal arteries, internal spermatic arteries, and lumbar arteries. All lymphatics are carefully identified and ligated to avoid the lymph leakage. The inferior mesenteric artery is not included in the graft, because of its very low junction to the aorta.

The infrahepatic IVC is dissected free from adjacent fatty connective tissue as low as the confluence of left renal vein. The right renal, the right adrenal, and the lumbar veins are ligated and divided. Ligaments around the liver, the falcifolm, and the right and left coronary ligaments, are separated. The left inferior phrenic vein is ligated at the entrance to the IVC. Splenectomy is performed by ligating the splenic arteries and veins at the splenic hilum. After injecting 300 U of heparin through the penal vein, intravenous catheter (20-gauge) is inserted into the end of the infrarenal aorta. The whole graft is perfused with 10 ml of iced lactated Ringer's solution after the thoracic aorta is clamped through thoracotomy. The IVC is transected in the chest and below the left renal vein. Soon after the graft is cooled down, the rectum and esophagus are transected. The abdominal aorta above the celiac axis is dissected towards the diaphragm and is ligated and divided. The cisterna chili and thoracic duct are also ligated and divided. The suprahepatic IVC is transected with a small cuff of diaphragm. The specimen is taken out from the donor and placed in the iced lactated Ringer's solution.

b. Graft Preparation

In the bath, the *en bloc* graft is perfused again via the aorta with cold lactated Ringer's solution to clean the vascular bed. Then, the contents of gastrointestinal tract (esophagus, stomach, small intestine, and colon) are flushed out with cold normal saline containing 1% neomycinphosfate (Figure 3). After the gastrointestinal lumen is cleaned, the esophagus and rectum are tied. The infrahepatic IVC and abdominal aorta are prepared for cuff anastomosis. For the infrahepatic IVC, PE 240 Intramedic polyethylene tubing, (I.D. 1.67, O.D. 2.42, Becton-Dickinson, Parsippany, NJ) and for the aorta, 16-gauge Intravenous Catheter (I.D. 1.2, O.D. 1.7 , Vycon, Ecouen, France) are used. Both cuffs consist of 0.2 cm length of cuff body and 0.2 cm of extension. The infrahepatic IVC is passed

Figure 3 Multivisceral graft. Arrows: arterial cuff, infrahepatic IVC cuff, and suprahepatic IVC.

through the lumen of the cuff. The cuff extension and vena cava is secured with a Satinsky clamp, which is rested on the bath wall. The end of the vein is spread and turned back over the cuff body. The circumferential suture secures the cuff. The arterial cuff is prepared in the same way. The suprahepatic IVC is prepared for suture anastomosis with small cuff of diaphragm. The whole graft is kept in the iced bath of lactated Ringer's solution, while the recipient is prepared.

c. Recipient Operation

After entering the abdomen through a complete midline incision, the infrarenal aorta is dissected and isolated, to which a 3 to 4 cm of piece of thoracic aorta from the recipient strain rat is anastomosed with 10-0 novafil suture. The other end of the aortic graft is tied. This two-step arterial anastomosis, recipient aorta to aortic graft, then aortic graft to graft aorta, is able to use the cuff technique for the graft arterial anastomosis. This aortic grafting can be done on the previous day.

The infrahepatic IVC is dissected free down to the right renal vein. Then the right adrenal and lumbar veins are ligated and divided. Ligaments around the liver are divided as described in the donor. The left inferior phrenic vein is ligated. After the celiac axis and superior mesenteric artery are exposed, the splenic artery is ligated and divided. The left gastric artery is isolated and is preserved with the esophagus and the small cuff of the stomach for the gastro-gastrostomy. Anesthesia is switched to pure oxygen, and the rectum is ligated and divided. The hepatic, and superior mesenteric arteries are ligated. The infrahepatic and suprahepatic IVC are cross-clamped with a mosquito and small Satinsky clamp, respectively. The stomach is resected leaving the small cuff for anastomosis. The whole visceral organs are taken out as an *en bloc* (Figure 4).

d. Graft Implantation

The graft is removed from the basin and placed orthotopically. The anastomosis between donor and recipient suprahepatic IVC is performed with 7-0 novafil suture. Infrahepatic IVC and aorta are anastomosed by simply inserting the cuffed donor vessels into the recipient vessels. For the aortic cuff, the thoracic aortic graft which has been placed before, is used for the recipient side of anastomosis. Using this aortic graft,

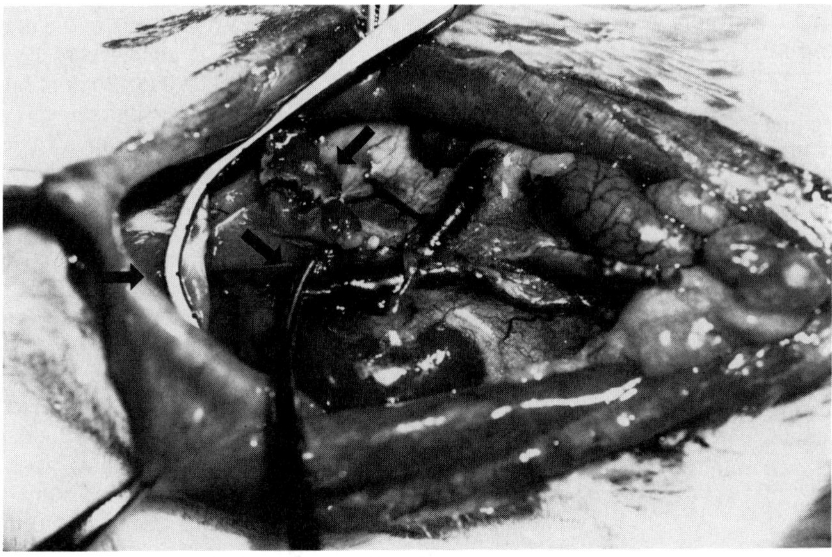

Figure 4 Recipient after removal of the visceral organs. Arrows: stomach, suprahepatic and infrahepatic IVC.

total implantation time can be reduced within 20 to 30 min. Soon after all three clamps are released, the whole graft is well revascularized in a satisfactory experiment.

e. Gastrointestinal Anastomosis

The graft stomach is incised in the cardiac portion (forestomach) and anastomosed with continuous 6-0 silk suture to a small cuff of recipient's stomach. A two-layer pyloroplasty is performed with 7-0 novafil suture. The graft rectum is anastomosed end-to-end with 6-0 silk suture to the residual rectum of the recipient. The peritoneal cavity was washed with warm irrigation fluid and the abdomen is closed in two layers with 4-0 dexon.

f. Recipient Care During and After Surgery

Because of the large amount of graft tissue, the recipient requires fluid and blood (whole blood with anticoagulant citrate dextrose [ACD]) during and after surgery. The right external jugular vein is used for a venous access. Lactated Ringer's solution with 10% dextrose is given as a continuous infusion with the flow rate of 1 to 2 ml/hour. Blood transfusion is added during and after the implantation of the graft with a total volume of 2 to 3 ml. Sodium bicarbonate (8.4%, 1.5 ml) and calcium chloride (10%, 0.1 ml) are given as bolus injections after the graft is revascularized. Continuous intravenous infusion is usually kept up for 1 to 2 hours in the postoperative period, and is disconnected. No more fluid is given postoperatively. The recipient animal receives a total 9 to 12 ml of fluid and 2 to 3 ml of whole blood.

Animals are placed under the warming light for few hours after surgery and placed back into a cage on the following day. Oral sugar and 10% dextrose are given freely for first 4 days, and then regular rat chow is started on 5th postoperative day. Cefamandole Nafate (20 mg/kg) is given intramuscularly for 5 days after surgery.

Total ischemic time of the graft is usually 60 to 90 min (Table 1), however, the graft preserved with lactated Ringer's solution for 150 min can function sufficiently. Implantation time (recipient vena cava clamping time) is between 20 to 30 min. During implantation, animals are stable and the arterial pressure is maintained in the normal range. At the moment of unclamping, the pressure goes down to 60 mmHg for a minute, and then

Table 1 **Surgical Time and Blood Pressure during MVTX**

Procedure	Time (min)	Blood pressure (mmHg)
Donor		
Dissection	30–45	
Ischemia *in situ*	10–20	
Recipient		
Dissection	30–45	100–130/60–90
Clamp vena cava		90/70
Implantation	20–30	100–140/90–120
Unclamp		60/40
Anastomosis in GI tract	45–60	90–150/70–100
Graft ischemic time	60–90	

quickly recovers to the normal range. Recipient surgery usually take 90 to 120 min. Immediately after the surgery, animals appear normal (Table 2).

Figure 5 shows the body weight gain after successful MVTX. Animals rapidly lose their weight for the first 3 days, up to 15 to 20% of initial weight by the 4th or 5th postoperative day. After starting regular rat chow on day five, the animals begin to gain weight and recover their initial body weight during the third postoperative week. By this time, they are healthy if rejection has been avoided.

Table 2 **Posttransplantation Status**

Na (mEq/L)	146 ± 2.7
K (mEq/L)	3.3 ± 0.67
Ca (mEq/L)	0.87 ± 0.12
Ht (%)	41.0 ± 2.9
pH	7.312 ± 0.114
PCO_2	40.5 ± 7.7
PO_2	197 ± 98.3
Base excess	–5.6 ± 5.8

g. Other Considerations

The use of nonabsorbable antibiotics, such as neomycinphosfate, for donor preparation and cleaning of all the gastrointestinal tract is effective. Without this manipulation hemorrhagic necrosis in the small bowel was common after revascularization. The role of bacteria indigenous to the gastrointestinal tract and effect of antibiotics upon animal survival after experimental intestinal ischemia has been reported in the nontransplanted model.[11]

The aortic graft for the graft arterialization is also useful. The principle of the cuff technique for the experimental organ transplantation has been well accepted.[12] The aortic grafting described here made it possible to use the cuff technique for the anastomosis which is basically side-to-end anastomosis. With this method the implantation time is reduced to 20 to 30 min, since the side to end arterial anastomosis with 10-0 novafil suture can be finished before implantation. The extra end-to-end arterial anastomosis using cuff takes only a few minutes to complete.

III. APPLICATIONS

In the recent advance of clinical visceral transplantation, the significance of multivisceral transplantation has been reappraised.[13] The whole graft can be used, by the same principle application for any intraabdominal grafting procedure which involves more than one organ, such as liver-duodenum-pancreas, liver-stomachduodenum-pancreas, liver-intestine, and other variations. These clinical transplantations of multiple abdominal viscera, derived from a multivisceral transplantation, have been performed more frequently with success.[13-16]

The key organs in multivisceral graft are the liver and small intestine. As an isolated graft, the liver graft has been known to tolerate rejection better than the other organs when transplanted individually, and the liver also appears to protect concomitantly transplanted organs, including the intestine. The small intestine graft is rejected easily when transplanted alone, and when transplanted alone or as a part of a multivisceral graft, it can

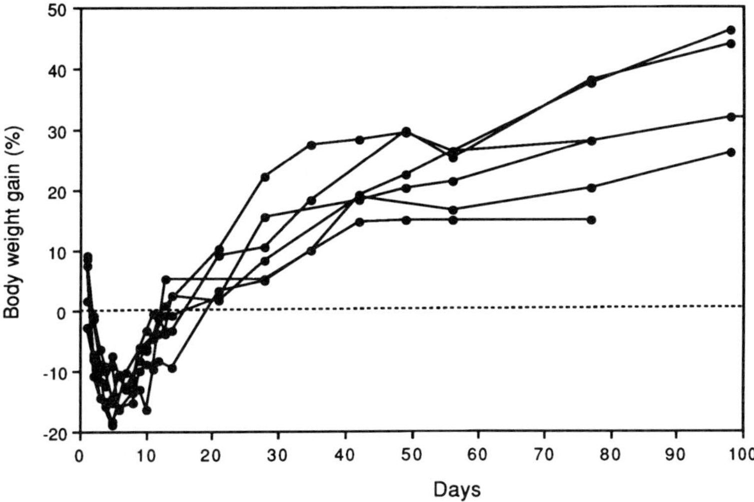

Figure 5 Body weight gain after syngeneic multivisceral transplantation.

cause GVHD because it contains a large amount of lymphoid tissue. Even in the earliest studies of untreated dog recipients of multivisceral grafts, evidence of GVHD was noticed.[1,2] In these early studies, the liver as the part of the multivisceral graft, was rejected later and less severely than after liver transplantation alone.[1,2] Whether this latter observation was a valid one remains to be confirmed or refuted by further studies in the more convenient rat model.

Some progress has been made. When multivisceral grafts from Brown Norway rats were transplanted to Lewis rats without immunosuppression, the intestinal component was rejected more aggressively than the companion liver and other organ components. This intestinal rejection was a lethal event and animal death occurred between 10 to 13 days after transplantation. In these untreated rats, the severity of the rejection of intestine in MVTX was as same as when intestine was transplanted alone, and animal survival was 9 to 14 days (Table 3). In either circumstance, the earliest and most intense intestinal rejection occurred in the intestinal lymphoid tissue, namely, the Peyer's patches and mesenteric lymph nodes. The T-cell dependent area of the Peyer's patches and mesenteric

Table 3 **Survival after Multivisceral Transplantation (MVTX), Small Bowel Transplantation (SBTX), or Liver Transplantation (LTX), Without Immunosuppressive Treatment**

Procedure	Strain combination		n	Survival (days)	Median (days)
	Donor	Recipient			
MVTX	LEW	LEW	8	7*, 72**, 81**, >100 × 5	>100
MVTX	BN	LEW	5	10, 10, 10, 10, 13	10.0
SBTX (caval drainage)	BN	LEW	7	9, 10, 11, 12, 13, 13, 14	12.0
SBTX (portal drainage)	BN	LEW	6	9, 9, 10, 11, 12, 14	10.5
LTX	BN	LEW	6	>100 × 6	>100

* Technical complication.

** Sacrificed in healthy condition as part of a colony depopulation program.

Table 4 **Survival after Multivisceral Transplantation (MVTX) and Small Bowel Transplantation (SBTX) under FK-506**

Transplantation	Treatment	*N*	Survival (days)
SBTX (caval drainage)	FK 0.64 mg/kg × 14 d	10	42, 105, (107), 112, 163, 167, 240, 286, >398, >400
SBTX (portal drainage)	FK 0.64 mg/kg × 14 d	9	(106), (112), 134, >146, >147, >148, >148, >166, >166
MVTX	FK 0.64 mg/kg × 14 d	7	12, 18, (114), (133), (134), >439, >473
SBTX (portal drainage)	FK 0.64 mg/kg × 14 d with splenectomy	6	10, (103), (110), >154, >160, >162

Parentheses: sacrificed for histopathology with healthy condition.

lymph nodes were markedly expanded by the proliferating population of blastic lymphoid cells 4 days after transplantation. This was associated or followed by cryptitis (at the base of the villi), epithelial cell necrosis, focal abscess formation, mural necrosis, and eventual perforation. Liver allografts transplanted alone or as part of multivisceral grafts had histological evidence of rejection, but this was self-limiting and spontaneously reversible when the liver was transplanted alone. Thus, the intestinal component was the Achilles heel of the multivisceral graft. The presence of the liver in the organ complex did not have protective effects in untreated rejection.[8]

However, when the immunosuppressive agent, FK 506, was used to treat Lewis recipients of multivisceral or isolated intestinal allografts of Brown Norway rats, the intestine was easier to protect from rejection if it was a part of a multivisceral graft than when it was transplanted alone (Table 4). The rejection of isolated small bowel graft was more difficult to control as judged by animal survival and by histopathologic examination. Histologically, the Peyer's patches and mesenteric lymph nodes are particularly affected, even though the epithelial intestinal components are still relatively intact; with chronic rejection, the lymphoid components were completely depleted and replaced by scar in isolated small intestine graft. This was less commonly seen after MVTX. Thus, the liver in the MVTX may have provided an advantage to the intestine and possibly to the other companion organs of the complex graft.[17]

REFERENCES

1. **Starzl, T.E. and Kaupp, H.A.,** Mass homotransplantation of abdominal organs in dogs, *Surg. Form.,* 11, 28-30, 1960.
2. **Starzl, T.E., Kaupp, H.A. Jr., Brock, D.R., Butz, G.W., Jr., and Linman, J.W.,** Homotransplantation of multiple visceral organs, *Am. J. Surg.,* 103, 219-229, 1962.
3. **Gridelli, B., Rossi, G., Colledan, L.R., Fassati, L.R., Ferla, G., Giacci, F., Gislon, M., Lucianetti, A., Andreoni, A., Doglia, M., and Galmarini, D.,** Organ procurement for multivisceral abdominal transplantation in pig, *Transplant. Proc.,* 20, 844-845, 1988.
4. **Starzl, T.E., Rowe, M., Todo, S., Jaffe, R., Tzakis, A., Hoffman, A., Esquival, C., Porter, K., Venkataramanan, R., Makowka, L., and Duquesnoy, R.,** Transplantation of multiple abdominal viscera, *JAMA,* 261, 1449-1457, 1989.
5. **Williams, J.W., Sankary, H.N., Foster, P.F., Lowe, J., and Goldman, G.M.,** Splancnic transplantation: an aproach to the infant dependent on parenteral nutrition who develops irreversible liver disease, *JAMA,* 261, 1458-1462, 1989.

6. **Jaffe, R., Trager, J.D.K., Zeevi, A., Sonmez-Alpan, E., Duquesnoy, R., Todo, S., Rowe, M., and Starzl, T.E.,** Multivisceral intestinal transplantation, surgical pathology, *Pediatr. Pathol.,* 9, 633-654, 1989.
7. **Margreiter, R.,** personal communication, Innsbruck.
8. **Murase, N., Demetris, A.J., Kim, D.G., Todo, S., Fung, J.J., and Starzl, T.E.,** Rejection of the multivisceral allografts in rats: a sequential analysis with comparison to isolated orthotopic small bowel and liver grafts, *Surgery,* 108, 880-889, 1990.
9. **Bivin, S.W., Crawford, M.P., and Brewer, N.R.,** Morphophysiology, in *The Laboratory Rat,* Baker, H.J., Lindsey, J.R., and Weisbroth, S.H., Eds., Academic Press, London, 1979.
10. **Donaldson, H.H.,** *Anatomy of the Rat,* Hafner Publishing, New York, 1935
11. **Jamieson, W.G., Pliagus, G., Marchuk, S., DeRose, G., Moffat, D., Stafford, L., Finley, R.J., Sibbald, W., Taylor, B.M., and Duff, J.,** Effect of antibiotic and fluid resuscitation upon survival time in experimental intestinal ischemia, *SGO,* 167, 103-108, 1988.
12. **Heron, I.,** A technique for accessory cervical heart transplantation in rabbits and rats, *Acta Pathol. Microbiol. Scand.,* 79(Suppl. A), 366, 1971.
13. **Starzl, T.E., Todo, S., Tzakis, A., Alessiani, M., Casavilla, A., Abu-Elmagd, K., and Fung, J.J.,** The many faces of multivisceral transplantation, *SGO,* 172, 335-344, 1991.
14. **Starzl, T.E., Todo, S., Tzakis, A., Podesta, L., Mieles, L., Demetris, A.J., Teperman, L., Selby, R., Stevenson, W., Gordon, R., and Iwatsuki, S.,** Abdominal organ cluster transplantation for the treatment of upper abdominal malignancies, *Ann. Surg.,* 210, 374-386, 1989.
15. **Tzakis, A.G., Todo, S., Madariaga, J., Tzoracoeleftherakis, E., Fung, J.J., and Starzl, T.E.,** Upper abdominal exenteration in transplantation for extensive malignancies of the upper abdomen: an update, *Transplantation,* 51, 727-728, 1991.
16. **Grant, D., Wall, W., Mimeault, R., Zhong, R., Ghent, C., Garcia, B., Stiller, C., and Duff, J.,** Successful small-bowel/liver transplantation, *Lancet,* 335, 181-184, 1990.
17. **Murase, N., Kim, D.G., Todo, S., Cramer, D.V., Fung, J.J., and Starzl, T.E.,** Induction of liver, heart, and multivisceral graft acceptance with a short course of FK 506, *Transplant. Proc.,* 22, 74-5, 1990.
18. **Murase, N., Demetris, A.J., Matsuzaki, T., Yagihashi, A., Todo, S., Fung, J.J., and Starzl, T.E.,** Long survival in rats after multivisceral versus isolated small bowel allotransplantation under FK 506, *Surgery,* 110, 87-98, 1991.
19. **Iwaki, Y., Starzl, T.E., Yagihashi, A., Taniwaki, S., Abu-Elmaged, K., Tzakis, A., Fung, J.J., and Todo, S.,** Replacement of donor lymphoid tissue in human small bowel transplantation under FK 506 immunosupression, *Lancet,* 337, 818-819, 1991.

Chapter 16

Intestinal Transplantation in the Rat

Robert Zhong, Robert Black, and David Grant

CONTENTS

I. INTRODUCTION

Small intestinal transplantation may eventually become the definitive treatment for short gut syndrome. Presently, however, this procedure remains highly experimental, in marked contrast to the great success of other clinical transplants such as the kidney, liver, and heart.[1,2] Intestinal grafting has been limited by problems with rejection, graft-versus-host disease, sepsis, and the development of post-transplant lymphoproliferative disorders. Researchers are now trying to develop protocols that will avoid these complications.

0-8493-3629-5/94/$0.00+$.50

Future progress will depend on our ability to achieve a better understanding of the immunology and the physiology of the small bowel allograft.

Inbred rats have been frequently used in intestinal transplant studies because of their low cost and the opportunity to manipulate immune responses by selecting certain combinations of donor and recipient strains. Heterotopic intestinal transplantation, developed 20 years ago by Monchik and Russell,[3] is a relatively simple surgical procedure with a low mortality rate. Because the graft is transplanted in an accessory position, this model is best suited for immunological studies. Orthotopic intestinal transplantation, developed by Kort et al.,[4] provides an appropriate model for physiological studies since the intestinal graft replaces the native jejunum and ileum, providing a normal environment for mucosal enterocytes. However, this model is technically demanding and mortality rates in most laboratories are higher with orthotopic transplantation than with heterotopic transplantation.[4]

Parallel to the development of our clinical intestinal transplant program, our laboratory has developed or modified several surgical models of intestinal transplantation in rats. Since 1985, more than 1200 intestinal transplants have been performed with high success rates.[5-16] Herein, we describe our surgical techniques and comment on the potential applications of each model.

II. MODELS

A. HETEROTOPIC (ACCESSORY) INTESTINAL TRANSPLANTATION

The intestinal graft is defunctioned with both ends or the proximal end of the graft exteriorized. The native gastrointestinal tract is left intact (Figures 1 and 2).

B. ORTHOTOPIC (IN-CONTINUITY) INTESTINAL TRANSPLANTATION

The intestinal graft is anastomosed to the remaining section of the native small bowel after resection of the recipient's jejunum and ileum (Figure 3). With this technique, host survival is dependent upon the provision of adequate nutrition by the intestinal graft.

C. SEGMENTAL INTESTINE AND COLON TRANSPLANTATION

The terminal ileum, ileocecal valve, cecum, and ascending colon are retrieved from the donor. The graft is then anastomosed to the native small bowel and colon after resection of the recipient's jejunum, ileum, and cecum (Figure 4). Again, with this technique, host survival is dependent upon the provision of adequate nutrition by the intestinal graft.

D. COMBINED LIVER AND INTESTINAL TRANSPLANTATION

The liver and intestinal grafts are harvested from separate donors and simultaneously transplanted (Figure 5). The liver is transplanted orthotopically. The intestine is transplanted heterotopically.

III. ANIMALS

A. SEX AND WEIGHT

Male rats, weighing between 250 and 300 g, are the ideal donors and recipients. Female rats are less tolerant of anesthesia and they lack a penile vein for intravenous access. Older rats have excessive fat making the surgical dissection more difficult. They are also more susceptible to respiratory diseases resulting in higher mortality rates.

B. STRAINS

Outbred Sprague-Dawley rats are ideal for practicing microsurgical procedures because they are hardy and inexpensive. Inbred Lewis rats are ideal for physiology studies since they are isogeneic and have a very gentle behavior. The selection of donor-recipient strain

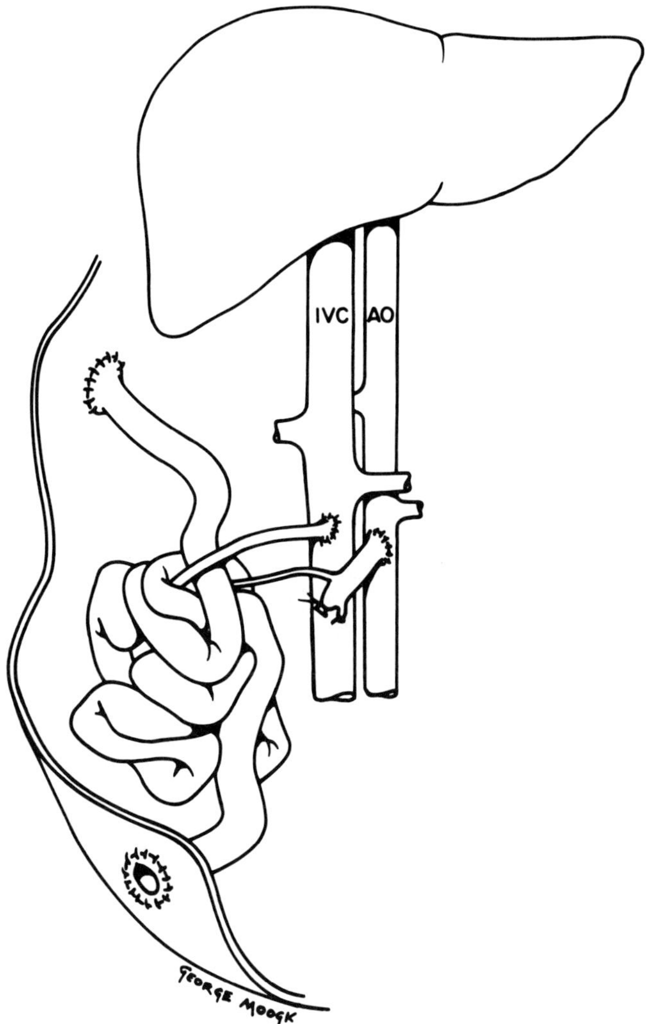

Figure 1 Heterotopic intestinal transplant model with both ends of the graft exteriorized as stomas. (IVC) Inferior vena cava and (AO) aorta.

combinations for immunology studies will depend on the type of immune response that is desired. Low-responder strain combinations, for example, have slightly longer graft survival times (Table 1).[16]

C. PREOPERATIVE CARE

The rats receive oral 5% glucose and 0.9% saline ad libitum for 24 hours before surgery. Atropine (0.04 mg/kg) is administered by subcutaneous injection prior to surgery. For the orthotopic intestinal and segmental intestinal/colon transplantation, cefoxitin (40 mg/kg) is given intramuscularly 1 hour preoperatively.

D. ANESTHESIA

All the donors and most recipients are anesthetized with an intraperitoneal injection of pentobarbital (40 mg/kg), supplemented with chloral hydrate (160 mg/kg, intraperitoneal)

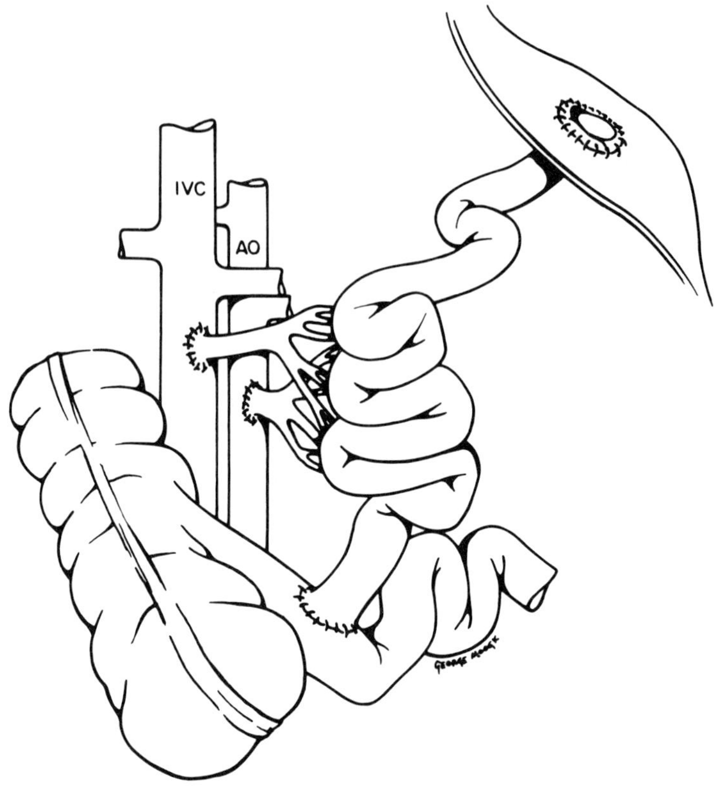

Figure 2 Heterotopic intestinal transplant model with exteriorization of the proximal end of the graft. (IVC) Inferior vena cava and (AO) aorta.

as required. The liver/intestinal transplant recipients are anesthetized by halothane inhalation using a small animal anesthetic machine (Med-Vet Anaesthetic Systems Incorporated, Toronto, Canada).

IV. HETEROTOPIC INTESTINAL TRANSPLANTATION

A. DONOR OPERATION

This procedure is modified from the technique developed by Monchik and Russell.[3] The abdomen is opened via a midline incision. An operating microscope is used to mobilize the graft. Manipulation of the intestine is minimal. The entire colon is removed and the portal vein is separated from the pancreas. A long segment of aorta, containing the superior mesenteric artery, is mobilized by ligating and dividing the renal and lumbar arteries (Figure 6). It is not necessary to ligate the left renal artery if its origin is far from the superior mesenteric artery. The lumbar arteries from the aorta are meticulously ligated with 8-0 silk sutures to minimize bleeding. The celiac trunk is ligated, then the pyloric and splenic veins are ligated separately using 8-0 silk sutures. The aorta is cannulated with a fine polyethylene catheter and the graft is perfused *in situ* with 2 to 3 ml of cold, heparinized lactated Ringer's solution. The intestine and its vascular supply are removed *en bloc* and stored in lactated Ringer's solution at 4°C. The donor surgery takes less than 45 min. During this time, approximately 10 ml of crystalloid is given intravenously to maintain normal blood pressure.

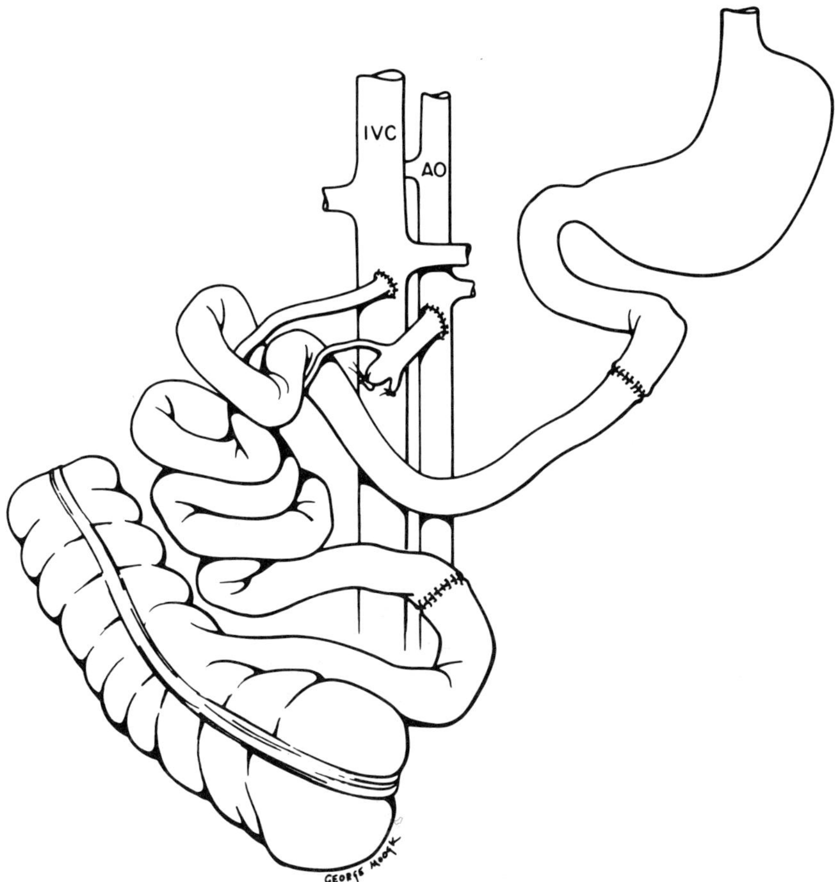

Figure 3 Orthotopic intestinal transplant model. (IVC) Inferior vena cava and (AO) aorta.

A good quality donor intestine is critical. Mechanical injury or ischemic injury can damage the graft.[17,18] We have recently abandoned the traditional practice of intra-luminal irrigation with an antibiotic solution because this procedure can sometimes damage the microcirculation of the graft. We also significantly reduced the volume of the intravascular perfusate (2 to 3 ml vs 12 to 20 ml in Monchik and Russell's study).[3] Early ligation of the pyloric and splenic veins causes splanchnic venous congestion leading to shock and ischemic injury to the graft. To avoid this complication, we ligate the celiac trunk first, followed by ligation of the pyloric and splenic veins just prior to graft removal. Ischemic injury is also minimized by (1) meticulous ligation of the lumbar vessels to avoid unnecessary blood loss, (2) early ligation of the distal abdominal aorta to improve perfusion of the superior mesenteric artery, and (3) the intravenous administration of lactated Ringer's solution during the retrieval.[4]

B. RECIPIENT OPERATION

The native intestine is left intact. The recipient's infrarenal aorta and inferior vena cava are carefully mobilized. A modified Lee's clamp is placed across these vessels (Figure 7). The aorta is punctured with a 30-gauge needle and opened via a longitudinal arteriotomy. The lumen is flushed with heparinized saline. Two 10-0 nylon stay sutures are placed at both apices of the aortotomy. A 7-0 nylon stay suture is placed at the midpoint of the

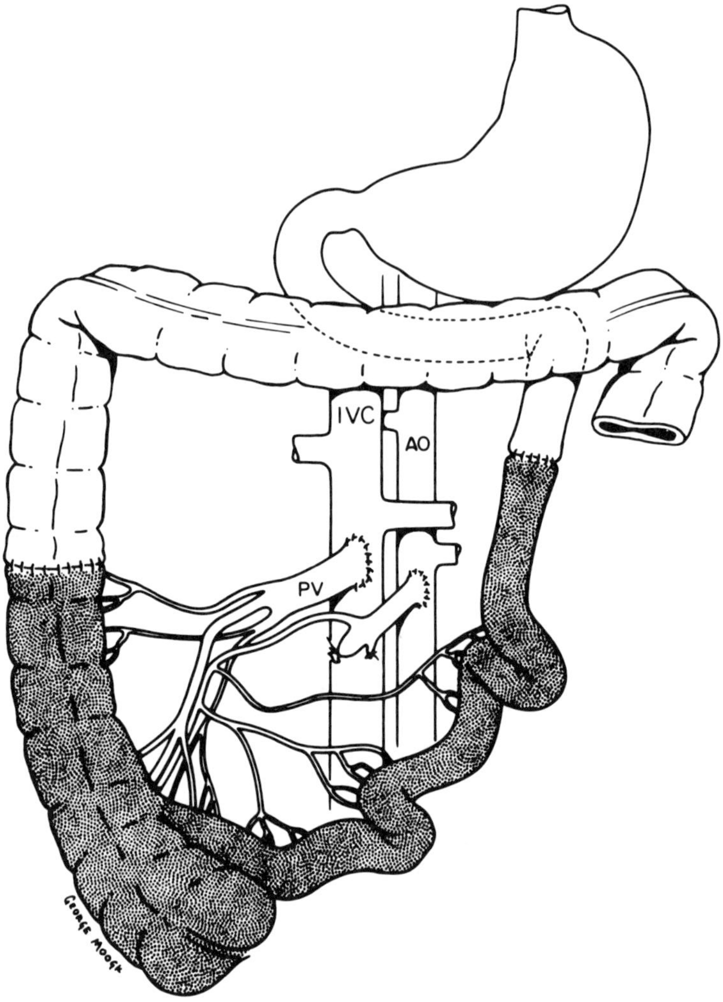

Figure 4 Segmental intestine and colon transplant model. (AO) Aorta, (IVC) inferior vena cava, and (PV) portal vein.

Table 1 **Intestinal Rejection in the Rat Using Different Strain Combinations**

Donor strain	Recipient strain	Mean graft survival +SD (days)
Low-responder combination:		
DA (RT1a)	PVG (RT1c)	7.7 ± 1
Intermediate-responder combination:		
Brown-Norway (RT1n)	Lewis (RT1l)	7.0 ± 1
High-responder combination:		
ACI (RT1a)	Lewis (RT1l)	6.0 ± 1

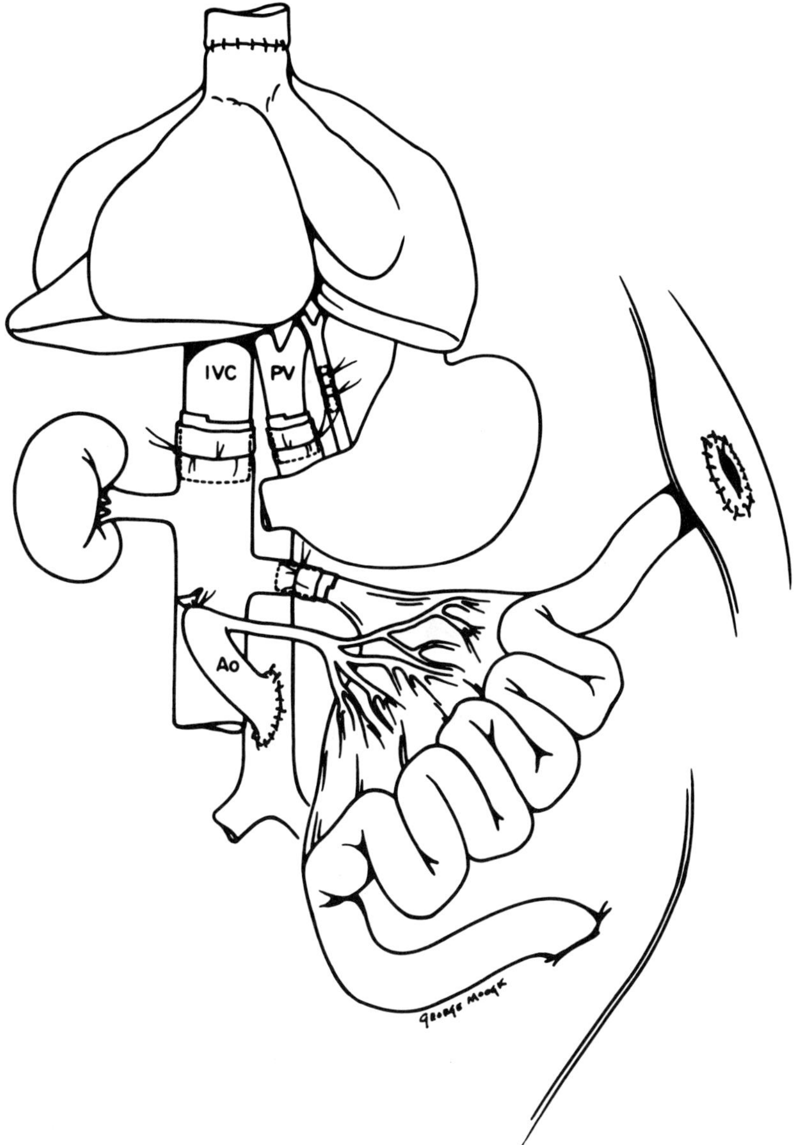

Figure 5 Combined liver/intestine transplant model. (PV) Portal vein, (IVC) inferior vena cava, and (AO) aorta.

aortotomy on the left side to act as a self-retaining retractor. A longitudinal venotomy is then made in the inferior vena cava slightly above the aortotomy. Two 10-0 nylon stay sutures are placed at both apices of the venotomy. The donor's small intestine, which has been surrounded by a gauze sponge packed with crushed ice, is removed from the ice water, and then placed on the right flank of the rat.

After ensuring that the donor portal vein is not twisted (Figure 8), an end-to-side anastomosis to the recipient inferior vena cava is performed first using a continuous 10-0 nylon suture. The posterior wall is anastomosed from the inside of the vessels without repositioning the graft. The anterior wall of the portal vein is anastomosed externally

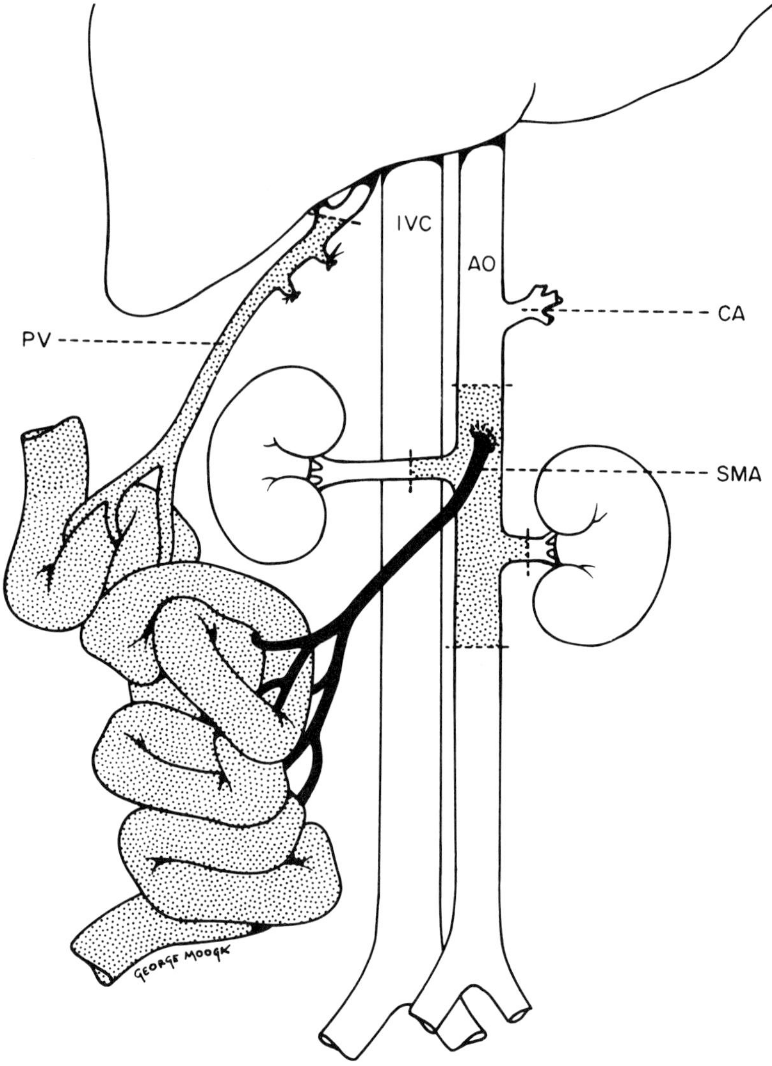

Figure 6 *En bloc* intestinal graft. (PV) Portal vein, (IVC) inferior vena cava, (AO) aorta, (CA) celiac artery, and (SMA) superior mesenteric artery.

using the same suture. The vein is irrigated with saline to keep the vessel walls apart during anastomosis. Before tying the sutures, the vein is gently pulled apart to avoid narrowing of the vessel at the anastomotic site. It takes less than 10 min to complete this anastomosis since only four or five stitches are needed for each side. Next, the front and back walls of the arterial anastomosis are sutured end-to-side. More stitches are required for this anastomosis. Care must be taken to ensure that each stitch passes through all layers of the arterial wall. The clamp is slowly released after completion of the anastomoses. The arterial anastomosis is compressed lightly with a dry sponge for 1 to 2 min after reperfusion. Persistent bleeding is easily controlled by placing a small quantity of Avitene R, a microfibrillar collagen, over the anastomotic site. The total time for anastomosis is less than 25 min.

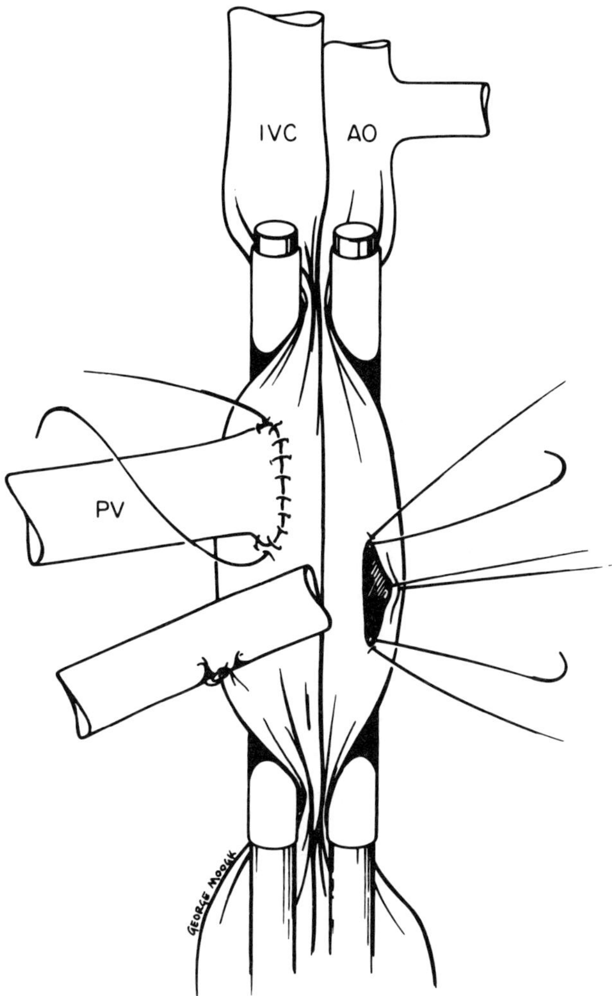

Figure 7 Cross-clamping aorta and inferior vena cava using a modified Lee's clamp. (IVC) Inferior vena cava, (AO) aorta, and (PV) portal vein.

Meticulous anastomotic technique with the aid of an operating microscope and 10-0 nylon sutures minimizes vascular complications. The incidence of vascular complications in our group is only 2.5%.[5] Portal thrombosis, the most common vascular complication, can be avoided by orienting the portal vein with the pyloric and splenic vein ligatures, and by widely separating the arterial and venous anastomoses to reduce torsion.

Both ends of the graft may be exteriorized as stomas. Alternatively, the proximal end of the graft can be exteriorized as a jejunostomy and the distal end can be attached to the native ileum. The stomas are secured with four 7-0 silk sutures between the host peritoneum and the seromuscular layer of the graft, and four 5-0 silk sutures between the skin and the everted mucosa of the graft.

The animals receive 10 ml/hour of lactated Ringer's solution by intermittent injection in the dorsal penile vein. Before and after vascular clamping, 3 ml of lactated Ringer's

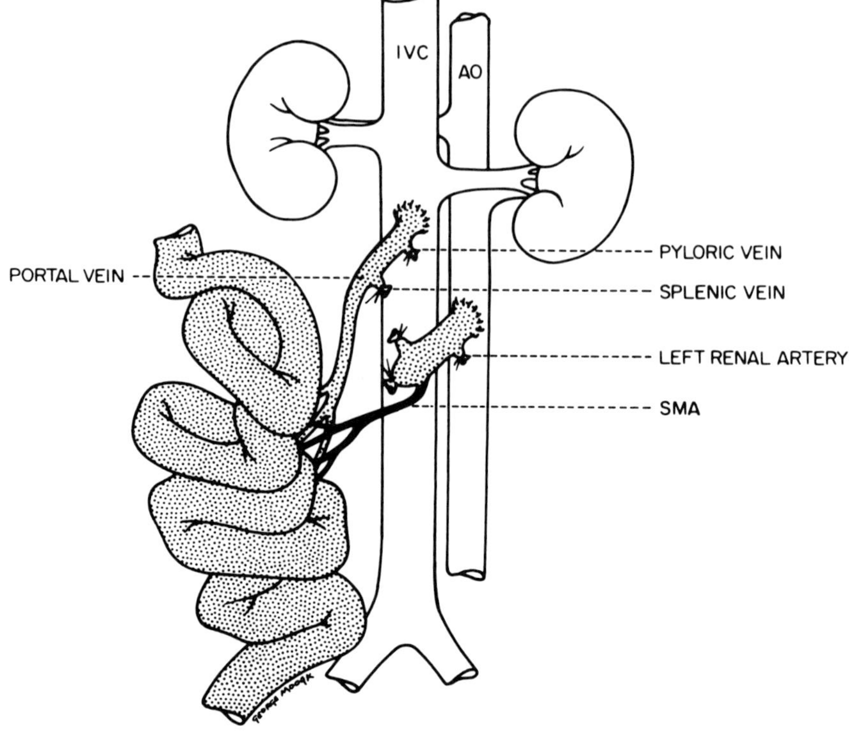

Figure 8 Vascular reconstruction. Note (1) the arterial and venous anastomoses are widely separated to prevent torsion; (2) the correct portal vein orientation is ensured by placing pyloric and splenic vein ligatures on the left of the anastomosis; and (3) the correct orientation of the artery is ensured by placing the ligature of left renal artery on the left side of the arterial anastomosis.

solution is administered intravenously. After abdominal closure, 20 ml of saline is given subcutaneously. No blood transfusion is required.

C. POSTOPERATIVE CARE

The rats are kept on a warming blanket and under a heat lamp for the first 24 hours. They usually recover from anesthesia within 1 hour after the operation. Postoperatively, the rats are given regular food and water *ad libitum.* No antibiotics or parenteral fluid are required after transplantation. The survival rate in 298 consecutive rats is 90%.[5]

D. APPLICATIONS

The heterotopic intestinal transplantation model is ideally suited for immunological studies. The relative ease of this surgery and the resulting low mortality rate are significant advantages. In addition, the exteriorized graft provides access for sequential mucosal biopsies to monitor rejection. However, this model may not be as physiologically relevant as orthotopic transplantation, since the graft is defunctioned. In contrast, the orthotopically transplanted gut is exposed to bacterial and food antigens and recipient survival depends entirely on the provision of adequate nutrition via the graft.[13] Rejection of the orthotopic graft is usually fatal whereas rejection of a heterotopic intestinal graft is surprisingly well tolerated by rats.[19] Intestinal obstruction may occur during the rejection process following orthotopic intestinal transplantation. Rats, which reject their heterotopic grafts, are able

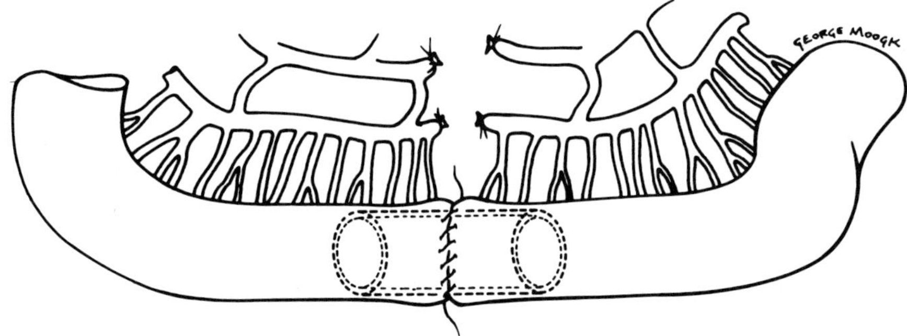

Figure 9 Macaroni noodle splint for the intestinal anastomosis.

to maintain intake of fluids and nutrients by their intact native gastrointestinal tract whereas rats with orthotopic grafts experience severe dehydration, malnutrition, and weight loss.

V. ORTHOTOPIC INTESTINAL TRANSPLANTATION

A. DONOR OPERATION

The donor procedure for orthotopic transplantation is the same as for heterotopic transplantation. However, the quality of the graft is more critical with orthotopic placement since the recipient's native small bowel is resected and the graft is placed in an immediately functional capacity. A damaged intestinal graft increases the incidence of leakage from the gut anastomosis and can cause hypovolemic shock following orthotopic intestinal transplantation.

B. RECIPIENT OPERATION

After the venous and arterial anastomoses, which are the same as described for heterotopic grafting, the recipient's jejunum and ileum are removed leaving only 2 cm each of the native jejunum and ileum to be anastomosed with the graft. The intestinal anastomoses are performed over a piece of dry macaroni using one layer of full thickness 7-0 silk continuous sutures (Figure 9). After anastomosis, the macaroni stent is pushed distally beyond the anastomotic site. This splinting method for rat intestinal anastomosis was developed by Kiernan[20] and modified by our group.[5] The macaroni is quickly absorbed thus avoiding problems with obstruction. Complications of gut anastomosis, such as leakage and stenosis, are less than 1%.[5]

Fluid management is the same as previously described for heterotopic transplantation, except 3 ml of anticoagulated isogeneic whole blood (3 U heparin per 1 ml blood) are transfused at the end of the operation to replace blood loss during enterectomy.

Hypovolemic shock is the most frequent cause of postoperative mortality after orthotopic grafting in rats.[4] Hemodynamic monitoring in our laboratory (Figure 10) demonstrates that significant reductions in blood pressure levels after revascularization are due to intra-luminal and extra-luminal isotonic fluid losses from the transplanted gut.[5] This complication can be avoided by infusing large volumes of crystalloid before and after vascular clamping and at regular intervals throughout surgery, as well as blood transfusion.

C. POSTOPERATIVE CARE

The animals receive subcutaneous injections of 5% dextrose and 0.45% sodium chloride solution (15 to 20 ml) three times daily for 2 days postoperatively. Cefoxitin (40 mg/kg)

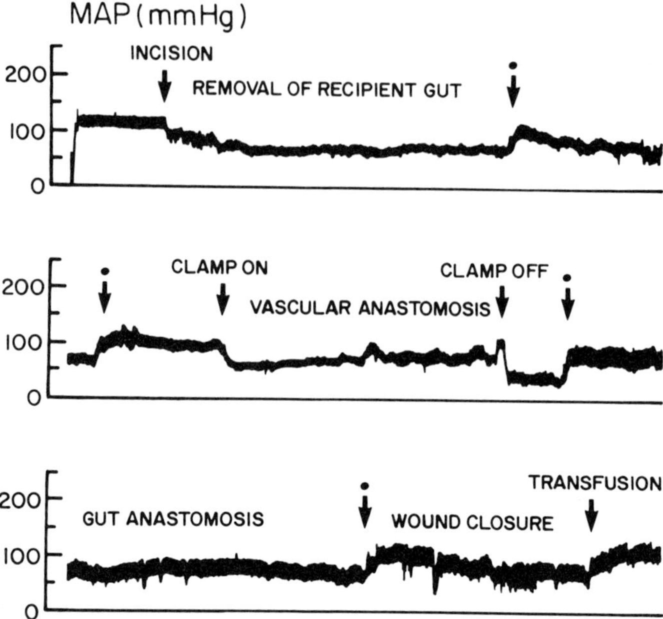

Figure 10 Changes in mean arterial pressure (MAP) during orthotopic intestinal transplantation. (* infusion of 3 to 5 ml of lactated Ringer's solution).

is given intramuscularly every 8 hours for the first 24 hours postoperatively. The animals have free access to sugar water 24 hours postoperatively, followed by normal food on the next day.

The orthotopic intestinal transplant model was first developed by Kort et al.[4] with a relatively high mortality rate and was later modified and improved by Deltz and Thiede[21] and Lee and Schraut.[22] Anastomotic complications can be avoided using techniques described above, none of the animals in our recent series of transplants have died from gut anastomotic complications. The survival rate in 102 consecutive rats is 86%.[5]

D. APPLICATIONS

Orthotopic placement is better than heterotopic placement for physiological studies since this technique provides a normal environment for mucosal enterocytes. Furthermore, graft failure leads to the animal's death, providing a well-defined, objective marker of impaired gut barrier and absorptive functions.[13,23] Orthotopic transplantation is more technically demanding than heterotopic transplantation.

VI. SEGMENTAL INTESTINE AND COLON TRANSPLANTATION

A. DONOR OPERATION

The graft consists of a short segment of terminal ileum (approximately 20 cm), ileocecal valve, cecum, and ascending colon (Figure 11). The terminal ileum is exposed through a midline abdominal incision. The proximal end of the segmental ileal graft is determined by measuring the small bowel starting at the ileocecal junction. The vessels adjacent to the mesenteric border of the small bowel are ligated and divided with 8-0 silk sutures. The mesenteric arcades, proximal to this point, are ligated in groups of four to six with 7-0 silk sutures. This technique is continued to the ligament of Treitz, thus devascularizing the entire jejunum and proximal segment of the ileum. The devascularized small bowel

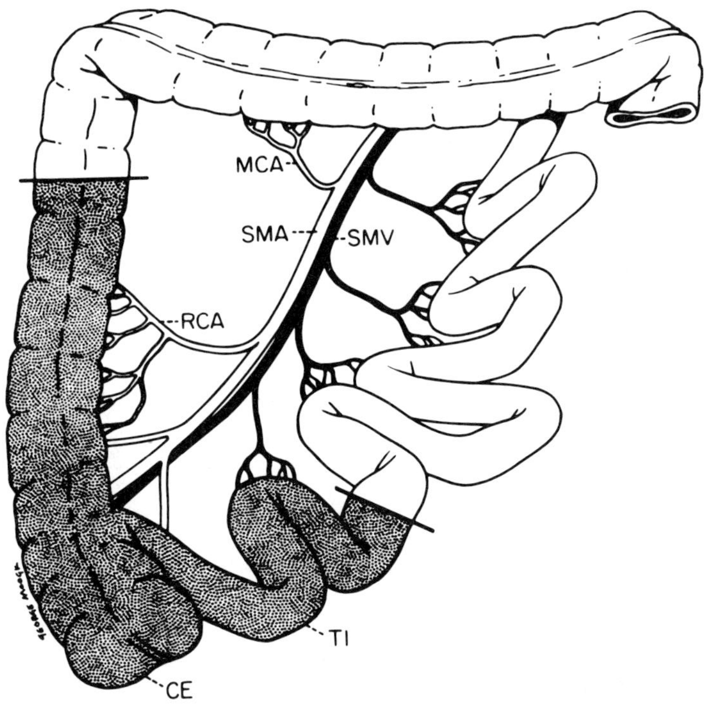

Figure 11 Donor procedure of segmental intestine and colon transplant. (RCA) Right colic artery, (MCA) middle colic artery, (SMA) superior mesenteric artery, (SMV) superior mesenteric artery, (SMV) superior mesenteric vein, (CE) cecum, and (TI) terminal ileum.

is removed. The colon, distal to the hepatic flexure, is devascularized by ligating and dividing the marginal vessels and the middle colic vessels with 8-0 silk sutures. The avascular colic mesentery is cut and the distal segment of the colon is discarded. Thereafter, dissection of the vascular pedicles and perfusion of the graft are performed as described for heterotopic transplantation.

B. RECIPIENT OPERATION

Revascularization is the same as described for heterotopic transplantation. After completion of the vascular anastomoses, a total small bowel enterectomy and cecectomy is performed by ligating and dividing the mesenteric arcades from the ligament of Treitz to the cecum, including the ileocolic vessels. The devascularized small bowel and cecum are removed.

The colo-colonic anastomosis between the distal end of the donor ascending colon and the proximal end of the recipient ascending colon is performed first to minimize blood loss. This anastomosis is performed rapidly using continuous 7-0 silk suture over a macaroni stent.

The donor ileum is anastomosed to the recipient duodenum in a similar fashion. The abdomen is closed and the recipient is given 3 cc of isogeneic whole blood intravenously via the dorsal penile vein. Postoperative care is the same as for orthotopically transplanted rats. With good surgical technique, attention to intraoperative fluid replacement, and broad-spectrum antibiotics, this procedure can be performed with less than 10% mortality.[24]

C. APPLICATIONS

Transplantation of a segment of the small intestine/colon is the most recent surgical model of intestinal transplantation developed by our laboratory.[24] This surgical model can be used to examine the role of the ileocecal valve, cecum, and colon in bowel transplantation. Successful segmental intestinal transplantation in rats is a precursor to performing living-related intestinal transplants in larger animals and, eventually, in humans.

VII. COMBINED LIVER/INTESTINE TRANSPLANTATION

A. DONOR OPERATION

The liver and intestine are retrieved from two different donors. These procedures are performed simultaneously by two surgeons. The donor liver is exposed through a cruciate abdominal incision. The liver graft is harvested with ligation of the hepatic artery according to Kamada's technique.[25] Heparin (200 U) is given intravenously before harvesting. The donor intestine is harvested from another rat as previously described in the heterotopic model. The portal vein is dissected at its bifurcation and a cuff of the vein is prepared in the ice-water basin using a 0.4-cm Teflon tube with a 0.2-cm outside diameter.

B. RECIPIENT OPERATION

Following recipient hepatectomy, a nonarterialized orthotopic liver transplant is performed. The suprahepatic portion of the recipient's inferior vena cava is anastomosed end-to-end to the donor suprahepatic inferior vena cava using a 7-0 prolene continuous suture. The donor portal vein and the infrahepatic portion of the vena cava are anastomosed to the corresponding native parts using the modified cuff technique described by Kamada and Calne.[25] A stent (polyethylene 50) is inserted into the donor and recipient bile ducts, which are attached with two circumferential 7-0 silk sutures. Liver circulation is reestablished with the removal of the clamps. Next, the left renal artery is isolated and ligated with 7-0 silk ligatures and the renal vein is cross-clamped near the inferior vena cava (Figure 12). The end of the donor aorta is sutured to the side of the recipient aorta as in the heterotopic model. A venotomy is made close to the kidney, following ligation of the left renal vein. The cuff of the donor portal vein is inserted into the recipient's left renal vein and secured with a circumferential 5-0 silk suture. Both ends of the intestinal graft are exteriorized as stomas. Whereas the liver is transplanted in an orthotopic position, the intestine is transplanted as an accessory graft. The venous clamp is released first. If the portal vein is properly oriented, the small intestine graft should become pink as a result of retrograde flow through the superior mesenteric vein. The intestinal graft is revascularized by releasing the arterial clamp. Warm ischemia times for liver and gut are less than 15 min.[9]

Both ends of the graft are brought to the surface of the abdomen and exteriorized as stomas. The stomas are secured with four 7-0 silk sutures between the host peritoneum and the seromuscular layer of the graft, and with four 5-0 silk sutures between the skin and the everted mucosa of the graft.

C. FLUID REPLACEMENT

The animals receive 10 ml/hour of lactated Ringer's solution during surgery by intermittent injection via the penile vein. Lactated Ringer's solution (3 ml) is given before and after the venous clamping during the liver grafting. After revascularization of the intestinal graft, 5 ml of Ringer's lactate are given and 3 ml of whole blood are transfused at the end of the surgery. Additional Ringer's solution is given as needed.

D. POSTOPERATIVE CARE

The rats are kept on a warming blanket and under a heat lamp for the first 24 hours. They do not receive any parenteral fluids and are allowed to take regular food and water *ad libitum.* The operative mortality is 10%. Survival 1 month after transplant is 71%.[9]

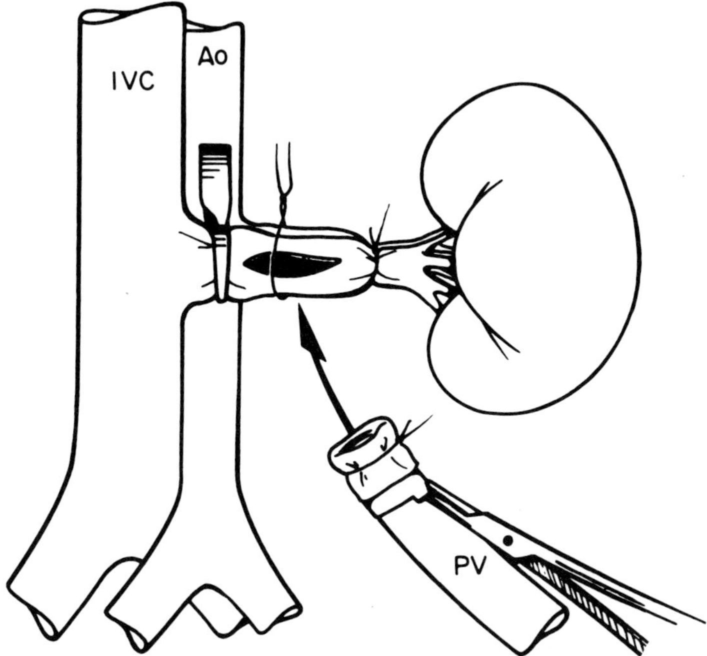

Figure 12 The cuff technique for an end-to-end anastomosis between donor portal vein and recipient's left renal vein.

E. APPLICATIONS

This transplant model permits the study of complex immune responses associated with liver/intestinal grafting. Recent experiments in our laboratory using this model have provided preliminary evidence that simultaneous liver grafting protects the intestinal allograft from rejection.[8] This model can be further used to study the mechanisms of this phenomenon.

ACKNOWLEDGMENTS

The authors acknowledge the technical assistance provided by Drs. Huifang Chen, Tetsuo Hashimoto, Gang He, Pengzhi Wang, Yoshiraru Sakai, Zheng Zhang and Linfu Zhu; Ms. Cate Abbott for reviewing the manuscript; and Mr. George Moogk for preparing the illustrations.

REFERENCES

1. **Grant, D.,** Intestinal transplantation: current status, *Transplant. Proc.,* 21, 2869, 1989.
2. **Grant, D., Wall, W., Mimeault, R., Zhong, R., Ghent, C., Garcia, B., Stiller, C., and Duff, J.,** Successful small-bowel/liver transplantation, *Lancet,* 335, 181, 1990.
3. **Monchik, G. J. and Russell, P. S.,** Transplantation of small bowel in the rat: technical and immunological considerations, *Surgery,* 70, 693,1971.
4. **Kort, W. J., Westbroeck, D. L., MacDicken, I., and Lameijer, L. D. F.,** Orthotopic total small-bowel transplantation in the rat, *Eur. Surg. Res.,* 5, 81, 1973.
5. **Zhong, R., Grant, D., Sutherland, F., Wang, P., Chen, H., Stiller, C., and Duff, J.,** A refined technique for intestinal transplantation in the rat, *Microsurgery,* 12, 268, 1991.

6. **Zhong, R., Grant, D., Black, R., Stiller, C., and Duff, J.,** Combined small bowel and kidney transplantation in the rat, *Transplant. Proc.,* 21, 2907, 1989.
7. **Zhong, R., Wang, P., Chen, H., Sutherland, F., Duff, J., and Grant, D.,** Surgical techniques for orthotopic intestinal transplantation in the rat, *Transplant. Proc.,* 22, 2473, 1990.
8. **Zhong, R., He, G., Sakai, Y., Li, X., Garcia, B., Wall, W., Duff, J., Stiller, C., and Grant, D.,** Combined small bowel and liver transplantation in the rat: possible role of the liver in preventing intestinal allograft rejection, *Transplantation,* 52, 550, 1991.
9. **Zhong, R., He, G., Sakai, Y., McAlister, V., Zhang, Z., Stiller, C., Duff, J., and Grant, D.,** Surgical model of combined liver/intestine transplant in rats, *Microsurgery,* 13, 126, 1992.
10. **Grant, D., Lamont, D., Zhong, R., Garcia, B., Wang, P., Stiller, C., and Duff, J.,** 51Cr-EDTA: A marker of early intestinal rejection in the rat, *J. Surg. Res.,* 46, 507, 1989.
11. **Grant, D., Zhong, R., Gunn, H., Stiller, C., Garcia, B., Keown, P., and Duff, J.,** Graft-versus-host disease associated with intestinal transplantation in the rat: host immune function and general histology, *Transplantation,* 48, 545, 1989.
12. **Garcia, B., Zhong, R., Wijsman, J. Wang, P., Chen, H., Sutherland, F., Duff, J., and Grant, D.,** Pathological changes following intestinal transplantation in the rat, *Transplant. Proc.,* 22, 2469, 1990.
13. **Grant, D., Zhong, R., Hurlbut, D., Duff, J., and Stiller, C.,** Comparison of heterotopic and orthotopic intestinal transplant in rats, *Transplantation,* 51, 948, 1991.
14. **Grant, D., Hurlbut, D., Zhong, R., Wang, P., Chen, H., Garcia, B., Behme, R., Stiller, C., and Duff, J.,** Intestinal permeability and bacterial translocation following small bowel transplantation in the rat, *Transplantation,* 52, 221, 1991.
15. Adam, P., Zhong, R., Haist, J., Flanagan, P., Grant, D., Mucosal iron in control of iron absorption in a rat intestinal transplant model, *Gastroenterology,* 100, 370, 1991.
16. **Zhong, R., He, G., Sakai, Y., Zhang, Z., Quan, D., Garcia, B., Duff, J., and Grant, D.,** The effect of donor-recipient strain combinations in combined liver/intestine transplantation in the rat, *Transplant. Proc.,* 24, 1208, 1992.
17. **Vega, R. G. and Toledo-Pereyra, L. H.,** Acute mesenteric small bowel ischemia in the rat, *Transplantation,* 49, 830, 1990.
18. **van Oosterhout, J. M., de Boer, H. H., and Jerusalem, C. R.,** Small bowel transplantation in the rat: the adverse effect of increased pressure during the flushing procedure of the graft, *J. Surg. Res.,* 36, 140, 1984.
19. **Lee, K. K. W. and Schraut, W. H.,** Small bowel transplantation in the rat. Graft survival with heterotopic vs. orthotopic position, in *Small Bowel Transplantation: Experimental and Clinical Foundations,* Deltz, E., Thiede, A., and Hamelmann, H., Eds., Springer-Verlag, Heidelberg, 1986, 7.
20. **Kiernan, J. A.,** Intestinal anastomosis in the rat facilitated by a rapidly digested internal splint and indigestible but absorbable sutures, *J. Surg. Res.,* 45, 427, 1988.
21. **Deltz, E. and Thiede, A.,** Microsurgical technique for small intestine transplantation, in *Microsurgical Models in Rats for Transplantation Research,* Thiede, A., Deltz, E., Engemann, R., and Hamelmann, H., Eds., Spring-Verlag, Berlin, 1985, 51.
22. **Lee, K. K. and Schraut, W. H.,** Structure and function of orthotopic small bowel allografts in rats treated with cyclosporine, *Am. J. Surg.,* 151, 55, 1986.
23. **Stangl, M. J., Schraut, W. H., Moynihan, H. L., and Lee, T.,** Rejection of ileal versus jejunal allografts, *Transplantation,* 47, 424, 1989.
24. **Black, R., Zhong, R., and Grant, D.,** unpublished data, 1992.
25. **Kamada, N. and Calne, R.,** A surgical experience with five hundred thirty liver transplants in the rat, *Surgery,* 93, 64, 1983.

Chapter 17

Multivisceral Transplantation in the Pig

Luis Podesta, Donald V. Cramer, Leonard Makowka, and Marcos Nores

CONTENTS

I. INTRODUCTION

Multivisceral transplantation (liver, stomach, duodenum, pancreas, small intestine, and a portion of the large intestine) has been recently employed in different combinations to treat patients with a variety of disease conditions.[1-5] These include biliary, pancreatic, gastric, and duodenal malignancies that have spread to the liver, as well as primary liver tumors with extrahepatic extension. A further application of combined liver and small intestinal transplantation has been performed in both adult and pediatric patients suffering from short bowel syndrome with superimposed liver failure secondary to parenteral nutrition. The transplantation of *en bloc* abdominal organs poses unusual problems and challenges which have not been encountered with single organ transplants. The pig is an excellent model to investigate this new therapeutic technique.

The earliest description of experiment multivisceral transplantation in large animals is attributed to Starzl and Kaupp in 1960.[6] These investigators reported the results of *en bloc* transplantation of multiple abdominal organs in 19 dogs. This early experience demonstrated the technical feasibility of the procedure and that functional recovery of the graft was possible. Many of the animals, however, died intraoperatively due to massive hemorrhage from the transplanted intestine. It soon became apparent that the pig was the better model to investigate this procedure due to greater tolerance to the operation and because of shared physiologic and metabolic characteristics between pigs and humans.[7] It was also felt that immunologic considerations could be better addressed in the pig.

0-8493-3629-5/94/$0.00+$.50

Garnier et al.[8] and Peacock and Terblanche[9] have explored the use of the pig for such procedures. In 1988, Gridelli et al. reported his experience with 35 multivisceral transplant operations performed in the pig to explore the technical aspects of multi-organ graft procurement, preservation, and implantation.[10] In 1991, Kobayashi et al. reported their experience with eight pigs undergoing abdominal organ cluster transplantation. The effect of the new immunosuppressant FK 506 on postoperative function of the organ cluster grafts was evaluated in this model.[11]

II. TECHNIQUES

A. PREOPERATIVE PREPARATION AND ANESTHESIA

Every effort must be made to maintain pigs under conditions with as minimal stress as possible in order to avoid the development of stress syndromes such as malignant hypothermia. Pigs weighing between 15 and 25 kg are the most appropriate size. Females are preferred as recipients and males as donors. It is also preferable for the recipient pig to weigh approximately 5 kg less than the donor. Pigs require at least 24 hours of complete deprivation of solid food to ensure an empty intestinal tract. Water is denied 4 to 6 hours prior to surgery in order to avoid passive regurgitation during anesthesia. The presence of food, either in the stomach or in the bowel of a donor can result in significant distension of the bowel, impaired respiration, and make abdominal closure difficult in the recipient. Selective bowel decontamination has been recommended by some authors in dogs[6] consisting of 2 g of Neomycin Sulfate per day and cathartics.

1. Induction of Anesthesia

We prefer to use Ketamine i.m. at 10 to 15 mg/kg for preoperative sedation only. During this time intravenous access in the ear can be established and intubation can be achieved. Prior to intubation, isoflurane (3%) is administered, via a face mask, at a rate of approximately 4 to 5 l/min for 2 to 3 min. This combination of Ketamine and isoflurane provides sufficient relaxation for intubation.

If inhalatory anesthetic machines are not available, i.v. anesthesia can be a useful alternative, particularly in the donor procedure. After induction with Ketamine, a muscle relaxant can be used. Boluses of Ketamine (10 mg/kg) combined with Pancuronium (.10 to .15 mg/kg) provides good anesthesia and relaxation for periods of 20 to 30 min.

Endotracheal intubation is performed with the animal lying on its back with a long-bladed laryngoscope; the larynx should be sprayed with lidocaine to decrease the risk of laryngospasm and vagally induced arrhythmias. After, a large stomach tube is passed to maintain decompression of the stomach. Anesthesia is maintained with isoflurane (3%) and oxygen at 4 to 5 l/min delivered via a conventional human anesthetic machine. A humidifier is needed if the room temperature is below 25°C.

2. Surgical Monitoring Techniques

During the preparation for surgery, the right carotid artery and the right jugular vein are catheterized for blood pressure monitoring, blood sampling, and for fluid and blood replacement. The external jugular vein should be preserved for bypass venous procedures if needed.

The animals are placed on heating pads and given intravenous fluids throughout the entire operation. Lactate Ringer's solution is administered at 10 ml/kg/hour. The depth of anesthesia is assessed by evaluation of the heart rate, blood pressure, respiratory rate, oxygen concentration, pH, and body temperature. The heart rate varies normally between 72 and 145 beats/min. Animals who receive Atropine tend to have tachycardia. Bradycardia (less than 70 beats/min) is seen with deep anesthesia. Body temperature is

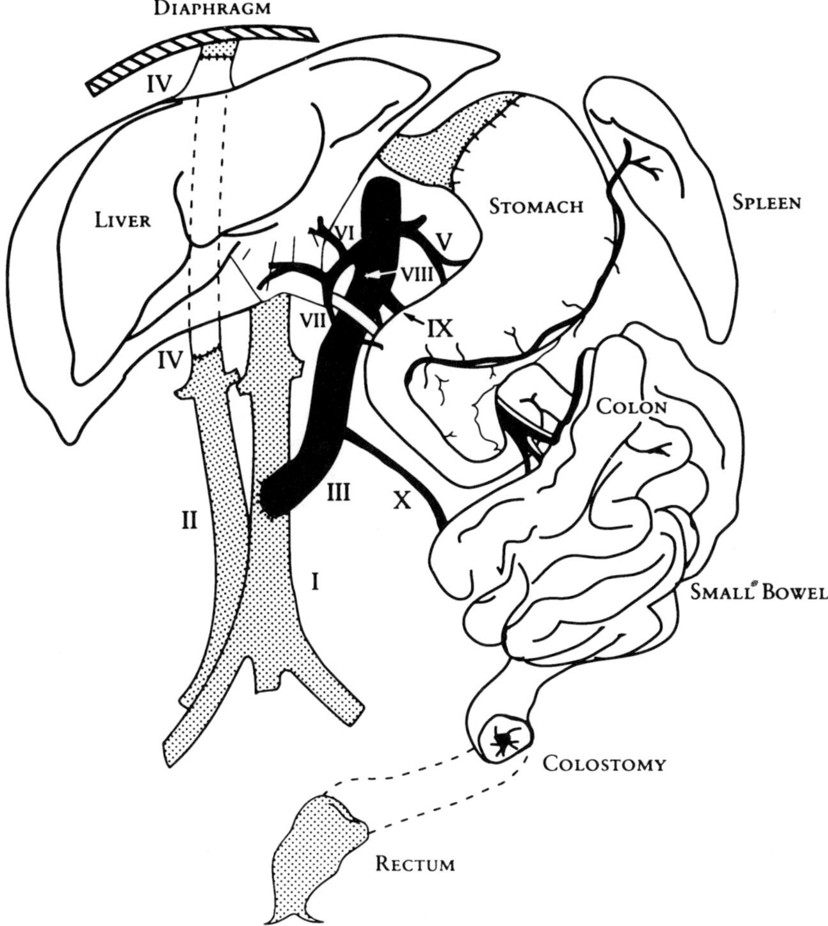

Figure 1 I. Recipient abdominal aorta. II. Recipient inferior vena cava. III. Donar abdominal aorta. IV. Donar inferior vena cava. V. Left gastric artery. VI. Common hepatic artery. VII. Right gastric artery. VIII. Gastroduodenal artery. IX. Superior mesenteric artery. X. Inferior mesenteric artery.

monitored with a rectal thermometer during the procedure. Antibiotic prophylaxis is achieved with 400 mgs of Keflex preoperatively and continued every 8 hours for 2 days.

B. DONOR OPERATION

The midline abdominal incision from the xiphi sternum to the pubis is the standard incision to remove the abdominal organs. The incision should deflect to the left to avoid the genito-urinary organs in the male. To ensure widest exposure, a self-retaining abdominal retractor is employed, and if necessary, two retractors can be placed, one in the upper abdomen and the other in the lower abdomen. The organs to be removed include the stomach, duodenum, small intestine, most of the large intestine, liver, pancreas, and spleen (see Figure 1). The kidneys are not included in this block since the recipient kidneys will be preserved. The spleen is ultimately removed at the time of the recipient operation, once the organs have been perfused. The specimen includes the aorta with its branching vessels to these organs, i.e., celiac trunk (CT), superior mesenteric (SMA), and inferior mesenteric arteries (IMA), and that portion of the inferior vena cava retrohepatic and the infrahepatic cava proximal to the renal veins. The proximal and distal enteric continuity is interrupted by dividing the distal esophagus at the junction with the stomach, and the left colon, distal to any of the important arcades of the inferior mesenteric artery. The GIA stapler is used for the bowel division. Mobilization of the retroperitoneum is carried out by retracting the root of the mesentery and dissecting on each side. The abdominal aorta is completely dissected, starting at the bifurcation of the common iliac arteries, ligating and dividing all the lumbar arteries. As dissection progress proximally, the renal arteries, including the left renal vein, adrenal and diaphragmatic arteries are divided. Only the CT, SMA, and IMA are preserved. Next, the liver is mobilized by dividing the left triangular, falciform, right triangular, and hepatorenal ligaments. The inferior vena cava is easily approached in its supra and infrahepatic portions. At this point, all the abdominal viscera should be entirely mobilized on the major vessels. Prior to cannulation, 5000 U of heparin are administered through a vascular access. A suitable plastic cannula is placed through an incision in the common iliac artery into the distal portion of the aorta, and connected to the cold perfusion system 2 min after heparin infusion. Next, the aorta proximal to the celiac trunk is clamped and an infusion of cold lactated Ringer's is initiated. All organs can be flushed with Ringer's and ultimately preserved with Viaspan R solution. If the timing between the donor procurement and recipient implantation is close enough (i.e., within 1 to hours of cold ischemia) then the preservation solution is not necessary. If required, 2 l of chilled Lactated Ringer's is infused, followed by l or 2 l of Viaspan solution (UW solution) flushed through the aorta.

As the initial infusion begins, the vena cava should be immediately transected. This can be done, either in the abdomen by transecting the infrahepatic cava, which is inconvenient, or by opening the diaphragm and transecting the right atrium. This is our preferred technique since a pool suction can be placed in the damaged area. Once approximately 50% of the infusion fluid has been administered, the evisceration is completed by transecting the diaphragm around the liver, the supraceliac aorta, and the cava above and below the liver. The organs are removed from the donor and placed into a basin with ice-cold Viaspan solution or Lactated Ringer's while the recipient is being prepared for implantation.

At the back table, the diaphragmatic attachments to the liver are excised and the cuffs of the supra- and infrahepatic cava are prepared for implantation. The aorta needs to be closed on one end. Usually the proximal aortic cuff is closed with a running 5-0 Prolene suture. The vascular integrity, both venous and arterial, is checked at the back table and the organs are ready for implantation.

C. RECIPIENT OPERATION

Once the recipient animal has been anesthetized, intubated, and placed on the operating table in supine position, the thorax, abdomen, groin, and neck areas are prepared using sterile techniques. As described previously, the internal jugular vein and the carotid artery on the right side are cannulated for fluid resuscitation, sampling, and blood pressure monitoring. Access for veno-venous bypass, consisting of isolation of the right external or left internal jugular vein and the left or right femoral vein, is undertaken before the laparotomy.

A xipho-pubic midline incision is used for abdominal access. Retractors are placed as with the donor procedure, one or two depending on the size of the pig. The "organectomy" is carried out by transecting the left colon with preservation of the inferior mesenteric artery, and proximally by a total gastrectomy, leaving the esophagus and 2 to 3 in. of the fundus of the stomach. If a larger portion of the stomach is preserved, venous congestion will occur. After the enteric divisions, the liver is entirely mobilized around the supra and infrahepatic cava. The gastrohepatic omentum is divided while performing the gastrectomy. The recipient's entire abdominal aorta and cava, except for the retrohepatic portion, and the kidneys, are preserved. The celiac trunk and SMA are encircled and prepared for division in order to complete the visceral resection while on veno-venous bypass. Cannulas are placed in the internal jugular vein and femoral veins. These cannulae (numbers 18, 20, or 22 French) vary in size depending on the vessels. No heparin is used and clotting of the cannulae is avoided by maintaining fluid flows consistently above 500 ml/min. If the flow decreases, the cannulas should be checked for correct positioning. This is a simple venous bypass that shunts the inferior vena cava to the superior vena cava while the supra and infrahepatic cava are clamped. Pigs can tolerate caval clamping without shunting for 10 to 15 min, but may require significant fluid volume to be infused in order to maintain hemodynamic stability. Once on veno-venous bypass, the celiac trunk and the superior mesenteric artery are ligated and divided. Clamps are placed on the infrahepatic cava and suprahepatic cava and the organs removed. Just prior to clamping, the liver can be squeezed gently to "pump" out residual blood. Hemostasis of the operative field should be checked and should be completely controlled at this time.

1. Implantation (see Figure 1)

Adequate cuffs should be prepared for both the recipient supra and infrahepatic vena cava. The suprahepatic cuff is fashioned by communicating all hepatic veins into one cuff with the vena cava.

The donor organs are brought to the field. In order to handle them without any rotation, the organs should be positioned over a towel at the back table. The first anastomosis to be performed is the suprahepatic vena cava. The sutures to be used depend on the surgeon's choice, but 4-0 or 5-0 Prolene are adequate for completing this anastomosis. A running anastomosis with no growth factor is standard. The pig suprahepatic cuff is not as long and easy to handle as the same cuff in humans. Sutures, especially on the back wall, should be relatively close to each other, avoiding the liver parenchyma if possible. Following the suprahepatic cava, the infrahepatic cava anastomosis is performed. The rest of the viscera require retraction towards the midline; they should all be contained within moist laporoscomy towels. The infrahepatic anastomosis can be done with a 5-0 Prolene running suture, and before completion, the organs should be flushed with lactated Ringer's if a preservation solution has been used. Viaspan R contains high levels of potassium and care should be taken to wash this out as well as any air. Chilled Lactated Ringer's (1 l, 4°C) should be adequate.

The last vascular anastomosis is that of the distal aorta of the donor to the recipient's infrarenal aorta. This is performed in an end-to-side fashion. The donor aorta is transected diagonally and anastomosis between the donor aorta and the recipient mid-abdominal aorta is completed with the use of 5-0 running prolene sutures. Care must be taken to avoid the lymphatics running adjacent to the aorta.

Once the arterial anastomosis is completed, the clamp on the aorta is removed, along with the infra and suprahepatic caval clamps. The reperfusion period is followed by a transient drop in blood pressure which should be managed with fluid, blood, and/or drug administration. Immediately after revascularization and after checking for hemostasis, the veno-venous bypass is interrupted. In order to save some of the blood within the lines of the bypass, the blood in the system can be re-infused into the pig. The cannulas in the jugular and femoral veins are then removed and the veins either ligated or repaired. At this time, hemostasis is checked and the organs are allowed to rest in an anatomically comfortable position for 15 to 30 min. After this period, the entire abdomen is reexamined for bleeding sites or lymphatic leaks. This portion of the operation is particularly important since a significant amount of bloody ascites can accumulate in the postoperative period resulting in a complication that is commonly fatal.

Finally, the gastrointestinal tract continuity is reestablished. The recipient's stomach cuff is checked for good perfusion and no congestion. An end-to-side anastomosis between the recipient fundal cuff and the donor's stomach is performed in two layers; an external seromuscular layer with interrupted 4-0 silk sutures and a continuous full thickness layer with 3-0 or 4-0 absorbable sutures (Maxon, Dexon, or Vicryl). The distal end of the graft large intestine and the recipient proximal sigmoid or rectum are checked for viability. An end-to-end anastomosis in one or two layers is then completed. Some surgeons exteriorize the graft's distal pouch of large intestine due to reasonable concerns about possible leaks. In this case, the recipient's distal sigmoid can be closed and a colostomy performed at the left lower quadrant. Another advantage of the colostomy is access to the intestinal mucosa if future biopsies are required. Once the anastomoses are completed, a final review of the entire abdomen is carried out to assess hemostasis, remove foreign objects, e.g., sponges, assess the viability of the graft, and to ensure appropriate positioning of the abdominal organs. Before closing the abdomen, 2 to 3 l of warm Ringer's lactate solution are used to wash and warm the abdominal cavity. We prefer a single-layer abdominal wall closure including the peritoneum and fascia with a running 0 Maxon or PDS. We use two sutures, starting one at the xiphoid and one at the pubic edges of the incision. This suture line needs to be tight as pigs tend to accumulate fluid after surgery. The subcutaneous tissues are usually not closed and the skin edges are approximated with interrupted mattress sutures.

D. POSTOPERATIVE CARE

The animal is maintained on the ventilator but can be removed from the operating table and placed on a warm blanket on the floor. This is to avoid falls from the table once the animal starts waking up. Maintaining adequate body temperature should be considered a high priority. Heating lamps can be placed on the floor and the animal should be rotated from one side to the other. The animal should be kept intubated, with or without the ventilator, and extubated only when they begin to strongly reject the tube. Before, and if possible, after extubation, hematocrit, blood gases and electrolytes can be monitored via femoral arterial punctures. If the pig is hypoxic, then it should remain intubated. Careful fluid management including adequate diureses, correction of the blood count, and acid-base or electrolytes abnormalities should be maintained. Once extubated, the animal is then placed in a warmed cage and the oral gastric tube removed. The pig should then be closely monitored for signs of laryngeal or bronchial spasms. An intravenous infusion of normal saline with 5 1/2 dextrose is maintained overnight at the rate of 60 ml/hour and for the next 2 or 3 days at 5 ml/kg/hour, depending on whether the pig can drink. Some animals recover surprisingly fast and are allowed to drink water with glucose the first day after surgery. Antibiotic therapy is maintained for at least 2 days with 400 mg of intravenous Keflex every 8 hours. A diet consisting of glucose water and baby food may be taken the second day after the transplant. For IV access, an angiocatheter or a broviac catheter can be placed into a jugular vein at the end of the surgery, exteriorized, and fixed securely to the dorsal aspect of the neck.

The use of FK 506 as an immunosuppressive agent has had an important impact on the use of the multivisceral procedure. The most clinical experience has been obtained at the University of Pittsburgh with intestinal transplantation alone or in combination with other visceral grafts.[5,12] FK 506 is administered at 0.1 to 0.15 mg/kg/day by continuous intravenous infusion begun immediately after revascularization. When enteral feeding is started, FK 506 is changed to a b.i.d. oral formulation with several days of overlap. Plasma levels of FK 506 can be measured on a daily basis (trough target levels should be 1 ng/ml). FK 506 can be used in combination with Methylprednisolone begun intraoperatively as a 200 mg bolus and rapidly weaned by decrements of 40 mg/day over the next 5 days to 20 mg/day. Prostaglandin E_1 at 0.6 to 0.8 μg/kg/hours may also be used for 7 to 14 days. Rejection is treated with an upward dose adjustment of FK 506, with an increase of steroids and OKT3 when necessary.

E. POSTOPERATIVE COMPLICATIONS

1. Hemorrhagic or Lymphatic Ascites

This is usually a technical complication. Bleeding can originate at the vascular anastomotic sites, or as a consequence of necrotic enteric tissue (i.e., significant preservation injury or rejection). Injury to lymphatic tissue can also result in unmanageable lymphatic ascites.

2. Enteric Anastomotic Leaks or Vascular Enteric Perforations

Inadequate circulation to the anastomotic sites, either because of ischemia, congestion, or poor surgical techniques, can cause leaks at the anastomoses. Enteric perforations may occur as a result of ischemia or rejection and can result in fatal complications.

a. Thrombosis of the Feeding Vessels

This is an unusual complication because of the size of the vessels involved in the anastomoses when it does occur, however, it results in massive necrosis, acidosis, and hemodynamic instability.

b. Infections

Enteric translocation due to disruption of the intestinal membrane, due to ischemia, preservation damage, or rejection is the usual source of infection. The use of a gut decontamination solution and prophylactic antibiotics has significantly altered this deadly complication. Other sources of infection may be intestinal leaks, either at anastomotic sites or gut perforations, lung infections, and other unexpected infections associated with excessive immunosuppression. Frequent cultures of the blood, stool, urine, sputum, wound exudate, and peritoneal discharge should be obtained. Anti-viral and anti-fungal prophylaxis are used in clinical practice, but not in the experimental setting. Nevertheless, acyclovir can be used as prophylaxis for CMV, Bactrim for *Pneumocystis carinii,* and amphotericin if there is concern of fungal infection.

3. Graft Versus Host Disease (GVHD)

As described earlier, the frequency and severity of GVHD is proportional to the amount of lymphoid tissue present in the donor organs. The following include conditions necessary for the development of GVHD: (1) engraftment of immunocompetent lymphoid cells capable of proliferating and attacking host tissues, (2) histoincompatibility difference between the host and donor effector cells, and (3) immunological incompetence of the host.[13,14]

GVHD is a frequent complication of bone-marrow transplantation, but may also be seen following solid organ transplants. Symptoms of GVHD include fever, diarrhea, skin rash, intestinal ulceration, and perforation. A diagnosis can be made by biopsy of the skin and detecting the presence of subepithelial lymphocytic inflammatory infiltrates in the biopsy. The treatment of GVHD is difficult and consists of increasing the level of immunosuppression.

4. Rejection

The intestine is probably the most sensitive organ to the rejection process, and as has been shown experimentally is difficult to protect from the rejection process. With the introduction of FK 506, better management of rejection has been achieved at the University of Pittsburgh.[5,12,15,16] The intestine is generally the most important target organ to follow postoperatively. Periodic intestinal biopsies should be performed, although technically difficult in animals if no jejunostomy has been created. The mucosal biopsy findings will exhibit variable degrees of cell infiltration, villus blunting, cryptitis, epithelial cell damage, and regeneration, and mucus cell rejection. Severe rejection is manifested by mucosal hemorrhage, sluffing of the intestinal epithelium, and the presence of micro-abscesses.

Chronic rejection has fewer inflammatory cells and cryptitis, evidence of epithelium regeneration, and submucosal fibrosis.

5. Malabsorption

Animals are usually allowed water 48 to 72 hours after the operation. Severe diarrhea can occur and is most commonly related to GVHD or rejection. Should these two processes be adequately controlled, normal absorption and weight increase should occur. D-xylose absorption tests and fecal lipid determinations can be used to monitor intestinal function.

III. CONCLUSION

Multivisceral transplantation in the pig is a very useful model to study the application of these techniques in humans. Technical success for both the animal and human surgical procedures has recently been achieved, but long-term survival remains a significant challenge.

REFERENCES

1. **Starzl, T.E., Rowe, M., Todo, S., Jaffe, R., Tzakis, A., Hoffman, A., Esquivel, C., Porter, K., Venkataramanan, R., Makowka, L., and Duquesnoy, R.,** Transplantation of multiple abdominal viscera, *JAMA,* 261, 1449-1457, 1989.
2. **Williams, J., Sankary, H., Foster, P., Lowe, J., and Goldman, G.,** Splachnic transplantation: an approach to the infant dependent on parental nutrition who develops irreversible liver disease, *JAMA,* 261, 1458-1462, 1989.
3. **Starzl, T., Todo, S., Tzakis, A., Podesta, L., Mieles, L., Demetris, A., Teperman, L., Selby, R., Stevenson, W., Stieber, A., Gordon, R., and Iwatsuki, S.,** Abdominal organ cluster transplantation for the treatment of upper abdominal malignancies, *Ann. Surg.,* 210, 374-386, 1989.
4. **Grant, D., Wall, W., Mimeault, R., Zhong, R., Ghent, C., Garcia, B., Stiller, C., and Duff, J.,** Successful small-bowel/liver transplantation, *Lancet,* 335, 181-184, 1990.
5. **Todo, S., Tzakis Andreas, G., Abu-Elmagd, K., Reyes, J., Nakamura, K., Casavilla, A., Selby, R., Nour, B.M., Wright, H., Fung, J.J., Demetris, A.J., Van Thiel, D.H., and Starzl, T.E.,** Intestinal transplantation in composite visceral grafts or alone, *Ann. Surg.,* 216, 223-234, 1992.
6. **Starzl, T.E. and Kaupp, H.,** Mass homotransplantation of abdominal organs in dogs, *Surg. Forum,* 11, 28-30, 1960.
7. **Eiseman, B. and Spencer, F.C.,** Man's best friend?, *Ann. Surg.,* 159, 159-160, 1964.
8. **Garnier, H., Clot, J.P., Bertrand, M., Camplez, P., Kunlin, A., Gorin, J.P., Le'Goaziou, F., Levy, R., and Cordier, G.,** Biologie experimentale: greffe de foie chez le porc: approche chirurgicale, *C.R. Acad. Sci. Paris,* 260, 5621, 1965.
9. **Peacock, J.H. and Terblanche, J.,** Orthotopic homotransplantation of the liver in the pig, in *The Liver,* Read, A.E., Ed., Butterworths, London, 1967, 333-336.
10. **Gridelli, B., Rossi, G., Colledan, M., Fassati, L., Ferla, G., Giacci, F., Gislon, M., Lucianerri, A., Andreoni, A., Doglia, M., and Galmarini, D.,** Organ procurement for multivisceral abdominal transplantation in the pig, *Transplant. Proc.,* 5, 844-845, 1988.
11. **Kobayashi, N., Sakagami, K., Takasu, S., Inagaki, M., Hasuoka, H., Yagi, T., Onoda, T., Matsuno, T., Saito, S., Kawamura, T., and Orita, K.,** Significance of pancreatic allograft rejection in swine abdominal organ cluster transplantation, *Transplant. Proc.,* 23, 635-639, 1991.

12. **Starzl, T.E., Todo, S., Tzakis, A., Alessiani, M., Casavilla, A., Abu-Elmagad, K., and Fung, J.,** The many faces of multivisceral transplantation, *Surg. Gynecol. Obstets.,* 172, 335-344, 1991.
13. **Elkins, W.L.,** Cellular immunology and the pathogenesis of graft versus host reactions, *Prog. Allergy,* 15, 78, 1971.
14. **Pirenne, J., Nakhleh, R.E., and Dunn, D.L.,** Graft-versus-host disease after multiorgan transplantation, *J. Surg. Res.,* 50, 622-628, 1991.
15. **Murase, N., Demetris, A.J., Matsuzaki, T., Yagihashi, A., Todo, S., Fung, J., and Starzl, T.E.,** Long survival in rats after multivisceral versus isolated small bowel allotransplantation under FK 506, *Surgery,* 110, 87-89, 1991.
16. **Hoffman, A.L., Makowka, L., Banner, B., Cai, X., Cramer, D.V., Pascualone, A., Todo, S., and Starzl, T.E.,** The use of FK 506 for small intestine transplantation: Inhibition of acute rejection and prevention of fatal graft versus host disease, *Transplantation,* 49, 483-490, 1990.

Section VI

Bone Marrow and Graft-Versus-Host Disease

Chapter 18

Bone Marrow Transplantation and Graft-Versus-Host Disease in Mice

Robert Korngold

CONTENTS

I. INTRODUCTION

Over the last two decades, bone marrow transplantation has developed as a primary therapeutic approach for the treatment of aplastic anemia, several forms of high-risk leukemia, and severe combined immunodeficiency.[1,2] In addition, this procedure has potential applicability to many other genetic, metabolic, and hematologic disorders, but only if the associated complications involved with transplantation can be significantly reduced. Graft-versus-host disease (GVHD) remains the most critical risk factor, particularly in allogeneic transplants for leukemia, where between 35 to 60% of patients exhibit moderate to severe acute GVHD and causes 12 to 20% fatality.[3,4] The skin, gut, liver, and lymphoid organs are the principal target organs for GVHD, and patients are usually immunodeficient and thereby susceptible to a wide range of opportunistic pathogenic infections. A significant number of transplant recipients (20 to 45%) surviving more than 6 months experience a chronic form of GVHD, which can cause severe debilitation.[5,6] This syndrome may involve several elements including an autoimmune-like response, suppressor cells, and a dysfunctional thymus as complicating factors. In addition to GVHD and prolonged immunodeficiency, the other major problems related to bone marrow transplantation are failure of marrow engraftment and the relapse of leukemia. The resolution of all of these complications in the future would greatly enhance the usefullness of marrow transplantation therapy. In order to develop these new approaches, the field will rely heavily upon animal studies and, as in the past, by far the most widely

0-8493-3629-5/94/$0.00+$.50

used model will be that of the mouse. The reasons for this popularity are quite practical in that the wide availability of inbred mouse strains has allowed for extensive immunogenetic analyses, aided by the multitude of known lymphocyte markers and reagents necessary to characterize GVHD and related responses. The following discussion is a brief review of some of the major aspects of bone marrow transplantation and GVHD in relation to murine studies and some of the basic procedures used to carry on these investigations.

A. EARLY MURINE STUDIES

Experimental studies in murine models for bone marrow transplantation and GVHD were initiated in the early 1950s when it was observed by Lorenz et al.[7] that lethally irradiated mice could be protected by the administration of syngeneic bone marrow cells. It was subsequently realized that allogeneic donor marrow could also avoid irradiation-related death but that the recipients often succumbed to what was originally described as "secondary disease", and what we now know as GVHD[8-11] (see reference 12 for a more complete review). Mice with this condition exhibited symptoms of drastic weight loss, diarrhea, and a hunched posture before their ultimate death. By 1967 it was realized that GVHD was the consequence of transferring immunocompetent cells in the marrow graft to an immunocompromised host expressing foreign transplantation antigens.[13] The major histocompatibility complex (MHC) antigens, and in particular the class I molecules, were recognized as important differences for generating lethal GVHD.[14,15] More recent studies with single MHC subregion differences has established that both class I and class II antigens can be effective stimulators of a GVHD response.[16-18] With regard to the nature of the cells responsible for MHC-directed GVHD induction, several investigations established that mature T cells contaminating the donor inoculum had this etiological role.[19-24]

B. GVHD ACROSS NON-MHC BARRIERS

In the 1970s, the combined results from both experimental and clinical studies convinced all in the field that an MHC-matched sibling marrow donor, if available, was the transplant of choice. Although these transplants exhibited significant improvement in survival rates, the incidence of GVHD, in either its acute or chronic form, still remained high and pointed to potential non-MHC antigens as targets for disease induction.[25,26] This possibility was verified experimentally in mice models where GVHD directed to non-MHC multiple minor histocompatibility (H) antigens could be prevented by elimination of contaminating T cells from the marrow inoculum.[27,28] The relative contributions of $CD4^+$ and $CD8^+$ T cell subsets in the etiology of GVHD across either MHC or minor H antigen barriers has been the subject of controversy over the years (reviewed in reference 29) but can be summarized as follows: $CD4^+$ T cells alone can cause GVHD directed to class II MHC antigens[30] and in some strain combinations can also be directed to minor H antigens;[31,32] $CD8^+$ T cells can cause disease when directed to class I MHC antigens[33,34] and are potent mediators of GVHD against minor H antigens, often doing so in apparent independence of CD4 "help".[31] The T cell responses of either subset to minor H antigens in GVHD is also complicated by the phenomenon of immunodominance, by which only a few antigens are actually driving the response out of the many that should actually be available for recognition.[35] The mechanisms controlling immunodominance and the actual nature and source of minor H antigens themselves is far from clear in murine systems and even less so in man.[36]

C. PATHOBIOLOGY OF GVHD

Although there is substantial evidence to support the necessity for appropriate T cells in the development of GVHD, little is understood about the mechanisms involved in actual target tissue destruction. This is a very complex issue that may involve multi-functionality of T cells (i.e., CD4+ cells may, in addition to producing lymphokines, also engage in direct cytotoxicity,[37] and CD8+ cells may also release critical lymphokines),[38] the activities of several different lymphokines (including IL-2,[39] TNF,[40] and IFN[41]), and recruitment of different types of effector cells (e.g., NK cells[42,43] and macrophages[40]). In regard to this topic, we have recently compared the sequential histopathology mediated by CD4+ vs. CD8+ T cells in minor H antigen-different strain combinations.[44] With CD8-cell-mediated GVHD, disease developed by the third week post-transplant and primarily involved the skin and liver, with only mild involvement of the gut. Dermal fibrosis was evident between the fourth and fifth week, along with bronchiolitis, and most of the mice died by the end of the sixth week. On the other hand, with CD4-cell-mediated GVHD, there were similar dermal and hepatic changes, but with much more pronounced intestinal involvement, beginning around the fourth week after transplantation. Furthermore, epididymal lesions were seen but there were no prominent pulmonary lesions. These findings do suggest that at least in GVHD directed to minor H antigens, CD8+ and CD4+ T cells mediate quite similar patterns of tissue damage, with the exception of gut damage being more apparent in recipients of CD4+ cells.

D. T CELL-DEPLETION OF DONOR MARROW

The implication of mature donor T cells in the marrow inoculum as the main cause of GVHD across either MHC or non-MHC barriers has stimulated attempts to purge the donor marrow inoculum of T cells before transplantation. This approach has been undertaken at several clinical transplant centers around the world and has been highly successful in reducing the incidence of GVHD.[45,46] The most common means of depleting T cells from marrow involve either techniques of anti-T cell-specific antibody and complement treatment or lectin-rosetting procedures (reviewed in reference 46), both of which can be quite effective. However it is now clear that T-cell depletion of marrow can cause a noticeable increase in the other serious complications of marrow transplantation, particularly graft failure, immunoincompetency, and relapse of leukemia.[46] Finding the means of avoiding GVHD and at the same time countering these other risk factors is currently an important goal of investigators working with all experimental models, and is obviously critical to the future success of bone marrow transplantation.

II. PROCEDURES

The procedures involved in setting up a bone marrow transplantation and lethal GVHD murine irradiation model will be discussed below largely in terms of the preparation of different types of donor cells that can be used and the optimum preconditioning of host animals. The techniques described are those used in our own laboratory and, of course, might vary considerably in other laboratories with similar outcomes.

A. MEDIA

Either Dulbecco's Phoshate Buffered Saline or Hank's Buffered Saline Solution supplemented with 0.1% bovine serum albumin (BSA) (Hyclone, Logan, UT) is suitable for use for all *in vitro* manipulations of donor bone marrow and lymphocytes. For injection of cells into recipient animals, cells are finally suspended in buffered saline solution, alone.

B. PREPARATION OF BONE MARROW CELLS

Femurae and tibiae bones are excised from donor mice and kept in media on ice while awaiting further processing. The bones are then scraped clean of all flesh with a scalpel (size #10 blade) and the joint ends clipped off to allow a 26-gauge needle insertion. Bone marrow cells are then obtained by flushing the bones with media from a 10-ml syringe, until marrow is no longer visible within the bone. The flushed cells are collected in a centrifuge tube, vigorously pipetted to break up cell clumps, and then passed over a nylon mesh filter to remove any remaining debris. The cells are then centrifuged at 1200 RPM, 4°C, resuspended, and counted. The yield of bone marrow cells will vary depending upon the strain and age of the donor mice, but usually falls within a range of 3 to 5 × 10^7 cells per mouse.

C. T CELL-DEPLETION OF BONE MARROW CELLS

For most types of bone marrow transplantation experiments, and of course depending upon intent, a control group for survival is included in which GVHD is to be avoided by the elimination of mature T cells in the donor marrow inoculum (the percentage of T cell contamination is usually less than 5%). This is accomplished by treating the bone marrow cells with anti-Thy-1 monoclonal antibodies (from the J1j hybridoma[47]) or an equivalent pan-T cell preparation in the presence of guinea pig complement for 50 min at 37°C. The marrow cells are then washed at least three times with 20 ml of media to sufficiently dilute out residual antibody, particularly if specially prepared T cell populations are to be mixed into the inoculum to test for GVHD potential. The cells are then counted and resuspended in saline solution for injection. Since T cells were only a small percentage of the starting population, the final cell numbers would only reflect a small drop in the cell yield.

D. PREPARATION OF DONOR T CELL POPULATIONS

1. Whole T Cells

For testing the capacity of different T cell populations to induce lethal GVHD, T cells are usually prepared from donor spleen and lymph nodes (LN) and then added to T-cell-depleted bone marrow in appropriate dosages. First, to figure out the number of donor mice needed, the number of whole T cells required for the experiment is doubled and then divided by 1.3 × 10^8 cells (the average yield, on the low side, of spleen and LN cells per mouse), and the result, of course, rounded up to the nearest mouse. Mice are sacrificed by CO_2 inhalation and the major LN sets excised (cervical, axial, brachial, inguinal, and mesenteric), along with the spleen, and kept in 50-ml centrifuge tubes with approximately 10 ml of media on ice. Pooled lymphoid collections from up to five mice per tube are then gently ground in glass Ten Broeck tissue grinders to release the cells, utilizing multiple washings with media. Harvested cells are then centrifuged at 1200 RPM, 8 min at 4°C, and the pellet resuspended in 10 ml of Gey's balanced salt lysing solution containing 0.7% NH_4CL for removal of red blood cells. The cells are kept on ice for 2 min, diluted with 20 ml of media and recentrifuged. To remove what usually is a substantial number of dead cells from the population, the pellet is resuspended in 10 ml of a low ionic buffer solution[48] and filtered over a cotton-packed funnel directly into a fresh tube containing normal media. The cells are centrifuged, resuspended in 10 ml media, and counted. To prepare what is essentially a restricted T-cell population (although not pure since it would still contain 5 to 10% macrophages), B cells would then be eliminated by either panning or cytolytic methods. For the former technique, lymphoid cells are incubated on Goat-anti-mouse Ig antibody-coated plates to which B cells will adhere.[49] The plates used are nontissue culture grade plastic petri dishes (150 × 15 mm) which are pre-incubated for 1 hour at room temperature with approximately 15 ml of a 1:200 dilution of affinity-purified G anti-MIg antibody (1 mg/ml, Cappel Laboratories, West Chester, PA). Plates are washed with media at least four times before cells are added, for up to a total of 2.2 × 10^8 cells in a volume of 10 ml. The plates are incubated in the refrigerator on a flat surface for 1 hour, after which the nonadherent cells (T cells) can be collected by gently swirling, pouring, and washing the plate. The T cells are then centrifuged, resuspended, counted, and are ready for use. Alternatively, lymphoid populations can be treated with an appropriate concentration of an anti-B cell specific antibody (e.g., J11d[47] or M1/69[50] mAbs) in the presence of guinea pig complement and incubated for 50 min at 37°C followed by two washes with centrifugation, counting, and final resuspension. Both methods of removing B cells are equally effective and a matter of personal preference in regard to whether there is a perceived problem in the presence of the dead B cells remaining in the suspension after antibody lysis. These of course could be removed if necessary by another round of low ionic buffer treatment.

2. T Cell Subsets

T cells of mice can be divided into two major subsets: the H-2 class II-restricted (helper-type) T cells bearing the CD4 phenotype; and the H-2 class I-restricted (cytotoxic T lymphocytes, CTL) cells expressing the CD8 marker.[51] The separation of T cells into highly purified subpopulations of CD4$^+$ or CD8$^+$ cells involves both negative and positive selection techniques[31,52] and is summarized in Figure 1.

For the preparation of CD4$^+$ cells, for example, a whole T cell population is first generated as described above. Cells are then treated with an anti-CD8 mAb (3.168)[53] and complement for 50 min at 37°C in a water bath. Cells are then washed and resuspended and panned over petri dishes (150 × 15 mm) pre-incubated for 1 hour at room temperature with 15 ml of anti-CD4 mAb (GK1.5)[54] at a dilution of 1:12000 of a 1 mg/ml concentration. CD4$^+$ T cells adhere to the plate and after incubation for 1 hour in the refrigerator, nonadherent cells are washed off, and the positively selected CD4$^+$ T cells are eluted from the plate by vigorous pipetting. This procedure can effectively yield a greater than 98% pure CD4$^+$ T cell population, as measured by flow cytometric analysis.

The method for purifying CD8$^+$ T cells is similar except that: the whole T cells are incubated with anti-CD4 mAb (RL172, a rat IgM antibody that is more effective than GK1.5 in lysing cells *in vitro*)[55] plus complement for 1 hour at 37°C; the cells are then washed and incubated with anti-CD8 mAb (3.168; on ice for 30 min); and the cells are then washed and panned over plates pre-incubated at room temperature for 1 hour with 15 ml of a 1:200 dilution of goat anti-rat IgM antibodies (1 mg/ml affinity purified from Cappel Laboratories). This positive selection is performed indirectly because the 3.168 mAb is of a rat IgM isotype and does not effectively bind to the plastic petri dish by itself. Accordingly, after 1 hour incubation in the refrigerator, nonadherent cells are gently removed by washing the plate, and the CD8$^+$ T cells are pipetted off and collected, with usually considerably greater ease than the CD4$^+$ cells described above.

E. IRRADIATION OF RECIPIENTS

Recipient mice are exposed to whole-body irradiation from a Gammacell[137] Cesium source (130 rad/min), with the dosage dependent upon the individual strain but usually falling between 800 to 900 rads. The object of the irradiation is to immunocompromise the host in order to avoid a host-versus-graft response directed against the donor marrow.[56] Therefore, the intention is usually to expose the recipients to as high as possible a dose of irradiation that will still allow them to survive when reconstituted with T cell-depleted bone marrow cells (see reference 57 for a review on the subject of engraftment). Mice are injected with the marrow and/or T cell inoculum with a minimum wait of 6 hours after irradiation. Alternatively, mice may be exposed to as high as 1100 rads of total irradiation if given over a split-dose separated by 3 hours, which minimizes gastrointestinal toxicity.[42]

F. MORTALITY ASSAY FOR GVHD

Donor T-cell-depleted bone marrow cells (2×10^6) alone as a negative control, or combined with an appropriate donor T cell population are injected intravenously in a volume of 0.5 ml. Mice are checked daily for morbidity and mortality until the experi-

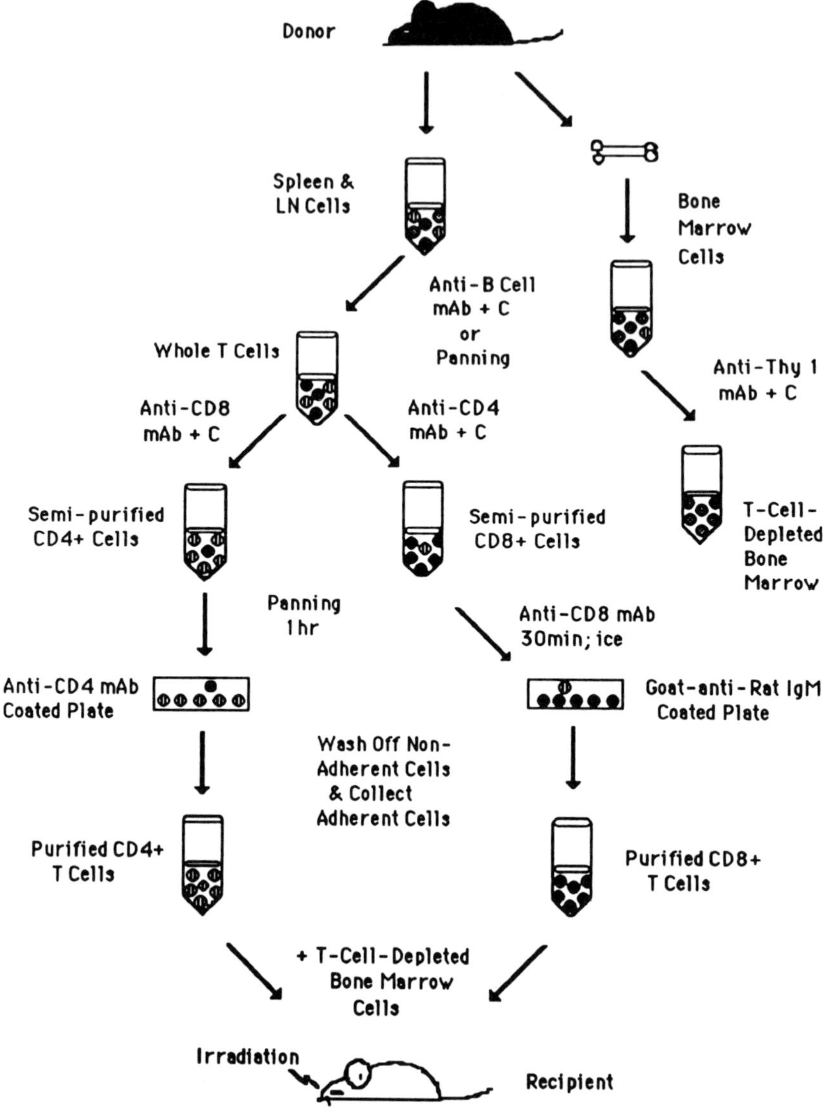

Figure 1 Protocol for T cell subset purification and GVHD assay. The cell preparation methods involve a combination of negative selection procedures utilizing mAb and complement treatment followed by positive panning on antibody-coated plastic plates. Highly purified T cell subset populations obtained in this way are then combined with T-cell-depleted donor bone marrow and injected intravenously into irradiated recipient mice for the GVHD mortality assay.

ments are terminated at 80 days post-transplantation. Data can be pooled from several separate experiments in order to calculate a median survival time (MST). Statistical comparisons between experimental groups for mortality curves can be performed by the nonparametric Wilcoxin signed rank analysis. In appropriate experiments recipient mice can also be weighed twice a week as an indicator of developing GVHD and the data expressed as the mean percentage of the initial weight for the group. Significance between the different weight groups can be determined using student's t-test for paired samples.

G. SURVIVAL OF NEGATIVE CONTROLS

A common problem with bone marrow transplantation models is that since irradiated mice are immunocompromised, the environmental health status of a particular mouse colony will often dictate the survivability of negative control mice receiving T-cell-depleted bone marrow cells alone. Therefore, to avoid costly complications to the results of a marrow or GVHD experiment, we have found that it is best to keep the mice under sterile conditions in microisolator cages from the time of their arrival into the colony. Although cumbersome, these procedures have virtually eliminated the loss of marrow controls, and thereby experiments along with them, due to endemic colony viral infections.

H. KINETICS FOR GVHD DEVELOPMENT

The dosage of donor T cells in an inoculum will affect the median survival times of the recipient mice, particularly in the case of GVHD across multiple minor H antigen barriers.[27] Higher dosages of T cells will usually cause a more rapid onset of GVHD (less than 30 days) with an increase in the severity of symptoms and mortality rates; lower dosages can lead to late onset GVHD (more than 50 days) and sometimes more chronic conditions like skin lesions, without fatality.[58] The exact relationship of cell dosage and GVHD kinetics will depend upon the strength of the antigen disparity in the particular strain combination being utilized.

I. OTHER TYPES OF LETHAL GVHD MODELS

GVHD can also be obtained in nonirradiated F_1 semi-allogeneic mice and is characterized by either a lymphoproliferative type of syndrome or an aplastic condition, depending upon which donor T cell subsets are utilized.[59-61] The procedures for setting up the model are otherwise very similar to those described above. Another interesting model for a GVHD-like syndrome is produced in syngeneic mice or rats in the presence of cyclosporin A administration and may resemble more the chronic form of GVHD, from a pathological point of view (reviewed in reference 62).

III. FUTURE PERSPECTIVES

In light of the recent observations in murine systems on the role of $CD4^+$ and $CD8^+$ T cell subsets in development of GVHD, new approaches to the current obstacles facing clinical bone marrow transplantation may utilize some form of selective T-cell-depletion of the marrow inoculum. In class I MHC-mismatched or MHC-identical, non-MHC-mismatched situations, the experimental results suggest that $CD8^+$ T cells are predominantly involved in GVHD onset, and therefore they should be completely eliminated, whereas leaving $CD4^+$ cells in the inoculum may be advantageous, if indeed, they are not involved in GVHD. Donor $CD4^+$ T cells can convey a level of immediate immune protection to the recipient, either directly or indirectly through lymphokine production, that can help counteract opportunistic infection. Hematopoietic stimulating factors secreted by $CD4^+$ cells may also enhance marrow engraftment, and, potentially, $CD4^+$ cells may be able to mount responses to prevent the relapse of residual leukemia cells. This latter effect, which is referred to as a graft-vs.-leukemia response may be possible if $CD4^+$ T cells can recognize tumor-specific antigens or even if they could recognize alloantigen at a level that would not lead to a serious GVHD response.[63,64] It may even be possible to enhance the activity of these $CD4^+$ cells with additional transfers to the recipient of CD8-cell-depleted donor peripheral blood lymphocytes. Ideally, $CD4^+$ cells may also be infused after prestimulation in culture to the specific leukemia cell antigens of the recipient, which would allow for a more rapid clearance of residual tumor cells. This scenario, again, would depend upon the relative inactivity of $CD4^+$ cells in GVHD directed to non-HLA antigens. Although very little is known about these human minor H antigens,[36] we are encouraged by: 1) a recent clinical study for CD8-cell-depletion which resulted in a low incidence of GVHD[65] and 2) data from analysis of anti-human minor-H antigen-specific T cells from GVHD patients which exhibited a dominant involvement of cytotoxic cells,[66] although proliferating T cells were also demonstrated.[67] The approach of selective T-cell depletion is currently being investigated in mice models and we expect that these studies will yield some insight into the problem.

REFERENCES

1. **Gale R.P. and Champlin R.E.,** *Bone Marrow Transplantation: Current Controversies,* UCLA Symposium on Molecular and Cellular Biology, Vol. 91, Alan R. Liss, New York, 1989.
2. **Burakoff, S.J., Deeg, H.J., Ferrara, J.L.M., and Atkinson K.,** *Graft-Versus-Host Disease: Immunology, Pathophysiology, and Treatment,* Marcel Dekker, New York, 1990.
3. **Gale, R.P.,** Graft-versus-host disease, *Immunol. Rev.,* 88, 193, 1985.
4. **Deeg, H.J. and Henslee-Downey, P.J.,** Management of acute graft-versus-host disease, *Bone Marrow Transplant.,* 6, 1, 1990.
5. **Atkinson, K.,** Chronic graft-versus-host disease, *Bone Marrow Transplant.,* 5, 69, 1990.
6. **Sullivan, K.M., Witherspoon, R.P., Storb,R., et al.,** Chronic graft-versus-host disease: recent advances in diagnosis and treatment, in *Bone Marrow Transplantation Current Controversies,* UCLA Symposium on Molecular and Cellular Biology, Vol. 91, Gale, R.P. and Champlin, R.E., Eds., Alan R. Liss, New York, 1989, 511.
7. **Lorenz, E., Uphoff, D. E., Reid, T. R., and Shelton, E.,** Modification of irradiation injury in mice and guinea pigs by bone marrow injections, *J. Natl. Cancer Inst.,* 12, 197, 1951.
8. **Lorenz, E. and Congdon, C. C.,** Modification of lethal irradiation injury in mice by injection of homologous and heterologous bone marrow, *J. Natl. Cancer Inst.,* 14, 955, 1954.
9. **Barnes, D.W.H. and Loutit, J.F.,** Spleen protection: the cellular hypothesis, in *Radiobiology Symposium, Liege,* Bacq, Z.M., Ed., Butterworths, London, 1955, 134.
10. **Cohen, J.A., Vos, O., and van Bekkum, D.W.,** The present status of radiation protection by chemical and biological agents in mammals, in *Advances in Radiobiology,* de Hersey, G.C., Forssberg, A.G., and Abbott, J.D., Eds., Oliver and Boyd, Edinburgh, 1957, 134.
11. **van Bekkum, D.W. and de Vries, J.J.,** *Radiation Chimaeras,* Logos, London, 1967.
12. **Santos, G.W.,** History of bone marrow transplantation, *Clin. Haematol.,* 12, 611, 1983.
13. **Billingham, R.E.,** The biology of graft-versus-host reactions, *Harvey Lect.,* 62, 21, 1967.
14. **Klein, J.,** Relative importance of H-2 regions in the development of graft-versus-host reactions, *Transplant. Proc.,* 8, 335, 1976.
15. **Elkins, W.L.,** Correlation of graft-versus-host mortality and positive CML assay in the mouse, *Transplant. Proc.,* 8, 343, 1976.
16. **Sprent, J., Schaefer, M., Lo, D., and Korngold, R.,** Functions of purified L3T4+ and Lyt-2+ cells in vitro and in vivo, *Immunol. Rev.,* 91, 195, 1986.
17. **Sprent, J., Schaefer, M., Lo, D., and Korngold, R.,** Properties of purified T cell subsets. II. In vivo responses to class I vs. class II H-2 differences, *J. Exp. Med.,* 163, 998, 1986.
18. **Mason, D.W.,** Subsets of T cells in the rat mediating lethal graft-versus-host disease, *Transplantation,* 32, 222, 1981.

19. **Cantor, H., Mandel, M.A., and Asofsky, R.,** Studies of thoracic duct lymphocytes of mice. II. A quantitative comparison of the capacity of thoracic duct lymphocytes and other lymphoid cells to induce graft-versus-host reactions, *J. Immunol.,* 104, 409, 1970.
20. **Cantor, H.,** The effects of anti-theta serum upon graft vs. host activity of spleen and lymph node cells, *Cell Immunol.,* 3, 461, 1972.
21. **Tyan, M.L.,** Modification of severe graft-versus-host disease with antisera to the theta antigen or to whole serum, *Transplantation,* 15, 601, 1973.
22. **Trentin, J.J. and Judd, K.P.,** Prevention of acute graft-versus-host (GVH) mortality with spleen-absorbed antithymocyte globulin (ATG), *Transplant. Proc.,* 5, 865, 1973.
23. **Rodt, H., Thierfelder, S., and Eulitz, E.,** Anti-lymphocytic antibodies and marrow transplantation. III. Effect of heterologous anti-brain antibodies on acute secondary disease in mice, *Eur. J. Immunol.,* 4, 25, 1974.
24. **Sprent, J., von Boehmer, H., and Nabholz, M.,** Association of immunity and tolerance of host H-2 determinants in irradiated F1 hybrid mice reconstituted with bone marrow cells from one parental strain, *J. Exp. Med.,* 142, 321, 1975.
25. **Bortin, M.,** A compendium of reported human bone marrow transplants, *Transplantation,* 9, 571, 1970.
26. **Thomas, E.D., Storb, R., Clift, R.A., et al.,** Bone marrow transplantation, *N. Engl. J. Med.,* 292, 832, 1975.
27. **Korngold, R. and Sprent, J.,** Lethal graft-versus-host disease following bone marrow transplantation across minor histocompatibility barriers in mice. Prevention by removing mature T cells from marrow, *J. Exp. Med.,* 148, 1687, 1978.
28. **Hamilton, B.L., Bevan, M.J., and Parkman, R.,** Anti-recipient cytotoxic T lymphocyte precursors are present in spleens of mice with acute graft-versus-host disease due to minor histocompatibility antigens, *J. Immunol.,* 126, 621, 1981.
29. **Korngold, R. and Sprent, J.,** Overview: T cell subsets and graft-versus-host disease, *Transplantation,* 44, 335, 1987.
30. **Sprent, J., Schaefer, M., and Korngold, R.,** Role of T cell subsets in lethal graft-versus-host disease (GVHD) directed to class I versus class II H-2 differences. II. Protective effects of L3T4$^+$ cells in anti-class II GVHD, *J. Immunol.,* 144, 2946, 1990.
31. **Korngold, R. and Sprent, J.,** Variable capacity of L3T4$^+$ T cells to cause graft-versus-host disease across minor histocompatibility barriers in mice, *J. Exp. Med.,* 165, 1552, 1987.
32. **Hamilton, B.L.,** L3T4-positive T cells participate in the induction of graft-vs-host disease in response to minor histocompatibility antigens, *J. Immunol.,* 139, 2511, 1987.
33. **Sprent, J., Schaeffer, M., Gao, E-K., and Korngold, R.,** Role of T cell subsets in lethal graft-versus-host disease (GVHD) directed to class I versus class II H-2 differences. I. L3T4$^+$ cells can either augment or retard GVHD elicited by Lyt-2$^+$ cells in class I-different hosts, *J. Exp. Med.,* 167, 56, 1988.
34. **Korngold, R. and Sprent, J.,** T-cell subsets in graft-vs-host disease, in *Graft-Versus-Host Disease: Immunology, Pathophysiology, and Treatment,* Burakoff, S.J., Deeg, H.J., Ferrara, J.L.M., and Atkinson, K., Eds., Marcel Dekker, New York, 1990, 31.
35. **Korngold, R. and Wettstein, P. J.,** Immunodominance in the graft-versus-host disease T cell response to minor histocompatibility antigens, *J. Immunol.,* 145, 4079, 1990.
36. **Perreault, C., Decary, F., Brochu, S., Gyger, M., Belanger, R., and Roy, D.,** Minor histocompatibility antigens, *Blood,* 76, 1269, 1990.
37. **Golding, H., Munitz, T.I., and Singer, A.,** Characterization of antigen-specific, Ia-restricted, L3T4+ cytolytic T lymphocytes and assessment of thymic influence on their self specificity, *J. Exp. Med.,* 162, 943, 1983.

38. **Roopenian, D.C., Widmer, M.B., Orosz, C.G., and Bach, F.H.,** Helper cell-independent cytolytic T lymphocytes specific for a minor histocompatibility antigen, *J. Immunol.,* 130, 542, 1983.
39. **Parkman, R.,** Clonal analysis of graft-vs-host disease, in **Graft-Versus-Host Disease in Immunology, Pathophysiology, and Treatment,** Burakoff, S.J., Deeg, H.J., Ferrara, J.L.M., and Atkinson, K., Eds., Marcel Dekker, New York, 1990, 51.
40. **Piguet, P.F.,** Tumor necrosis factor and graft-vs-host disease, in *Graft-Versus-Host Disease in Immunology, Pathophysiology, and Treatment,* Burakoff, S.J., Deeg, H.J., Ferrara, J.L.M., and Atkinson, K., Eds., Marcel Dekker, New York, 1990, 255.
41. **Cleveland, M.G., Annable, C.R., and Klimpel, G.R.,** In vivo and in vitro production of IFN-β and IFN-γ during graft-vs host disease, *J. Immunol.,* 141, 3349, 1988.
42. **Ferrara, J. and Burakoff, S.J.,** The pathophysiology of acute graft-vs.-host disease in a mice bone marrow transplant model, in *Graft-Versus-Host Disease: Immunology, Pathophysiology, and Treatment,* Burakoff, S.J., Deeg, H.J., Ferrara, J.L.M., and Atkinson, K., Eds., Marcel Dekker, New York, 1990, 9.
43. **Ghayur, T., Seemayer, T.A., and Lapp, W.S.,** Histologic correlates of immune functional deficits in graft-vs.-host disease, in *Graft-Versus-Host Disease: Immunology, Pathophysiology, and Treatment,* Burakoff, S.J., Deeg, H.J., Ferrara, J.L.M., and Atkinson, K., Eds., Marcel Dekker, New York, 1990, 109.
44. **Murphy, G.F., Whitaker, D., Sprent, J., and Korngold, R.,** Characterization of target injury of mice acute graft-versus-host disease directed to multiple minor histocompatibility antigens elicited by either CD4+ or CD8+ effector cells, *Am. J. Pathol.,* 138, 983, 1991.
45. **Butturini, A. and Gale, R.P.,** Clinical trials of T-cell depletion: current controversies, future directions, in *Bone Marrow Transplantation: Current Controversies,* UCLA Symposium of Molecular and Cellular Biology, Vol. 91, Gale, R.P. and Champlin, R.E., Eds., Alan R. Liss, New York, 1989, 495.
46. **Martin, P.J. and Kernan, N.A.,** T-cell depletion for the prevention of graft-vs-host disease, in *Graft-Versus-Host Disease: Immunology, Pathophysiology, and Treatment,* Burakoff, S.J., Deeg, H.J., Ferrara, J.L.M., and Atkinson, K., Eds., Marcel Dekker, New York, 1990, 371.
47. **Bruce, J., Symington, F.W., McKearn, T.J., and Sprent, J.,** A monoclonal antibody discriminating between subsets of T and B cells, *J. Immunol.,* 127, 2496, 1981.
48. **Von Boehmer, H. and Shortman, K.,** The separation of different cell classes from lymphoid organs. IX. A simple and rapid method for removal of damaged cells from lymphoid cell suspensions, *J. Immunol. Methods,* 2, 293, 1973.
49. **Mage, M.G., McHugh, L.L., and Rothstein, T.L.,** Mouse lymphocytes with and without surface immunoglobulin: preparative scale separation in polystyrene tissue culture dishes coated with specifically purified anti-immunoglobulin, *J. Immunol. Methods,* 15, 47, 1977.
50. **Springer, T., Galfre, G., Secher, D.S., and Milstein, C.,** Monoclonal xenogeneic antibodies to mice cell surface antigens: identification of novel leukocyte differentiation antigens, *Eur. J. Immunol.,* 8, 539, 1978.
51. **Sprent, J. and Webb, S.,** Function and specificity of T cell subsets in the mouse, *Adv. Immunol.,* 41, 39, 1987.
52. **Sprent, J. and Schaefer, M.,** Properties of purified T cell subsets. I. In vitro responses to class I vs. class II H-2 alloantigens, *J. Exp. Med.,* 162, 2068, 1985.
53. **Sarmiento, M., Glasebrook, A.L., and Fitch, F.W.,** IgG or IgM monoclonal antibodies reactive with different determninants on the molecular complex bearing Lyt 2 antigen block T cell-mediated cytolysis in the absence of complement, *J. Immunol.,* 125, 2665, 1980.

54. **Dialynas, D.P., Wilde, D.B., Marrack, P., et al.,** Characterization of the mice antigenic determinant, designated L3T4a, recognized by monoclonal antibody GK1.5: expression of L3T4a by functional T cell clones appears to correlate primarily with Class II MHC antigen-reactivity, *Immunol. Rev.,* 74, 29, 1983.
55. **Ceredig, R., Lowenthal, J.W., Nabholz, M., and MacDonald, H.R.,** Expression of interleukin-2 receptors as a differentiation marker on intrathymic stem cells, *Nature (London),* 314, 98, 1985.
56. **Nakamura, H. and Gress, R.E.,** Graft rejection by cytolytic T cells. Specificity of the effector mechanism in the rejection of allogeneic marrow, *Transplantation,* 49, 453, 1990.
57. **Vallera, D.A. and Blazar, B.R.,** T cell depletion for graft-versus-host-disease prophylaxis. A perspective on engraftment in mice and humans, *Transplanation,* 47, 751, 1989.
58. **Hamilton, B.L. and Parkman, R.,** Acute and chronic graft-versus-host disease induced by minor histocompatibility antigens in mice, *Transplantation*, 36, 150, 1983.
59. **Rolink, A.G., Pals, S.T., and Gleichman, E.,** Allosuppressor and allohelper T cells in acute and chronic graft-versus-host disease. III. Different Lyt subsets of donor T cells induce different pathological syndromes, *J. Exp. Med.,* 158, 546, 1983.
60. **Rolink, A.G., Strasser, A., and Melchers, F.,** Autoimmune diseases induced by graft-vs-host disease, in *Graft-Versus-Host Disease: Immunology, Pathophysiology, and Treatment,* Burakoff, S.J., Deeg, H.J., Ferrara, J.L.M., and Atkinson, K., Eds., Marcel Dekker, New York, 1990, 161.
61. **Hakim, F.T. and Shearer, G.M.,** Immunologic and hematopoietic deficiencies of graft-vs-host disease, in *Graft-Versus-Host Disease: Immunology, Pathophysiology, and Treatment,* Burakoff, S.J., Deeg, H.J., Ferrara, J.L.M., and Atkinson, K., Eds., Marcel Dekker, New York, 1990, 133.
62. **Hess, A.D.,** Syngeneic graft-vs-host disease, in *Graft-Versus-Host Disease: Immunology, Pathophysiology, and Treatment,* Burakoff, S.J., Deeg, H.J., Ferrara, J.L.M., and Atkinson, K., Eds., Marcel Dekker, New York, 1990, 95.
63. **Truitt, R.L., LeFever, A.V., Shih, C.C.-Y., Jeske, J.M., and Martin, T.M.,** Graft-vs-leukemia effect, in *Graft-Versus-Host Disease: Immunology, Pathophysiology, and Treatment,* Burakoff, S.J., Deeg, H.J., Ferrara, J.L.M., and Atkinson, K., Eds., Marcel Dekker, New York, 1990, 177.
64. **Slavin, S., Ackerstein, A., Naparstek, E., Or, R., and Weiss, L.,** The graft-versus-leukemia (GVL) phenomenon: is GVL separable from GVHD?, *Bone Marrow Transplant.,* 6, 155, 1990.
65. **Champlin, R., Gajewski, J., Feig, S., et al.,** Selective depletion of CD8 positive T-lymphocytes for prevention of graft-versus-host disease following allogeneic bone marrow transplantation, *Transplant. Proc.,* 21, 2947, 1989.
66. **Van Els, C.A.C.M., Bakker, A., Zwinderman, A.H., Zwaan, F.E., van Rood, J.J., and Goulmy, E.,** Effector mechanisms in graft-versus-host disease in response to minor histocompatibility antigens. I. Absence of correlation with cytotoxic effector cells, *Transplantation,* 50, 62, 1990.
67. **Van Els, C.A.C.M., Bakker, A., Zwinderman, A.H., Zwaan, F.E., van Rood, J.J., and Goulmy, E.,** Effector mechanisms in graft-versus-host disease in response to minor histocompatibility antigens. II. Evidence of a possible involvement of proliferative T cells, *Transplantation,* 50, 67, 1990.

Chapter 19

Bone Marrow Transplantation in Dogs

H. Joachim Deeg and Rainer Storb

CONTENTS

0-8493-3629-5/94/$0.00+$.50

I. INTRODUCTION

Randomly bred animals such as the dog and non-human primates, although genetically less well-defined than murine models, are attractive for other reasons.[1,2] They allow, for example, sequential *in vivo* and *in vitro* studies in individual animals. Furthermore, data obtained in these models seem to be more directly applicable to man than data obtained in rodents. While primates are phylogenetically closer to man, dogs are easier to handle, less susceptible to infections, and less expensive to maintain; they also have larger litters at more frequent intervals, thus greatly facilitating genetic studies and intrafamilial transplantation experiments.[3] In this chapter we will provide a brief description of canine hematology, immunology, and immunogenetics as currently defined, and then describe the use of the dog as a bone-marrow transplantation model.

II. IMMUNOLOGY

The development, anatomy, and function of the canine immune system parallel those described in other species.[4] Lymphocyte responsiveness to mitogens increases with fetal age.[5] Fetuses become competent for antibody production close to term,[6] although at about 45 days of gestation skin allograft survival is still prolonged as compared to newborns or adults.[7] Lymphocyte responses to phytohemagglutinin (PHA) peak between the age of 6 weeks and 6 months and then decline;[5] consistent allogeneic *in vitro* responses are present at 4 months of age. Generally we do not use dogs for transplantation experiments until 6 months of age.

We have recently reviewed differentiation antigens and functional capacities of canine lymphohemopoietic cells and tissues.[8] Identifiable cells include T-lymphocytes, B-lymphocytes, large granular lymphocytes, dendritic cells, and those of myeloid lineage such as granulocytes, monocytes/macrophages, eosinophils, basophils, etc. The development of monoclonal antibodies (mabs) against canine cell subpopulations has been slow, in part because no reference marker for T cells was available.[9] It was also of note that most canine T cells expressed major histocompatibility complex (MHC) class II antigens even in the resting state.[10] The observation that canine thoracic duct lymph contained neither hemopoietic precursor cells[11] nor cells with antigen-presenting or accessory functions[12] has proven helpful in defining functions of "clean" cell populations. Several mabs directed at differentiation antigens have been described over the past decade (Table 1). Also, several cytokines, including Il-1, Il-2, and interferon, either cross-species reactive or isolated from the dog, have been investigated.[13,14]

III. HEMOPOIESIS

In the dog, as in man, hemopoiesis evolves through three embryological stages: mesodermal (yolk sac), hepatosplenic, and medullary.[15] Erythropoiesis is identifiable first, fol-

Table 1 **Monoclonal Antibodies Generated Against Canine Cells or Cells from other Species but Reactive with Canine Cells**

mab	IG Isotype	Specificity	Antigen	Comments	Ref.
DLy-1	IgG2b	Leukocytes	p213	T200-like	125
A-5	IgG2b	Leukocytes	p96	Suppressor cells	126
6C6	IaG2b	Lymphocytes	ND[1]	On medullary but not cortical thymocytes	127
DLy-6	IgM	Lymphocytes, dendritic cells	p60	Mature cells only	125
F3-20-7	IgG1	Thy-1	P24	Pan T cell, NK cells	128
1A1	IgM	T cells	ND	Pan T cell, NK cells, LGL[a]	129
DT-2	IgG2a	T-cell subset	P70	Helper/inducer cells	130
E11	IgG3	T-cell subset	ND	Suppressor cells, CTL	126
W1D10	IgG1	T-cell subset	p39	Inhibit mitogen activation	131
W1H3	IgG2a	T-cell subset	p26		131
T83[b]	IgG3	T-cell subset	ND	Inhibits MLC	132
Wig4	IgG1	Monocytes, platelets	p24	Activates C′	131
60.3	IaG2a	Leukocytes	p97,170,180	51% PBMC[c]	133
T811	IgG1	T suppressor, (CD8), Tp32	ND	CTL	134
J2	IgM	Activated T cells, malignant B cells, human gp 24	ND	15% PBMC[c]	135
CA13.164	IgG1	T cell subset (CD4)	p55	45% PBMC	Moore, unpublished
CA 4.1D3	ND	CD 45 RA	p205,220		Moore, unpublished
CA9.J03	IgG2a	T cell subset (CD8)	p32,38	20—25% PBMC	Moore, unpublished
DM5	IgG1	Myeloid, mature	p19,21,23	Not expressed on lymphoid cells	136
5F1	IgM	Monocytes, platelets, human g[85]	ND	16% PBMC[c]	137

[a] (ND) Not done; [b]We have not studied mab T83 in our laboratory; [c]Percent reactivity as determined by indirect cytofluorescence (FACS 440).

lowed by myelopoiesis, thrombopoiesis, and finally lymphopoiesis. At the time of birth, the bone marrow has become the only site of hemopoiesis, although the spleen can "store" a considerable amount of blood. Accordingly, hemopoietic stem cells can be obtained from fetal (but not adult) livers and spleens,[16] while adult donors can donate stem cells in the form of bone marrow[17,18] or from peripheral blood which contains circulating stem cells.[19,20] In fact, the dog was the first randomly bred species in which the presence of circulating stem cells in blood was proven by transplantation of cells exclusively obtained from peripheral blood.[21]

Calvo et al.[22] determined that marrow cellularity in young adult (1 to 2 year) beagles is highest in centrally located bones such as ribs, vertebrae, and pelvis, with values ranging from 8500 to 12,000 nucleated cells per mm^3. Cellularity is decreased in bones of peripheral body parts such as carpal, tarsal, and tail segments (150-300 cells/mm^3). Canine marrow contains fewer T lymphocytes than human marrow but substantially more than found in the mouse; generally about 10 to 15% of canine marrow cells are identified as T cells.[10]

In vitro assays for hemopoietic precursor cells including CFU-E,[23] BFU-E,[24] CFU-GM,[25] and a long-term culture system have been described.[26] These assays have been employed to determine precursor frequencies in marrow or peripheral blood cells used for transplantation, to quantitate recovery of hemopoiesis after transplantation, and to monitor effects of marrow purging, and, more recently, the efficacy of retroviral gene transfer into hemopoietic cells (see below).

Several recombinant hemopoietic growth factors, including granulocyte-colony stimulating factor (G-CSF),[27] granulocyte-macrophage colony stimulating factor (GM-CSF),[28] stem cell factor (SCF),[29] and Il1, either with cross-species reactivity or produced from the cloned canine genes, are currently being investigated. RhG-CSF, for example, if administered early after irradiation, prevents fatal marrow aplasia otherwise induced by 4 Gy of total body irradiation (TBI).[31] If delayed until day 7 after TBI, rhG-CSF is no longer effective. In dogs given 9.2 Gy of TBI and marrow grafts from DLA identical littermates, the administration of rhG-CSF at 100 μg/kg (or rcG-CSF at 10 μg/kg) for 10 days after transplantation significantly accelerates recovery of granulocytes, macrophages, and lymphocytes without affecting platelet recovery.[30] RcGM-CSF given subcutaneously induces a steep rise in blood granulocytes and monocytes and to a lesser extent eosinophils; interestingly there appears to be a concurrent decline in platelets to less than the baseline values.[28] SCF given subcutaneously induces a striking leukocytoses of 50 to 100 $\times 10^9$ WBC/l; intravenous administration may cause anaphylactic reactions (unpublished observations). Results from numerous ongoing studies have not been reported yet.

IV. IMMUNOGENETICS

The dog was the first randomly bred animal in which MHC identity of donor and recipient was shown to be of predictive value for successful bone marrow transplantation.[32] MHC as well as nonMHC (or minor) antigens are important both for host-versus-graft (HVG) and graft-versus-host (GVH) reactions (see below). While characterization of nonMHC antigens continues to be rather elusive, three international workshops have begun to standardize canine MHC antigens.[33,34] Three loci, DLA-A, B, and C, defined serologically,[34] were thought to be encoded for by class I genes; however, the existence of a class I DLA-B locus has been questioned,[35] and Doxiadis et al.[36] have presented data indicating that DLA-B antigens have, in fact, the structure of class II antigens. This notion is supported by preliminary studies by Sarmiento and Storb[37] who found molecular evidence for only one expressed class I gene. The canine class I gene has approximately an 85% homology with the human gene in regards to both DNA and predicted amino acid sequences.

In addition to the serologically defined gene products, there is at least one locus, DLA-D,[33,38] and possibly a second locus, DLA-E[39] encoding for antigens that are responsible for the proliferation of lymphocytes in mixed leukocyte culture (MLC). In light of the evidence discussed above, it is possible that DLA-B, in close linkage with DLA-D, also contributes to MLC reactivity. Recent work from this laboratory has provided evidence for at least four class II genes, composed of alpha and beta chains, that show a high degree of homology with human and bovine class II genes.[40,41] There is an excellent correlation between DLA-D type as determined by homozygous cell typing and DNA sequences as determined by restriction fragment length polymorphism (RFLP),[42] but a strict correlation between a given class II locus and MLC reactivity has, so far, not been established. The gene products described here are expressed codominantly; antigens comparable to the recessively inherited murine hemopoietic histocompatibility (Hh) antigens have not been described in the dog.

V. SELECTION OF A MODEL

Early clinical investigations indicated that much of the information obtained in murine models was difficult to transfer to man and almost all marrow transplants in man failed at that time. Subsequent work in dogs and primates led to renewed enthusiasm for the application of marrow grafting to a variety of human diseases.[43] Many seminal observations have been made in the canine model, and results from these studies have formed the basis for clinical protocols in transplantation centers worldwide. Numerous problems in clinical marrow grafting remain and need to be approached through preclinical investigations, ideally in a random-bred species such as the dog.

A. AUTOLOGOUS TRANSPLANTATION

Treatment with high-dose cytotoxic therapy can cure some patients with malignant disease, provided that following therapy marrow function can be secured. Autologous bone marrow transplantation is one means of rescuing patients treated with otherwise lethal marrow suppressive regimens. The dog has been used as a model to investigate parameters relevant for clinical application.[44]

The marrow harvesting procedure in dogs consists of administering a general anesthetic and aspirating marrow from both humeri and femora, as described below. A marrow-ablative regime such as total body irradiation (TBI) or high-dose chemotherapy is then applied. Approximately 0.5×10^8 unmanipulated marrow cells/kg are required for hemopoietic reconstitution. Cell counts are corrected for peripheral blood cell contamination. Marrow cells can be cryopreserved,[45] infused immediately after irradiation or after 24 to 72 hour following chemotherapy to allow clearing of parent substance and metabolites to nontoxic levels.

Generally, after irradiation, white blood cell counts reach a nadir at 6 to 7 days and begin to rise at about day 10. Normal levels in surviving dogs are usually attained on days 20 to 25. Platelets reach a nadir around day 9 or 10; their recovery is more gradual and normal levels are usually not reached until days 35 to 40. Reticulocyte counts generally fall to undetectable levels by day 7 and do not reappear until days 14 to 15; they usually remain elevated until day 30, at which time hemoglobin values also return to normal.

The autologous model allows for the study of the toxicity of conditioning regimens, cell dose requirements, kinetics of reconstitution, efficiency, and toxicity of cytokines and growth factors and gene transfer experiments without the potential interference of histocompatibility barriers.[1,2]

B. ALLOGENEIC TRANSPLANTATION

The allogeneic model has generally been used to study the impact of major and minor histocompatibility barriers on engraftment and graft-versus-host disease (GVHD), and to investigate immunological reconstitution and mechanisms of tolerance.[1,2] In principle, four donor/recipient combinations have been studied:

- DLA genotypically identical littermates
- DLA haploidentical littermates
- DLA incompatible unrelated dogs
- DLA phenotypically matched unrelated dogs

Some studies have also investigated the impact of nonidentity for selected (class I or II) MHC antigens.[46]

The transplant procedure is similar to that with autologous marrow; however, generally the marrow is **not** cryopreserved. Dependent upon the question under investigation the conditioning regimen and the marrow cell dose will vary. TBI at a dose of 920 cGy and a donor cell dose of 4×10^8 cells/kg are often used.[47] Frequently an arterio-venous shunt is placed in the donor at the time of marrow harvest to allow for subsequent leukapheresis and buffy coat cell transfusion to the recipient.[3]

C. SYNGENEIC TRANSPLANTATION

A syngeneic canine model is not available.

VI. BONE-MARROW HARVEST

A. DIAGNOSTIC BONE-MARROW ASPIRATION

For diagnostic marrow aspirates dogs are given an anesthetic (e.g., ketamine) i.v., and a selected site is prepared with alcohol, surgical soap or both. Generally, small samples are aspirated from the humerus. A 16-gauge, 2-in. marrow-aspiration needle is inserted into the proximal humerus just distal to the joint. A loss of resistance is felt when the marrow cavity is entered. A marrow sample is obtained by applying suction with a syringe. Some anticoagulant (e.g., heparin 1:1000 at a dose of 0.1 ml/5 ml marrow) must be used if not only smears are to be evaluated. Marrow can also be aspirated from the femur using the trochanteric fossa as the entry point, but this is more difficult.[3]

B. COLLECTION AND CRYOPRESERVATION OF BONE MARROW FOR TRANSPLANTATION

Dogs are anesthetized and prepared as for diagnostic marrow aspiration. Large volumes of bone marrow can be collected aseptically from humeri and femora using 4- to 6-in. 14- or 16-gauge needles (Figures 1a and 1b). Bone marrow is more readily collected by means of suction through polyvinyl tubing attached to the needle and originating from a 1-l Erlenmeyer flask connected to a vacuum pump (Figure 1c). The marrow is suspended in tissue culture medium containing heparin, and passed through stainless steel screens of 300- and 200-μm pore size (Figure 1d) and placed in plastic blood-administration bags. Since marrow samples are contaminated by peripheral blood, the bone-marrow cell count can be determined by subtracting from the total cell count the number of peripheral blood leukocytes expected to be contained in a volume of blood equivalent to the amount aspirated.

Bone marrow can be cryopreserved for days, months or years, as reviewed before.[48] An equal volume of a mixture of 80% TC199 medium and 20% dimethylsulfoxide (DMSO) is added to bone marrow in plastic blood-administration bags. The bags are then placed between copper plates to secure a uniform thin layer and frozen to –40°C at a rate of 1°C/min in a controlled rate freezer. Subsequently, the bags are frozen at a rate of 4°C/min to –80°C and then stored at –100°C.

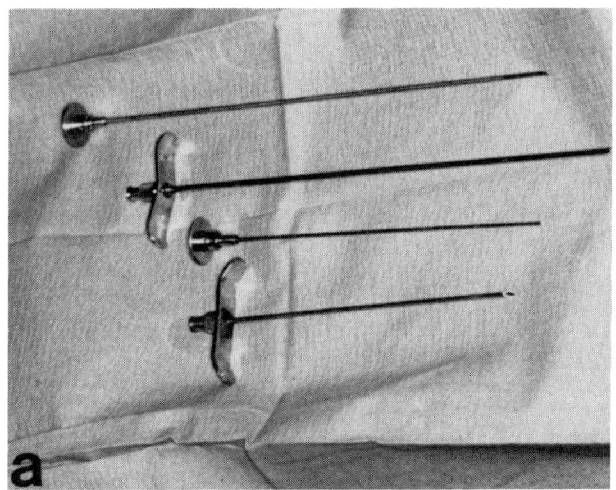

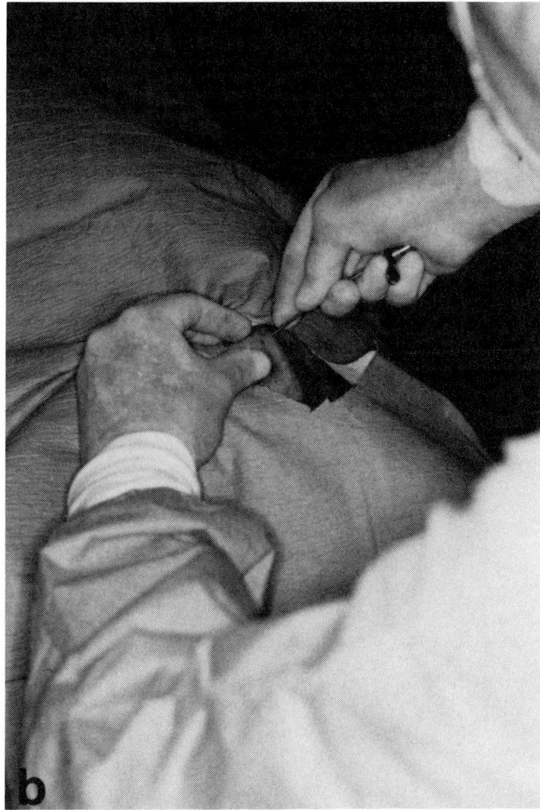

Figure 1 (a) Two different-sized needles (4-in. 16-gauge and 6-in. 14-gauge) used for aspirating bone marrow from the long bones of dogs. (b) Insertion of the needle into the trochanteric fossa of the femur, and aspiration of bone marrow. The femur is being supported by the palm of the left hand, while the greater trochanter is held with the thumb. The distal aspect of the femur projects to the left of the picture. (c) Bone marrow is being aspirated into the holding flask by means of a vacuum pump. (d) The bone marrow is placed into a syringe fitted with screens to remove clumps or bone particles. (From Ladiges, W.C., Storb, R., Graham, T., and Thomas, E.D., in *Methods of Animal Experimentation, Volume 7, Part C,* Academic Press, New York, 1989. With permission.)

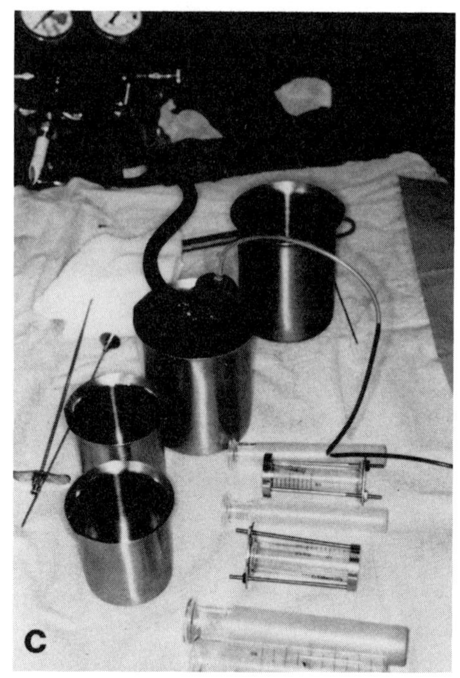

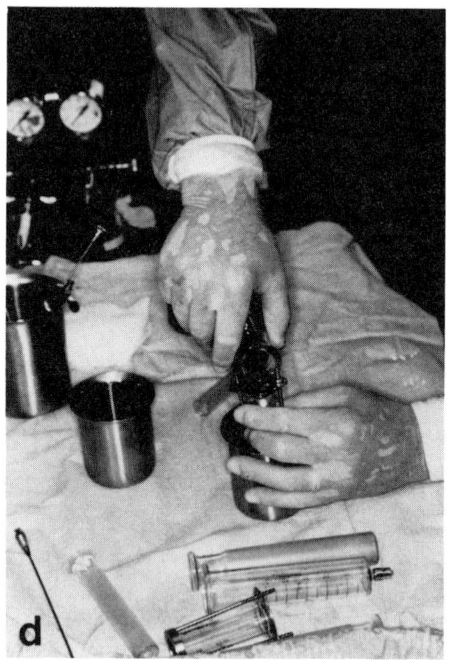

Figures 1c and 1d.

VII. COLLECTION OF PERIPHERAL BLOOD LEUKOCYTES AND STEM CELLS

A. VASCULAR ACCESS

To obtain blood samples for *in vitro* studies or to administer small amounts of medication intravenously, it is most practical to access the external jugular vein on either side of the neck or the cephalic vein on the front leg. If large amounts of blood need to be processed or if a medication or a biological is to be administered repeatedly or continuously, it is more practical to insert indwelling catheters.

1. Arterio-Venous Shunts

An arteriovenous (AV) shunt is useful for collecting large volumes of blood, e.g., in the form of leukapheresis. The shunt, most frequently between the carotid artery and jugular vein, is placed under general anesthesia.[3,49] Two incisions through the skin expose the right carotid artery and left jugular vein (or vice versa), allowing insertion of Teflon®-Silastic® tubes 2.5 mm in diameter. The tubes are tied into the vessels and the incisions closed. The external position of the shunt can be stabilized by taping it to a piece of cardboard. The dog's neck is then wrapped with gauze, protected by a collar (e.g., made of cardboard), and wrapped with tape. Alternatively, femoro-femoral AV shunts can be placed in the groin. Generally these shunts can be kept functional for 3 to 5 days. Upon completion of the experiment the arterial and venous limbs of the shunt are ligated. After 8 days the sutures holding the shunt in place are severed and the shunt is removed under general anesthesia. The procedure is well tolerated without long-term sequelae.

2. Indwelling Venous Catheter

An indwelling venous catheter is useful for continuous infusion of growth factors, immunosuppressive agents, or other substances. The catheter is inserted basically as described above for the venous branch of an AV shunt. As shown in Figure 2, the tubing is then connected to the infusion pump which is carried by the dog in a custom-fit jacket. The pump reservoir can be refilled at intervals as determined by the infusion rate.

B. LEUKAPHERESIS

Intermittent leukapheresis can be carried out by manually withdrawing blood from an AV shunt using 50-ml syringes. The dog receives an infusion of 200 to 400 ml of Ringer's solution into the vein to compensate for the 200 to 400 ml of blood subsequently removed from the artery. The blood is placed into heparinized 50-ml glass tubes and centrifuged at 1500 rpm for 15 min. The leukocyte layer (buffy coat) is removed and the plasma and red blood cells are returned to the dog through the venous limb of the shunt.[3] Leukapheresis can also be carried out using a continuous-flow blood-cell separator.[50,51] Blood flow through the separator is maintained at 40 to 60 ml/min, the bowl speed is kept at 1200 rpm, and buffy coat cells are collected at 3 ml/min. During collection, animals are anticoagulated with heparin. Dogs are maintained on the cell separator for 3 hour or until 500 ml of buffy coat is obtained. This cell preparation can be further concentrated by centrifugation. The procedure can be repeated. Peripheral blood leukocytes have been used in transfusion studies, to facilitate engraftment across histocompatibility barriers, and as a source of hemopoietic stem cells.

C. PERIPHERAL BLOOD STEM CELLS

Cells capable of lymphohemopoietic reconstitution circulate in canine peripheral blood.[52,53] More peripheral blood mononuclear cells (PBMC) than marrow cells are needed to rescue

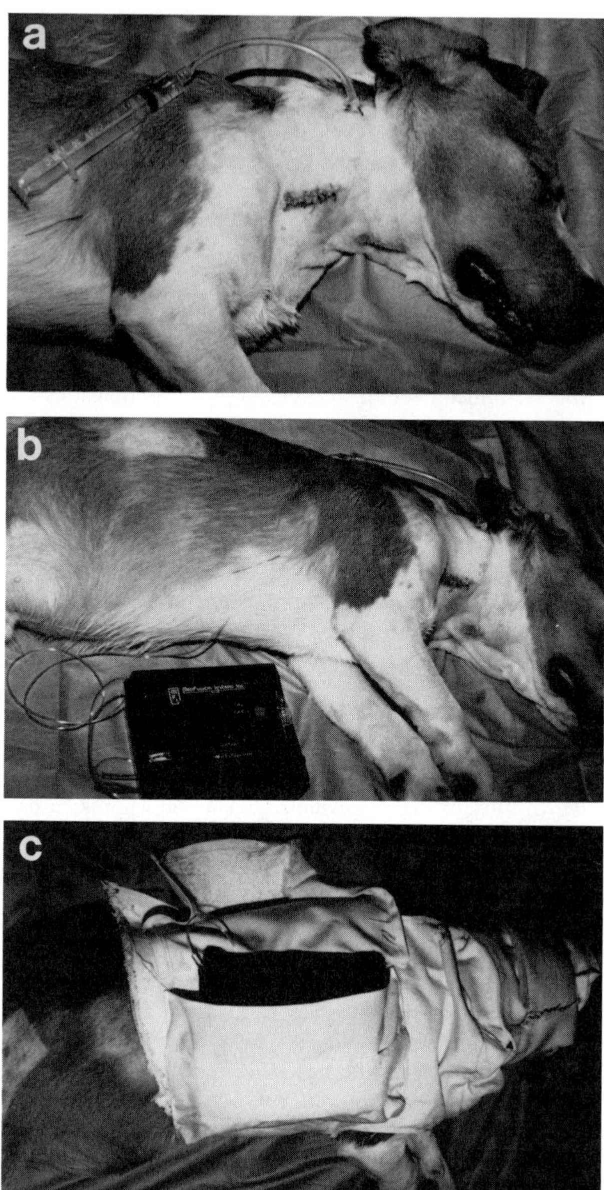

Figure 2 Indwelling catheter inserted into jugular vein, connected to continuous infusion pump which is carried by the dog in a jacket. (a) Catheter has been inserted into the vein and brought out through a separate skin incision at the lateral neck; (b) catheter connected to the infusion pump; (c) the dog has been equipped with a jacket, closed around the neck and the chest, and the infusion pump placed in a pocket of the jacket.

a dog after marrow ablative TBI: 5 to 10×10^8 PBMC/kg are necessary to rescue 100% of dogs given 9.2 Gy TBI, as compared to only 0.5×10^8 marrow cells.[51]

The procedure is the same as described above for leukapheresis. Peripheral blood stem cells can be cryopreserved the same way as marrow cells. Appelbaum et al.[54] have transplanted peripheral blood stem cells in dogs with spontaneous lymphoma using $10.43 \pm 9.75 \times 10^8$ mononuclear cells/kg collected in three leukapheresis on three consecutive days. All evaluable dogs had hemopoietic reconstitution.

Gerhartz et al.[20] performed allogeneic transplants of canine peripheral blood stem cells further purified by discontinuous albumin gradients. Dogs received an average of 20.9×10^6 mononuclear cells/kg, and hemopoiesis was restored in 9 of 12 dogs; tolerance developed in recipients of allogeneic transplants.

D. FETAL LIVER CELLS

The liver is an active site of hemopoiesis during fetal development and a rich source of hemopoietic precursor cells.[15,16,55] Since the fetal immune system is immature and the risk of GVHD low, fetal liver cells may offer an attractive alternative to marrow cells.

Fetuses are obtained via hysterotomy at about 50 days of gestation, and exsanguinated from the brachial or carotid artery. The liver is removed surgically. A single-cell suspension is prepared by irrigating the liver with tissue-culture medium while gently massaging it at the same time. The suspension is then passed though wire mesh screens or cotton gauze to remove debris. Viability and numbers of colony forming units are determined as with other cell populations. Recipient dogs are irradiated and cells are given i.v.

One problem with fetal liver cells has been failure of engraftment. Engraftment has generally been achieved only with cells obtained from selectively bred DLA homozygous donors and after high dose (e.g., 2×8 Gy) TBI.[55] Under those conditions hemopoietic reconstitution has been satisfactory, although immunoreconstitution may have occurred more slowly than with marrow cells. Interestingly, GVHD has not been a problem.

VIII. MODIFICATION OF HEMOPOIETIC STEM CELLS

A. MANIPULATION OF MARROW *IN VITRO*

Several methods have been applied to fractionate canine marrow cells such that hemopoietic stem cells were retained, while T lymphocytes with the potential to trigger GVHD were eliminated. Discontinuous albumin gradients have been used to successfully enrich, in fraction 3, hemopoietic stem cells capable of allogeneic marrow reconstitution with a reduced incidence of GVHD.[20]

Kolb et al. prepared a rabbit anti-dog anti-thymocyte serum (ATS) for cytolytic T cell elimination from the marrow *in vitro*.[56] With this approach haploidentical marrow was transplanted successfully, and half the dogs survived without GVHD. Mabs have also been used for marrow treatment; success has been limited in as far as T-cell depleted marrow failed to engraft consistently (unpublished observations).

Chemicals such as 4 hydroperoxycyclophosphamide have been used in an attempt to purge canine marrow of lymphoma cells.[57]

B. RETROVIRAL GENE TRANSFER

Retroviral vectors have been used to transfer genes into several canine cell populations including hemopoietic precursor cells.[58-60] For example, amphotropic helper virus-free vectors containing the neomycin phosphotransferase (neo) gene, a mutated dihydrofolate reductase (DHFR) gene, or a human adenosine deaminase (ADA) gene have been used. The approach has usually involved cocultivation of virus-producing cell lines with hemopoietic cells,[58] and CFU-GM have been tested to determine the fraction of cells carrying the reporter gene (e.g., resistance to G418 or methotrexate [MTX]). The efficiency of transfection has been improved by coculturing marrow and virus-producing cell lines followed by maintenance in long-term culture and feeding with virus containing supernatant. The average rate of gene expression was about 50%.[59]

In some experiments transfected hemopoietic cells were transplanted into the autologous recipient and gene expression, albeit at low levels, has been observed for more than 1 year (Schüening, F., unpublished observations).

IX. CONDITIONING FOR MARROW TRANSPLANTATION

A. CHEMOTHERAPY

1. Cyclophosphamide (Cy)

Cy is a potent inhibitor of the immune system in man and animals. Storb et al.[61] have shown that at a dose of 100 mg/kg administered as a single 1-hour i.v. infusion Cy is lethal to dogs; dogs can be rescued by autologous or allogeneic marrow transplantation. Allogeneic recipients generally are mixed chimeras. Aside from marrow suppression and pancytopenia, acute side effects include anorexia, vomiting, diarrhea, and hematuria as early as day 2 and lasting 4 to 7 days. In contrast to other models, postgrafting Cy at doses of 5 to 10 mg/kg i.v. does not prevent or delay lethal GVHD in dogs given marrow grafts from histoincompatible unrelated donors.

2. Dimethyl Busulfan (DMB)

In rodents, busulfan has profound marrow toxicity which can be overcome by infusion of syngeneic marrow.[62] In dogs, a similar compound, DMB, 3 to 5 mg/kg causes pancytopenia lasting more than 40 days.[63] Dogs given 7.5 mg/kg develop fatal marrow aplasia within 17 days; however, with autologous marrow infusion they survive with complete hemopoietic recovery. DMB is only moderately immunosuppressive at these doses as determined by antibody formation to sheep red blood cells and bacteriophage ϕX174, and by the development of lymphocytotoxic antibodies after transfusion of allogeneic blood products. The level of immunosuppression is not sufficient for successful engraftment of DLA incompatible bone marrow, although engraftment can be achieved with the addition of ATS. In DLA identical littermate transplants DMB at a dose of 10 mg/kg allowed for engraftment in half of the recipients; consistent engraftment was achieved when ATS or a combination of ATS and procarbazine was added.[64]

3. Procarbazine

As indicated above, procarbazine in combination with ATS and DMB has proven effective in securing engraftment of DLA-identical marrow. It is of note that a similar regimen using procarbazine and ATS has been used with success in patients transplanted for nonmalignant acquired or congenital disorders.[65,66]

B. TOTAL BODY IRRADIATION (TBI)

Radiation has been used extensively in preparation for bone marrow transplantation. Numerous TBI regimens have been tested. Thomas et al.[67] described a technique using single-dose exposure from two mobile opposing cobalt-60 (^{60}Co) sources. Dogs are placed unanesthetized in plastic transport cages midway between the ^{60}Co sources. The exposure rate can be modified by changing the distance between the sources; rates from less than 1 to more than 20 cGy/min have been investigated. The total in vivo exposure is monitored by means of lithium fluoride radioluminescence dosimeters. In a 10-kg beagle the midplane tissue exposure corresponds to approximately 75% of the midline exposure in air measured in Roentgen.[48,68]

A TBI dose of approximately 300 cGy without subsequent marrow infusion is lethal to about 70% of dogs (LD_{70}), and the LD_{99} is about 400 cGy[69] (see Section III). These doses are insufficient, however, for allogeneic engraftment. Allogeneic grafts from DLA identical littermates require at least 500 to 600 cGy, and histoincompatible grafts from littermate or unrelated donors require doses of 1500 to 1800 cGy for sustained engraft-

Figure 3 Dose/response curves for acute radiation morality (day 7) in dogs given single dose or fractionated TBI followed by autologous marrow infusion. Curves correspond to logistic regression model fit assuming common slope parameter for all four dose rates. Different symbols indicate different exposure rates (= 2.1; = 5; = 10; = 20 cGy/min). For single dose TBI the calculated LD 50/7 values (95% confidence intervals) for 2.1, 5, 10, and 20 cGy/min were 1692 (1540, 1846), 1499 (1362, 1636), 1261 (1143, 1388), and 1056 (939, 1174) cGy, respectively. The corresponding values for fractionated TBI at 2.1, 5, 10, and 20 cGy/min were 1628 (1423, 1834), 1470 (1265, 1675), 1184 (1008, 1361), and 1320 (1136, 1505) cGy, respectively. (From Deeg, H.J., Storb, R., Longton, G., Graham, T.C., Shulman, H.M., Appelbaum, F., and Thomas, E.D., *Int. J. Radiat. Oncol. Biol. Phys.*, 15, 647, 1988. With permission.)

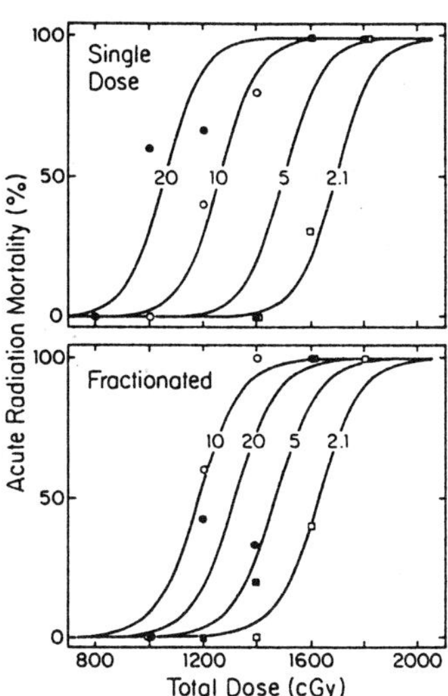

ment if not other graft-facilitating means are additionally applied.[69,70] These doses are tolerated only if administered in smaller increments (see below). The current standard regimen in our laboratory consists of 920 cGy delivered at a rate of 7 cGy/min. This dose assures sustained engraftment of DLA-identical littermate marrow; for sustained engraftment of histoincompatible marrow additional manipulations such as the infusion of donor buffy coat cells are needed.[71]

We have shown[48,72] that higher doses of TBI can be given if irradiation is fractionated into smaller increments. However, the analysis of these data is complex since toxicity, efficacy, and maximum tolerated dose depend upon several parameters including exposure rate, increment size, fractionation interval and others. At 48 hour intervals, 1800 cGy given in 3 fraction of 600 cGy and delivered at 2.1 cGy/min were well tolerated and sufficiently immunosuppressive to allow for sustained engraftment of DLA-incompatible marrow from unrelated donors.[70] In a study using increments of 200 cGy given at 6-hour intervals at exposure rates of 2.1, 5, 10, or 20 cGy/min up to 2400 cGy could be given at the lowest rate; an advantage of fractionation in regards to acute toxicity was noted only at the highest rate of 20 cGy/min (Figure 3). However, at total doses of 1000 cGy or more, fractionation provided a significant advantage for long-term survival at all exposure rates.[72] Additional fractionation regimes have been described by Kolb et al.[73]

Recent experiments by Storb et al. show that the immunosuppressive efficacy of fractionated TBI may differ from that of single-dose TBI;[74] in dogs prepared with single-dose TBI and given a marrow graft from a DLA-identical littermate no autologous recovery was observed as doses greater than 600 cGy and sustained engraftment was achieved in 95% of dogs at 920 cGy. In contrast, with fractionated TBI (increments of 150 to 200 cGy) both graft rejection and autologous recovery were observed even after 920 cGy.

C. POLYCLONAL AND MONOCLONAL ANTIBODIES

1. Antithymocyte Serum

Methods for the preparation of anti-dog ATS by means of immunization of rabbits with puppy thymocytes have been described by several investigators.[56,75] The immunosuppressive properties of ATS have generally been assessed by its ability to prolong skin graft survival. ATS has been found to be useful for the treatment of established GVHD but not as a prophylactic agent. These studies in dogs provided the basis for the use of anti-human ATS for the treatment of GVHD in clinical marrow transplantation.

ATS has also been incorporated into conditioning regimens, particularly in combination with procarbazine or DMB (see above). In these combinations it provides powerful immunosuppression; its efficacy with DLA genotypically identical but not with nonidentical transplants has been interpreted as being in support of the notion that with DLA nonidentical transplants recipient cells other than T lymphocytes mediate resistance to engraftment.

2. Monoclonal Antibodies

A panel of mabs has been developed by this and by other laboratories (reviewed in Reference 8; see Table 1). Mabs are being used as part of the conditioning regimen, for bone marrow purging, for postgrafting immunosuppression, and other purposes.

One protocol involves the use of a mab (7.2) specific for nonpolymorphic class II determinants in addition to TBI to overcome DLA barriers and facilitate engraftment.[76] Recipients are prepared by 920 cGy single-dose TBI, and mab 7.2 is given at daily doses of 0.2 mg/kg i.v. in the pre- (starting day –5) or peritransplant period (day –1 to +3). Approximately 50% of dogs given DLA nonidentical unrelated marrow engraft. Increasing the mab dose to 0.4 mg/kg does not increase the percentage of dogs which engraft, nor does the addition of a second mab, HB10a, given at the same dose and schedule. Interestingly the addition of i.v. methotrexate at a dose of 0.4 mg/kg on days 1, 3, 6, and 11 posttransplant increases the fraction of dogs with sustained engraftment to about 70%. More recently a mab, S5, directed at the CD44 antigen, when given at appropriate doses pretransplant was shown to allow for engraftment in about 75% of dogs as compared to 5 to 10% in dogs given an irrelevant mab or no mab at all.[77,78] The mechanism of action of these mabs is not clear; conceivably mab 7.2 inhibits recipient cytotoxic cells, while mab S5 may affect cell homing.

Additional studies have been carried out with mabs to which radioisotopes such as ^{131}I are conjugated.[79,80] This approach allows to readily define the biodistribution of antigens recognized by a given mab, and provides a means of delivering high doses of in situ irradiation without systemic toxicity. For example, with an ^{131}I/S5 conjugate a dose of 5 µg/kg resulted in lethal marrow suppression; however, dogs so treated could be rescued by subsequent bone marrow infusion. Hence it appears possible to use this approach for pretransplant conditioning.[81]

For biodistribution studies, dogs are anesthetized and given an injection of the radiolabeled mab. While under general anesthesia, the animals are scanned with a gamma scintillation camera for 3 hour or longer. The animals are scanned again at 24 hour and at 48 hour. At an appropriate time point animals are sacrificed and individual organs are biopsied, weighed, and counted for the amount of isotope within each separate organ. In addition, samples are snap-frozen for immunoperoxidase staining.

X. GVHD PROPHYLAXIS

A. CHEMOPROPHYLAXIS

Based on the observation by Uphof in the early 1950s that aminopterin was capable of preventing or ameliorating secondary disease in mice,[82] E. D. Thomas and colleagues in Cooperstown, and subsequently Storb et al. in Seattle investigated the usefulness of MTX to prevent GVHD in dogs. These investigators first reported that MTX had a beneficial effect both on engraftment and suppression of GVHD in dogs given marrow grafts from random donors.[83] Subsequently, systematic studies were carried out between DLA-typed recipients and donors.[84,85] Among dogs given transplants from DLA identical littermates and MTX on days 1, 3, 6, 11, and then weekly until day 102 after transplantation, none developed GVHD, and more than 90% became long-term survivors.[86] In contrast, among

DLA identical transplant recipients not given postgrafting immunosuppression more than half developed GVHD and only approximately 40% became long-term survivors.[87]

Results with DLA nonidentical transplants are summarized in Table 2. Without postgrafting administration of MTX, the median survival of dogs given marrow grafts from DLA nonidentical unrelated donors is approximately 10 days, and all die with GVHD. With the intermittent administration of MTX over a period of 102 days, median survival increases to 3 weeks, and 10 to 15% become healthy long-term survivors. These results are not substantially improved by other single agents such as 6-mercaptopurine, azathioprine, procarbazine, or cytosine arabinoside. Cyclosporine (CSP) on the other hand, although quite effective in preventing GVHD, interferes with engraftment and does not improve overall survival.[88]

It was of interest, therefore, to test combinations of immunosuppressive drugs.[84,85,89,90] Results are summarized in Table 3. Although median survival was prolonged as compared to single agents, there was a decrease in the overall likelihood of long-term survival usually due to infection. The only exception so far has been a combination of MTX and CSP: median survival is improved to more than 4 months, and 35% of dogs become long-term survivors. This combination is also effective in preventing GVHD in the majority of dogs transplanted with marrow from a haploidentical littermate, and 75% of these chimeras become healthy long-term survivors.[90] Similar regimens are now used clinically.[91,92]

B. T-CELL DEPLETION

T-cell depletion of transplanted lymphohemopoietic cells was first shown in murine models to be an effective means of GVHD prophylaxis. Based on those observations Kolb et al. investigated the efficacy of bone-marrow treatment with multiply absorbed antithymocyte globulin (ATG) and complement for GVHD prevention in a canine model.[56] DLA heterozygous dogs were given lethal doses of TBI and marrow grafts from DLA homozygous, i.e., haploidentical littermates. Among dogs with engraftment, 50% became healthy survivors.

There is still a paucity of mabs against canine cell subpopulations (Table 1). No studies on the use of those antibodies for GVHD prophylaxis have been reported. In our model of DLA incompatible marrow transplantation from unrelated donors we gave donor marrow depleted of various lymphocyte populations by treatment with mabs such as DLy6 or DT2 and complement after recipient conditioning with 3 × 600 cGy of TBI. Almost uniformly dogs failed to achieve sustained engraftment in contrast to consistent engraftment in dogs given unmanipulated marrow[70] (unpublished observations). Instead of using mabs, investigators have used physical methods of cell fractionation from peripheral blood or bone marrow. Gerhartz et al.[20,93] used density gradients to obtain cell fractions largely depleted of lymphocytes and enriched for hemopoietic precursors (e.g., CFU-C). Dogs transplanted with those cells generally did not develop GVHD.[93] A similar approach has been taken by Vriesendorp et al.,[71] who observed a quantitative relationship between the number of lymphocytes transplanted and the probability of developing GVHD.[94] In their model the removal or omission of lymphocytes was associated with an increased risk of engraftment failure. This was overcome by the use of increased doses of (fractionated) TBI.[70,71,95] Some dogs in those experiments died from a poorly understood wasting syndrome, possibly related to incomplete immunoreconstitution.

XI. POSTGRAFTING CARE AND EVALUATION

Preparative regimens necessary for immunosuppression and marrow ablation are sufficiently toxic to nonlymphohemopoietic organs to require intensive postgrafting support.

Table 2 **GVHD and Survival in Dogs Given 920 cGy of TBI and Hemopoietic Grafts from DLA Nonidentical Unrelated Donors. Single Agent GVHD Prophylaxis**

GVHD Prophylaxis post grafting	Number of dogs: Studied	Number of dogs: With sustained engraftment	Number of dogs: With GVHD	Survival (days): Range	Survival (days): Median	No. of dogs surviving >100 days
None	10	10	10	9—15	10	—
MTX[a]	40	36	25	11—>100	21	7
6-MP[b]	8	7	7	9—22	14.5	—
Azathioprine[c]	2	2	2	19, 21	20	—
Procarbazine[d]	6	6	5	7—11	9.5	—
Ara C[e]	6	6	5	9—13	10	—
CSP[f] (early)	13	7	3	5—120	17	1
CSP[g] (late)	5	3	—	9—51	10	—

Note: The day of TBI and marrow infusion is designated day '0'. All drugs except cyclosporine (CSP), were given i.v.
[a] Methotrexate, 0.4 to 0.5 mg/kg on days 1, 3, 6, 11 and once weekly until day 102; [b] 6-Mercaptopurine, 5 mg/kg on days 1, 3, and 6; 7.5 mg/kg on day 11, and 10 mg/kg once weekly thereafter; [c] Azathioprine, 1 mg/kg/day starting 24 hours after marrow infusion; [d] Procarbazine, 20 mg/kg on days 1, 3, 6, and 11; [e] Cytosine arabinoside, 2.5 mg/kg twice daily on days 1, 3, 6, and 11; [f] CSP was given at doses of 20 to 25 mg/kg/day, i.m. on days 0 to 7, and p.o. on days 8 to 25; [g] CSP, 15 mg/kg/day (i.m. on days 0 to 7, p.o. subsequently) on days 5 to 25, and at gradually reduced doses on days 26 to 100.

Table 3 **GVHD and Survival in Dogs Given 920 cGy of TBI and Hemopoietic Grafts from DLA-Nonidentical Unrelated Donors. Combination Regimens of GVHD Prophylaxis**

GVHD Prophylaxis post grafting	Number of dogs			Survival (days)		No. of dogs surviving >100 days
	Studied	With sustained engraftment	With GVHD	Range	Median	
MTX[a] + CY[b]	8	7	7	17—38	19.5	—
MTX[a] + 6-MP[c]	8	8	6	18—>100	36.5	1
Azathioprine[d] + CSP[e]	10	5	4	15—>100	22	2
MTX[a] + 6-MP (higher dose)[f] + CY[b] + Prednisone[g]	8	6	0	23—>100	40	1
MTX[a] + 6-MP (lower dose)[h] + CY[b] + Prednisone[g]	8	7	6	16—87	23	—
MTX[i] + Procarbazine[j]	9	7	6	19—>100	27	1
MTX[i] + Procarbazine[j] + ATS[k] (higher dose)	9	8	7	17—49	32	—
MTX[i] + Procarbazine[l] + ATS[m] (lower dose)	8	7	7	21—67	29	—
MTX[n] + CSP (short)[o]	17	17	4	6—105	18	1
MTX[n] + CSP (11-100)[p]	6	5	4	13—51	26	—
MTX[n] + CSP (0-100)[p]	10	10	5	11—>100	122	6

Note: The day of TBI and marrow infusion is designated day '0'. All drugs were given i.v. except cyclosporine (CSP) which was given i.m. on days 0 to 7 and then switched to p.o. ATS (antithymocyte serum) was given s.c.; [a] MTX (methotrexate), 0.2 mg/kg on days 1, 3, 6, 11, and once weekly until day 102; [b] CY (cyclophosphamide), 2 mg/kg on days 4, 6, 8, 10, 12, 15, 18, 21, 24, 27, and 30; [c] 6-MP (6-Mercaptopurine) was given 2 mg/kg/day on days 1 to 6, 1 mg/kg/day on days 7 to 14, and 5 mg/kg once weekly until day 98; [d] Azathioprine, 1 mg/kg/day on days 1 to 11; [e] Cyclosporine, 7.5 mg/kg i.m. twice daily on days 0 to 7, 7.5 mg/kg p.o. twice daily on days 8 to 25, 10 mg/kg/day p.o. days 26 to 50, 5 mg/kg/day p.o. days 51 to 75, and 5 mg/kg every other day p.o. days 76 to 100. Details of this cyclosporine regimen have been reported previously;[9,10] [f] 6-MP, 1mg/kg/day on days 1 to 20, and 0.5 mg/kd/day on days 21-30; [g] Prednisone, 2 mg/kg/day in divided doses on days 0 to 4; [h] 6-MP, 1 mg/kg/day on days 1 to 14, and 1 mg/kg every other day until day 30; [i] MTX, 0.4 mg/kg on days 1, 3, 6, 11, and then weekly until day 102; [j] Procarbazine, 10 mg/kg every other day from day 10 to 30; [k] ATS, 7 mg IgG/kg s.c. from day 11 to 31; [l] Procarbazine, 7.5 mg/kg every other day from day 10 to 30; [m] ATS, 5.25 mg IgG/kg s.c. from day 11 to 31; [n] MTX, 0.4 mg/kg on days 1, 3, 6 and 11; [o] CSP, 25 to 30 mg/kg/day on days 5 to 25 or 11 to 25 (i.m. until day 7, then p.o.); [p] CSP, 15 mg/kg/day starting on day 11 or day 0, and continued at gradually reduced doses until day 100.

Postgrafting supportive care involves administration of fluids such as Ringer's solution during periods of vomiting and diarrhea, antibiotics such as ampicillin and gentamicin during periods of granulocytopenia, and transfusion of whole blood or platelet preparations (irradiated with 15 Gy) when platelet counts are below 10×10^9/l.

We have found it useful to give, additionally, nonabsorbable oral antibiotics usually polymyxin B and neomycin, starting 1 week before transplantation and continued until 1×10^9 granulocytes/l are reached. Established infections, especially pneumonia, are difficult to treat in the dog, and the emphasis must be on prophylaxis.

The most relevant question after transplantation usually is that of marrow engraftment and hemopoietic reconstitution. Marrow engraftment can be assessed by promptly rising granulocyte and platelet counts after the postirradiation or chemotherapy induced decline, the clinical development of GVHD, the presence of genetic markers of donor origin in marrow and peripheral blood cells including karyotype (sex chromosomes), isoenzyme patterns, and the histologic features of the marrow at autopsy. Long-term observations have focused on the development of secondary, probably radiation-induced malignancies.[96]

XII. IMPORTANCE OF DLA AND NON-DLA ANTIGENS IN GRAFT-HOST INTERACTIONS

Available data are insufficient to determine whether class I or class II histocompatibility antigens are more relevant for the development of GVHD in dogs. It is certain, however, that antigens other than those determined within MHC, participate in graft-host interactions.[86,92] For example, among dogs given 9.2 Gy of TBI and grafts of unmanipulated marrow from DLA identical littermates about 50% develop acute GVHD and die within 5 months of transplantation;[64] the remaining dogs do not develop GVHD and become healthy long-term survivors. Dogs given marrow grafts from either DLA nonidentical littermates or unrelated donors develop acute GVHD more rapidly than dogs given grafts from DLA identical littermates, and mortality is virtually 100% if no postgrafting immunosuppression is administered.[69,97,98] Dogs transplanted from phenotypically DLA-matched donors also fare less well than those given transplants from DLA genotypically identical littermates; even with the administration of MTX after transplant only one third of dogs given unrelated transplants will survive compared to 90% of dogs given grafts from DLA identical littermates.[87,98] Similarly, dogs prepared with 9.2 Gy of TBI and transplanted with marrow from a DLA-identical littermate very consistently achieve sustained engraftment. In contrast, dogs given marrow from DLA-incompatible donors or from donors phenotypically matched for DLA antigens but presumably differing for non-DLA antigens have a probability of achieving sustained engraftment of only 90 and 50%, respectively.[47,64,72,86,94,95,99] NonMHC antigens are generally not detected by cell mediated lympholysis or MLC,[99] and have not been characterized by serologic means.

As reviewed elsewhere, non-DLA antigens play a major role in transfusion-induced sensitization of transplant recipients, leading to rejection of marrow grafts from DLA identical littermates.[1,2] Sensitization appears to be mediated by antigen-presenting cells in the transfusion product and can be prevented by UV or gamma irradiation.[138,139]

XIII. CLINICAL SPECTRUM OF GVHD

Both acute and chronic forms of GVHD have been described in dogs.[63,102,103] Incidence and kinetics depend upon the extent of histoincompatibility between donor and recipient, and the efficacy of GVHD prophylaxis regimens applied after transplantation. Without GVHD prophylaxis usually only acute GVHD is observed. However, with the administration of immunosuppressive drugs after transplantation, some dogs develop chronic GVHD.[100]

A. ACUTE GVHD

Acute GVHD generally manifests itself as an erythematous skin rash, prominent in ears, axillae, and abdominal skin.[101,102] This can progress to skin ulcerations and denudation often complicated by bacterial superinfections. Histological features include atrophy with hyperkeratosis, marked basilar inflammation with necrotic keratinocytes and eosinophilic body formation, and overall basal layer irregularity. An inflammation of pilar units is less frequent. There is often glandular inflammation rather typically involving the sweat glands in the foot pads. In addition, there is often injection of the conjunctivae and inflammation of the oral mucosa. Frequently, dogs have bloody diarrhea. Histologically, there is crypt cell degeneration with dilatation and abscess formation, often more prominent in the small bowel than in the colon. There is a good correlation between clinical manifestation and histological involvement.

Many dogs develop an icterus and show elevations of serum transaminases and alkaline phosphatase. Histological examination of the liver reveals triaditis, injured bile ducts, evidence of cholestasis and hepatocyte degeneration. Often the central vein is surrounded by lymphoplasmacytic infiltrates.[102] There may also be involvement of the pancreas with mild lymphocytic exocytosis and apoptosis of ductal epithelium.[101] Lymphoid organs usually show moderate to marked depletion of the germinal cords. However, this may at times also be observed in dogs given autologous grafts. Occasionally, a dog can die suddenly without having developed the characteristic clinical syndrome, although autopsy changes are those of GVHD. The most frequent cause of death is septicemia or pneumonitis; an occasional dog may die from hepatic failure.

With DLA incompatible transplants and without immunosuppression, clinical signs of GVHD may develop as early as day 5 or 6 after transplantation. With minor antigen discrepant transplants and with the administration of postgrafting immunosuppression, signs of GVHD may not develop until 2 to 5 months posttransplant.

B. CHRONIC GVHD

Chronic GVHD in dogs presents with patchy hair loss, skin ulcerations, at times ascites and associated gram-positive infections.[100] These changes may develop 100 to 200 days after transplantation.[103,104] Atrophy of the skin and hyperkeratosis are more marked than with acute GVHD and may represent an isolated finding. There may also be marked basilar inflammation, necrotic keratinocytes and basal layer irregularity, along with glandular inflammation. Eventually, dermatofibrosis will develop.

The liver shows a decrease in bile ducts, and proliferation of small bile ductules. There can be marked triaditis with plasmacytic, lymphocytic, or plasmalymphocytic infiltration. In contrast to acute GVHD, there is piecemeal necrosis and bridging with portal fibrosis. Dogs with other signs of chronic GVHD generally also have histological evidence of GVHD in the liver. Similar to findings in man, there may be associated evidence of veno-occlusive disease.[105] Generally, the gut shows only minimal histological changes.

XIV. MECHANISMS OF TOLERANCE

A. BLOCKING FACTORS

Early experiments in dogs conditioned with TBI, and given DLA incompatible hemopoietic grafts, suggested a role for blocking factors identified in the serum of tolerant chimeras.[106] These results, however, could not be substantiated in subsequent experiments.[107]

B. CLONAL ABORTION AND SUPPRESSOR CELLS

In more recent investigations in DLA incompatible littermates given MTX postgrafting, lymphocytes from chimeras without active GVHD were found to be unresponsive or to show only minimal reactivity in MLC against cryopreserved host lymphocytes, but responded well to lymphocytes from third-party donors.[108] In dogs with GVHD, however, lymphocytes showed strong MLC reactivity against host cells. These observations are consistent with clonal deletion or abortion as a mechanism of tolerance. However, GVHD could not be induced by the infusion of large numbers of viable peripheral blood lymphocytes from the marrow donor,[109] an observation arguing against clonal deletion and more compatible with a suppressor mechanism.

Substantial differences were observed dependent upon donor/recipient combination (littermate vs. unrelated donors) and postgrafting immunosuppression (MTX vs. MTX + CSP).

1. Haploidentical Littermate Chimeras

Among 20 MTX treated chimeras ten developed acute GVHD, and chimera lymphocytes consistently showed proliferative responses to stimulation by host cells in MLC.[108] In six dogs, the onset of GVHD was delayed beyond 2 months, and in four of these chimera, lymphocyte responses to host cells were negative postgrafting. However, proliferative responses recurred coincidentally with the development of GVHD and persisted until the animals' deaths. Four dogs without GVHD had nonreactive MLC, although borderline stimulation index (SI) values of 3.5 to 5 were observed. Unresponsiveness in MLC was specific for host DLA-D alleles. Mixing experiments between chimera and donor cells failed to reveal evidence for circulating suppressor cells.[108,110] In CML assays, chimera cells failed to generate cytotoxic effector cells against host targets or unrelated cells homozygous for one of the host DLA haplotypes. Responses against third party targets were comparable to those of donor cells.[108] Chimera serum had no effect on either MLC or CML responses. These results were consistent with the hypothesis of clonal deletion or abortion being mediated by a central mechanism.

In dogs treated with MTX and CSP after transplantation, lymphoproliferative responses also correlated with the presence of GVHD,[103] and MLC unresponsiveness was specific for the host DLA-D alleles.[110] In contrast to MTX treated dogs, the addition of chimera cells to MLCs early postgrafting resulted in suppression of proliferation of donor cells against host and third party stimulators, both in dogs with and without GVHD;[111] after 3 to 4 months, however, cells from chimeras without GVHD showed specific suppression. Suppressor cells were nylon-wool adherent, showed the suppressor/cytotoxic T-cell phenotype, and were generally radiation resistant (2000 cGy). The disappearance of suppressor cells later after transplant is consistent with the concept of clonal reduction as described in other models.[112]

In CML assays, chimera lymphocytes, regardless of the presence of GVHD, failed to generate cytotoxic effector cells against host targets or unrelated dogs homozygous for one of the host DLA haplotypes. However, chimera cells showed cytotoxicity against third-party targets. Regardless of GVHD, marrow-donor skin grafts were permanently accepted, whereas third-party skin grafts were rejected.[110]

Thus a large proportion of dogs given DLA haploidentical marrow grafts can develop host-specific tolerance. Although it is likely that suppressor cells do not represent the only mechanism, it is conceivable that suppressor cells abort host reactive donor cells *in vivo* which would prevent their recognition *in vitro*. As the number of host reactive cells declines with time after grafting, a reduction of suppressor cells follows, and tolerance is achieved by establishing a balance of aggressor and suppressor cells. Reexposure to host lymphoid cells, as shown in a rat model, leads to reexpansion of the suppressor clone.[113] Such a concept also accommodates earlier canine data: the infusion of donor lymphocytes (see above) may trigger expansion of suppressor cells (requiring a finite length of time) and readjustment of the aggressor/suppressor balance. If specifically **sensitized** donor cells are infused, kinetics are altered and the pace of alloaggression is out of proportion to the rate of suppressor cell expansion, thus leading to the development of GVHD.[109]

2. Completely Allogeneic Chimeras

Among dogs given MTX + CSP postgrafting that developed GVHD,[89] proliferative responses of chimera lymphocytes to host stimulator cells showed virtually no increase.[110] The addition of chimera cells, in particular nylon-wool adherent cells, suppressed proliferative responses of donor cells against host cells to 10 to 25% of baseline values. However, in contrast to results in haploidentical littermates, suppression remained nonspecific throughout the entire period of observation, i.e., chimera cells suppressed donor-cell responses to any stimulator cells. There was no detectable difference between dogs with and without GVHD.

Regardless of GVHD, lymphocytes of these dogs did not generate cytotoxic effector cells against host targets[110] nor against DLA nonidentical third party target cells. Nevertheless, skin grafts from the respective marrow donors survived indefinitely, while grafts from DLA incompatible third party donors were uniformly rejected, albeit with some delay when compared to haploidentical chimeras. Thus, the mechanism of tolerance or GVHD appears to be more complex in completely allogeneic chimeras.[114] As in haploidentical littermates there was evidence for clonal abortion. Specific suppressor cells, however, were not detected; rather, suppression was nonspecific. For several years after transplantation chimeras remained unable to generate cytotoxic effector cells even against completely allogeneic third-party target cells.

The thymuses of completely allogeneic chimeras even in the absence of clinical GVHD, remained severely lymphocyte-depleted for years after transplantation.[110] On the other hand, the thymuses of DLA haploidentical chimeras that did not develop GVHD showed good reconstitution, albeit markedly delayed as compared to DLA identical or autologous transplants. These observations suggest a role for a restrictive genetic element in the development of GVHD and immunocompetence.

C. STUDIES ON TRANSFER OF TOLERANCE

A series of experiments has been carried out in dogs given marrow grafts from DLA-identical littermates. The basic approach was to transplant (1 to 2 months after hemopoietic grafts) full thickness skin grafts from the marrow chimeras to the original donor who was also injected with various populations of chimera cells.[114,115] The scheme is shown in Figure 4. The most potent effect on skin graft survival was provided by chimera peripheral blood leukocytes, given either along with the marrow graft or at the time of skin graft placement. In several instances the grafts survived permanently. Bone marrow by itself was ineffective, as was the addition of chimeric spleen cells.

The fact that transplantation of chimera marrow by itself did not allow for significant prolongation of skin graft survival argues against clonal deletion of alloreactive cells in DLA identical chimeras. However, the results provide strong support for the hypothesis that a population of circulating suppressor cells is involved in maintaining tolerance in DLA identical (minor antigen different) chimeras.[114]

XV. MARROW TRANSPLANTATION FOR THE TREATMENT OF CANINE DISEASES

Along with many experiments related to transplantation biology and immunology, several studies have been carried out applying the knowledge obtained in healthy dogs to dogs with spontaneously occurring diseases.[1,2]

It was shown, for instance, that cyclic neutropenia was due to a stem cell defect rather than to a defect of marrow regulation.[116] Marrow grafting studies in dogs with factor VIII

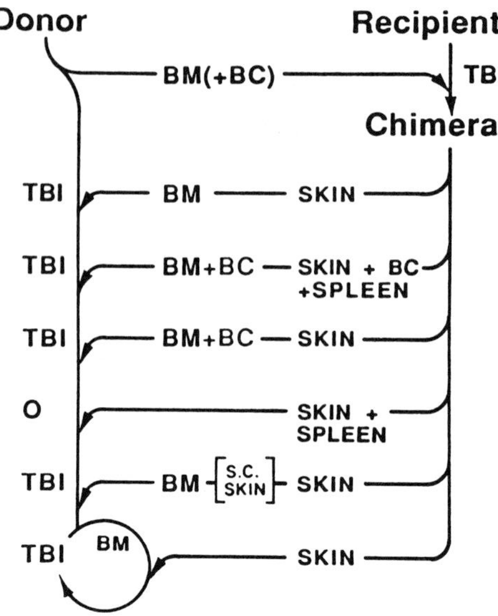

Figure 4 Tolerance transfer experiments. Recipient dogs (right) were given 920 cGy of TBI and hemopoietic grafts from a DLA identical littermate donor (left). Once the recipient had become a stable chimera, the original donor was given 920 cGy of TBI and hemopoietic graft from the chimera (except in experiment 4, where **no** marrow transplant was performed and in experiment 6, in which donors were given autologous marrow). The transplants consisted of bone marrow only (BM) or bone marrow and peripheral blood buffy coat cells (BM + BC). After 1 to 2 months the original donor was given skin grafts from the chimera, either without additional manipulation (experiments 1 and 3), along with buffy coat (BC) and spleen cells (experiment 2), spleen cells only (experiment 4), following repeated subcutaneous skin implants from the chimera (experiment 5), or skin alone (after autologous transplantation; experiment 6). Only dogs given BC along with BM showed indefinite skin graft survival.

deficiency ruled out the hemopoietic and lymphoid systems as relevant sources of factor VIII.[118] Severe life-threatening hemolytic anemia secondary to pyruvatekinase deficiency in Basenji dogs was corrected by marrow transplantation from DLA-identical littermates not affected by the disease.[117] Follow-up studies have shown that hemosiderosis and organ damage associated with hemolytic disease were phenomena secondary to hemolysis rather than a separate genetic defect.[119] Iron storage in the liver decreased with time after transplantation. Bone-marrow transplant attempts in dogs with ceroid lipofuscinosis and GM, gangliosidosis, although technically successful, showed that these genetic defects are not correctable by marrow transplantation.[121,122] Fucosidosis and mucopolysacchasidosis, on the other hand have been shown to be correctable or at least are markedly attenuated.[123,124]

In several laboratories, spontaneously occurring canine lymphoma has been treated by autologous or allogeneic marrow transplantation following TBI and chemotherapy. Among dogs in chemotherapy induced complete remission that were given autologous marrow grafts after consolidation with TBI, 25% have become healthy long-term survivors,[120] a result very encouraging for studies in man. Peripheral-blood stem cells have been transplanted with similar success in dogs with lymphoma.[54]

XVI. CONCLUSIONS

The dog is an excellent model for many questions related to bone-marrow transplantation. Important data on irradiation and chemotherapy conditioning regimens have been obtained and extrapolated to the clinical setting. Pivotal observations on the role of histocompatibility antigens on transplant outcome in a random bred species have been made in the dogs. Numerous regimens of GVHD prophylaxis and treatment have been tested in this model. The application of molecular technology has further refined the model particularly in the area of hemopoietic growth factors and histocompatibility. Finally, the availability of dogs with genetically determined or acquired diseases has permitted the investigation of the efficacy of bone-marrow transplantation in several disease categories before its clinical application.

ACKNOWLEDGMENT

We thank Deborah Gayle for helping compile the cited literature and with manuscript preparation.

REFERENCES

1. **Storb, R. and Deeg, H.J.,** Contributions of the dog model in marrow transplantation, *Plasma Ther. Transfus. Technol.,* 6, 303, 1985.
2. **Deeg, H.J., Storb, R., and Thomas, E.D.,** The dog as a preclinical model of marrow transplantation, in *Recent Advances in Bone Marrow Transplantation,* Gale, R.P., Ed., Alan R. Liss, 1983, 527.
3. **Ladiges, W.C., Storb, R., Graham, T., and Thomas, E.D.,** Experimental techniques used to study the immune system of dogs and other large animals, in *Methods of Animal Experimentation, Volume 7, Part C,* Gay, W.I. and Heavener, J.E., Eds., Academic Press, New York, 1989, 103.
4. **Clark, W.,** *The Experimental Foundations of Modern Immunology, 3rd ed.,* John Wiley & Sons, New York, 1986.
5. **Gerber, J.D. and Brown, A.L.,** Effect of development and aging on the response of canine lymphocytes to phytohemagglutinin, *Infect. Immun.,* 10, 695, 1974.
6. **Shifrine, M., Smith, J.B., Bulgin, M.S., Bryant, B.J., Zee, Y.C., and Osburn, B.I.,** Response of canine fetuses and neonates to antigenic stimulation, *J. Immunol.,* 107, 965, 1971.
7. **Dennis, R.A., Jacoby, R.O., and Griesemer, R.A.,** Development of immunity in fetal dogs: skin allograft rejection, *Am. J. Vet. Res.,* 30, 1511, 1969.
8. **Ladiges, W., Deeg, H.J., Aprile, J., Raff, R., Schuening, F., and Storb, R.,** Differentiation and function of lymphohemopoietic cells in the dog, in *Differentiation Antigens in Lymphohemopoietic Tissues,* Trnka, Z. and Miyasaka, M., Eds., Marcel Dekker, New York, 1988, 307.
9. **Atkinson, K., Deeg, H.J., Storb, R., Weiden, P.L., Gerhard-Miller, L., Torok-Storb, B.J., Seigneuret, M., and Thomas, E.D.,** Canine lymphocyte subpopulations, *Exp. Hematol.,* 8, 821, 1980.
10. **Deeg, H.J., Wulff, J.C., DeRose, S., Sale, G.E., Braun, M., Brown, M.A., Springmeyer, S.C., Martin, P.J., and Storb, R.,** Unusual distribution of Ia-like antigens on canine lymphocytes, *Immunogenetics,* 16, 445, 1982.
11. **Storb, R., Epstein, R.B., and Thomas, E.D.,** Marrow repopulating ability of peripheral blood cells compared to thoracic duct cells, *Blood,* 32, 662, 1968.
12. **Deeg, H.J. and Storb, R.,** Functional dichotomy of canine thoracic duct lymphocytes: proliferation in mixed leukocyte culture and unresponsiveness to plant lectins, *J. Reticuloendothel. Soc.,* 29, 395, 1981.
13. **Daemen, A.J., Buurman, W.A., Linden, C.J., Groenewegen, G., and Kootstra, G.,** Canine Il-2: characterization and optimal conditions for production, *Vet. Immunol. Immunopathol.,* 5, 247, 1984.

14. **Aprile, J., Gerhard-Miller, L., and Deeg, H.J.,** Cluster formation of canine dendritic cells and lymphocytes is calcium dependent and not inhibited by cyclosporine, *Exp. Hematol.,* 18, 32, 1990.
15. **Clara, M.,** *Entwicklungsgeschichte des Menschen, Auflage 6,* Edition Leipzig, Leipzig, 1966.
16. **Salzstein, E.C., Bortin, M.M., and Rimm, A.A.,** Long lived canine allogeneic radiation chimera produced with combined fetal liver and thymus cells, *Transplantation,* 18, 461, 1974.
17. **Thomas, E.D., Collins, J.A., Herman, E.C., Jr., Korner, B., and Ferrebee, J.W.,** Radiation and marrow transplantation in disease-free dogs, *Radiat. Res.,* 14, 168, 1961.
18. **Thomas, E.D., Collins, J.A., Kasakura, S., and Ferrebee, J.W.,** Lethally irradiated dogs given infusions of fetal and adult hematopoietic tissue, *Transplantation,* 1, 514, 1963.
19. **Epstein, R.B., Graham, T.C., Buckner, C.D., and Thomas, E.D.,** Allogeneic marrow engraftment by cross circulation in lethally irradiated dogs, *Blood,* 28, 692, 1966.
20. **Gerhartz, H.H., Nothdurft, W., Carbonell, F., and Fliedner, T.M.,** Allogeneic transplantation of blood stem cells concentrated by density gradients, *Exp. Hematol.,* 13, 136, 1985.
21. **Cavins, J.A., Scheer, S.C., Thomas, E.D., and Ferrebee, J.W.,** The recovery of lethally irradiated dogs given infusions of autologous leukocytes preserved at –80°C. *Blood,* 23, 38, 1964.
22. **Calvo, W., Fliedner, T.M., Herbst, E.W., and Fache, I.,** Regeneration of blood-forming organs after autologous leukocyte transfusion in lethally irradiated dogs. I. Distribution and cellularity of the bone marrow in normal dogs, *Blood,* 46, 453, 1975.
23. **Torok-Storb, B., Deeg, H.J., Atkinson, K., Weiden, P.L., Adamson, J.W., and Storb, R.,** Erythroid colony stimulating and inhibiting cells in peripheral blood of transfused dogs: separation of function by velocity sedimentation, *Blood,* 54, 955, 1979.
24. **Erickson, V. and Torok-Storb, B.,** Erythroid burst forming units (BFU_E) grown from canine marrow and peripheral blood, *Exp. Hematol.,* 9, 468, 1981.
25. **Schuening, F., Emde, C., and Schaefer, U.W.,** Improved culture conditions for granulocyte-macrophage progenitor cells, *Exp. Hematol.,* 11(Abstr.), 205, 1983.
26. **Schuening, F.G., Storb, R., Meyer, J., and Goehle, S.,** Long-term culture of canine bone marrow cells. *Exp. Hematol.,* 17, 411, 1989.
27. **Lothrop, C.D., Jr., Warren, D.J., Souza, L.M., Jones, J.B., and Moore, M.A.S.,** Correction of canine cyclic hematopoiesis with recombinant human granulocyte colony-stimulating factor, *Blood,* 72, 1324, 1988.
28. **Nash, R.A., Schuening, F., Appelbaum, F., Boone, T., Morris, C.F., Slichter, S.J., and Storb, R.,** Molecular cloning and in vivo evaluation of canine GM-CSF, *Blood,* 78, 930, 1991.
29. **Zsebo, K.M., Wypych, J., McNiece, I.K., Lu, H.S., Smith, K.A., Karkare, S.B., Sachdev, R.K., Yuschenkoff, V.N., Birkett, N.C., Williams, L.R., Satyagal, V.N., Tung, W., Bosselman, R.A., Mendiaz, E.A., and Langley, K.E.,** Identification, purification, and biological characterization of hematopoietic stem cell factor from buffalo rat liver-conditioned medium, *Cell,* 63, 195, 1990.
30. **Schuening, F.G., Storb, R., Goehle, S., Graham, T.C., Hackman, R., Mori, M., Sousa, L.M., and Appelbaum, F.R.,** Recombinant human granulocyte colony-stimulating factor accelerates after DLA-identical littermate marrow transplants in dogs, *Blood,* 76, 636, 1990.
31. **Schuening, F.G., Storb, R., Goehle, S., Graham, T.C., Appelbaum, F.R., Hackman, R., and Souza, L.M.,** Effect of recombinant human granulocyte colony-stimulating factor on hematopoiesis of normal dogs and on hematopoietic recovery after otherwise lethal total body irradiation, *Blood,* 74, 1308, 1989.

32. **Epstein, R.B., Storb, R., Ragde, H., and Thomas, E.D.,** Cytotoxic typing antisera for marrow grafting in littermate dogs, *Transplantation,* 6, 45, 1968.
33. **Deeg, H.J., Raff, R.F., Grosse-Wilde, H., Bijma, A.M., Buurman, W., Doxiadis, I., Kolb, H.J., Krumbacher, K., Ladiges, W., Losslein, K.L., Schoch, G., Westbroek, D.L., Bull, R.W., and Storb, R.,** Joint report of the third international workshop on canine immunogenetics. I. Analysis of homozygous typing cells (HTCs), *Transplantation,* 41, 111, 1986.
34. **Bull, R.W., Vriesendorp, H.M., Cech, R., Grosse-Wilde, H., Bijma, A.M., Ladiges, W.L., Krumbacher, K., Doxiadis, I., Ejima, H., Templeton, J., Albert, E.D., Storb, R., and Deeg, H.J.,** Joint report of the third international workshop on canine immunogenetics. II. Analysis of the serological typing of cells, *Transplantation,* 43, 154, 1987.
35. **Krumbacher, K., van der Feltz, M.J.M., Happel, M., Gerlach, C., Lösslein, L.K., and Grosse-Wilde, H.,** Revised classification of the DLA loci by serological studies, *Tissue Antigens,* 27, 262, 1986.
36. **Doxiadis, I., Krumbacher K., Neffjes, J.J., Pleogh, H.L., and Grosse-Wilde, H.,** Biochemical evidence that the DLA-B locus codes for a class II determinant expressed on all canine peripheral blood lymphocytes, *Exp. Clin. Immunogenet.,* 6, 219, 1989.
37. **Sarmiento, U.M. and Storb, R.,** Nucleotide sequence of a dog class-1 cDNA clone, *Immunogenetics,* 31, 400, 1990.
38. **Raff, R.F., Deeg, H.J., Farewell, V.T., DeRose, S., and Storb, R.,** The canine major histocompatibility complex. Population study of DLA-D alleles using a panel of homozygous typing cells, *Tissue Antigens,* 21, 360, 1983.
39. **Bijnen, A.B., Vriesendorp, H.M., Grosse-Wilde, H., and Westbroek, D.L.,** Polygenic control of mixed lymphocyte reactions in dogs, *Tissue Antigens,* 9, 187, 1977.
40. **Sarmiento, U.M. and Storb, R.,** Nucleotide sequence of a dog DRB cDNA clone, *Immunogenetics,* 31, 396, 1990.
41. **Sarmiento, U.M., Sarmiento, J.I., and Storb, R.,** Allelic variation in the DR subregion of the canine major histocompatibility complex, *Immunogenetics,* 32, 13, 1990.
42. **Sarmiento, U.M. and Storb, R.F.,** Characterization of class II alpha genes and DLA-D region allelic associations in the dog, *Tissue Antigens,* 32, 224, 1988.
43. **Thomas, E.D., Storb, R., and Epstein, R.B.,** Bone marrow chimeras in man and animals. Plenary sessions scientific contributions. XIII. International Congress of Hematology, Munchen, *Hematolund. Bluttrans.,* 9, 86, 1970.
44. **Thomas, E.D., Plain, G.L., Graham, T.C., and Ferrebee, J.W.,** Long-term survival of lethally irradiated dogs given homografts of bone marrow (Brief Note), *Blood,* 23, 488, 1964.
45. **Epstein, R.B., Storb, R., Clift, R.A., and Thomas, E.D.,** Transplantation of stored allogeneic bone marrow in dogs selected by histocompatibility typing, *Transplantation,* 8, 496, 1969.
46. **Abb, J., Grosse-Wilde, H., Scholz, S., and Albert, E.D.,** Matching for DLA-A, DLA-B, and DLA-D antigens and skin allograft survival in unrelated beagle dogs, *Eur. Surg. Res.,* 10, 142, 1978.
47. **Storb, R., Weiden, P.L., Graham, T.C., and Thomas, E.D.,** Studies of marrow transplantation in dogs, *Transplant. Proc.,* 8, 545, 1976.
48. **Deeg, H.J., Storb, R., Weiden, P.L., Schumacher, D., Shulman, H., Graham, T., and Thomas, E.D.,** High-dose total-body irradiation and autologous marrow reconstitution in dogs: dose-rate-related acute toxicity and fractionation-dependent long-term survival, *Radiat. Res.,* 88, 385, 1981.
49. **Thomas, E.D., Plain, G.L., and Thomas, D.,** Leukocyte kinetics in the dog studied by cross circulation, *J. Lab. Clin. Med.,* 66, 64, 1965.

50. **Buckner, D., Eisel, R., and Perry, S.,** Blood cell separation in the dog by continuous flow centrifugation, *Blood,* 31, 653, 1968.
51. **Zander, A.R., Gray, K.N., Hester, J.P., Johnston, D.A., Spitzer, G., Raulston, G.L., McCredie, K.B., Jardine, J.H., Wu, J., Gleiser, C., Cardiff, J., and Dicke, K.A.,** Rescue by peripheral blood mononuclear cells in dogs from bone marrow failure after total-body irradiation, *Transfusion,* 24, 42, 1984.
52. **Cavins, J.A., Kasakura, S., Thomas, E.D., and Ferrebee, J.W.,** Recovery of lethally irradiated dogs following infusion of autologous marrow stored at low temperature in dimethyl-sulphoxide, *Blood,* 20, 730, 1962.
53. **Storb, R., Epstein, R.B., Ragde, H., and Thomas, E.D.,** Marrow engraftment by allogeneic leukocytes in lethally irradiated dogs, *Blood,* 30, 805, 1967.
54. **Appelbaum, F.R., Deeg, H.J., Storb, R., Graham, T.C., Charrier, K., and Bensinger, W.,** Cure of malignant lymphoma in dogs with peripheral blood stem cell Transplantation, *Transplantation,* 42, 19, 1986.
55. **Stitzel, K.A., Champlin, R., and Gale, R.P.,** Fetal liver cell transplantation in dogs: a possible alternative source of hematopoietic cells for transplantation, in *Recent Advances in Bone Marrow Transplantation,* Alan R. Liss, New York, 1983, 831.
56. **Kolb, H.J., Rieder, I., Rodt, B., Netzel, B., Grosse-Wilde, H., Scholz, S., Schäffer, E., Kolb, H., and Thierfelder, S.,** Antilymphocytic antibodies and marrow Transplantation, *Transplantation,* 27, 242, 1979.
57. **Appelbaum, F.R., Brown, P.A., Graham, T.C., Sandmaier, B.M., Schuening, F.W., and Storb, R.,** Characterization of malignant lymphoma in dogs and use as a model for the development of treatment strategies, in *Recent Advances and Future Directions in Bone Marrow Transplantation, Experimental Hematology Today—1987,* Baum, S.J., Santos, G.W., and Takaku, F., Eds., Springer-Verlag, New York, 1988, 31.
58. **Schuening, F.G., Storb, R., Stead, R.B., Goehle, S., Nash, R., and Miller, A.D.,** Improved retroviral transfer of genes into canine hematopoietic progenitor cells kept in long-term marrow culture, *Blood,* 74, 152, 1989.
59. **Schuening, F.G. and Miller, A.D.,** Gene transfer into hematopoietic stem cells, in *Hematopoietic Stem Cells,* Müller, Sieburg, Torok-Storb, Visser, and Storb, Eds., Springer-Verlag, 1992, 237.
60. **Stockschlaeder, M.A.R., Storb, R., Osborne, W.R.A., and Miller, A.D.,** L-histidinol provides effective selection of retrovirus-vector-infected keratinocytes without impairing their proliferative potential, *Gene Ther.,* Berlin, 2, 33, 1991.
61. **Storb, R., Epstein, R.B., Rudolph, R.H., and Thomas, E.D.,** Allogeneic canine bone marrow transplantation following cyclophosphamide, *Transplantation,* 7, 378, 1969.
62. **Floersheim, G.L. and Ruszkiewicz, M.,** Bone-marrow transplantation after antilymphocytic serum and lethal chemotherapy, *Nature,* 222, 854, 1969.
63. **Kolb, H.J., Storb, R., Weiden, P.L., Ochs, H.D., Kolb, H., Graham, T.C., Floersheim, G.L., and Thomas, E.D.,** Immunologic, toxicologic and marrow transplantation studies in dogs given dimethyl myleran, *Biomedicine,* 20, 341, 1974.
64. **Storb, R., Weiden, P.L., Graham, T.C., Lerner, K.G., Nelson, N., and Thomas, E.D.,** Hemopoietic grafts between DLA-identical canine littermates following dimethyl myleran, evidence for resistance to grafts not associated with DLA and abrogated by antithymocyte serum, *Transplantation,* 24, 349, 1977.
65. **Parkman, R.,** Preparation for bone marrow transplantation, *Springer Semin. Immunopathol.,* 7, 59, 1984.
66. **Storb, R., Thomas, E.D., Buckner, C.D., Appelbaum, F.R., Clift, R.A., Deeg, H.J., Doney, K., Hansen, J.A., Prentice, R.L., Sanders, J.E., Stewart, P., Sullivan, K.M., and Witherspoon, R.P.,** Marrow transplantation for aplastic anemia, *Sem. Hematol.,* 21, 27, 1984.

67. **Thomas, E.D., Ashley, C.A., Lochte, H.L., Jr., Jaretzki, A.,III, Sahler, O.D., and Ferrebee, J.W.,** Homografts of bone marrow in dogs after lethal total-body radiation, *Blood,* 14, 720, 1959.
68. **Thomas, E.D., LeBlond, R., Graham, T., and Storb, R.,** Marrow infusions in dogs given midlethal or lethal irradiation, *Radiat. Res.,* 41, 113, 1970.
69. **Deeg, H.J., Storb, R., Shulman, H.M., Weiden, P.L., Graham, T.C., and Thomas, E.D.,** Engraftment of DLA-nonidentical unrelated canine marrow after high-dose fractionated total body irradiation, *Transplantation,* 33, 443, 1982.
70. **Vriesendorp, H.M., Klapwijk, W.M., van Kessel, A.M.C., Zurcher, C., and van Bekkum, D.W.,** Lasting engraftment of histoincompatible bone marrow cells in dogs, *Transplantation,* 31, 347, 1981.
71. **Deeg, H.J., Storb, R., and Thomas, E.D.,** Marrow graft studies in dogs: Factors influencing resistance to engraftment and graft-versus-host disease, *Surv. Immunol. Res.,* 1, 148, 1982.
72. **Deeg, H.J., Storb, R., Longton, G., Graham, T.C., Shulman, H.M., Appelbaum, F., and Thomas, E.D.,** Single dose or fractionated total body irradiation and autologous marrow transplantation in dogs: effects of exposure rate, fraction size and fractionation interval on acute and delayed toxicity, *Int. J. Radiat. Oncol. Biol. Phys.,* 15, 647, 1988.
73. **Kolb, H.J., Rieder, I., Bodenberger, U., Netzel, B., Schaffer, E., Kolb, H., and Thierfelder, S.,** Dose rate and dose fractionation studies in total body irradiation of dogs, *Pathol. Biol.,* 27, 370, 1979.
74. **Storb, R., Raff, R.F., Appelbaum, F.R., Graham, T.C., Schuening, F.G., Sale, G., and Pepe, M.,** Comparison of fractionated to single-dose total body irradiation in conditioning canine littermates for DLA-identical marrow grafts, *Blood,* 74, 1139, 1989.
75. **Storb, R., Kolb, H.J., Graham, T.C., Kolb, H., Weiden, P.L., and Thomas, E.D.,** Treatment of established graft-versus-host disease in dogs by antithymocyte serum or prednisone, *Blood,* 42, 601, 1973.
76. **Deeg, H.J., Sale, G.E., Storb, R., Graham, T.C., Schuening, F., Appelbaum, F.R., and Thomas, E.D.,** Engraftment of DLA-nonidentical bone marrow facilitated by recipient treatment with anti-class II monoclonal antibody and methotrexate, *Transplantation,* 44, 340, 1987.
77. **Schuening, F., Storb, R., Goehle, S., Meyer, J., Graham, T.C., Deeg, H.J., Appelbaum, F.R., Sale, G.E., Graf, L., and Loughran, T.P., Jr.,** Facilitation of engraftment of DLA-nonidentical marrow by treatment of recipients with monoclonal antibody directed against marrow cells surviving radiation, *Transplantation,* 44, 607, 1987.
78. **Sandmaier, B.M., Storb, R., Appelbaum, F.R., and Gallatin, W.M.,** An antibody that facilitates hematopoietic engraftment recognizes CD44, *Blood,* 76, 630, 1990.
79. **Appelbaum, F.R., Brown, P., Sandmaier, B., Badger, C., Schuening, F., Graham, T.C., and Storb, R.,** Antibody-radionuclide conjugates as part of a myeloblative preparative regimen for marrow transplantation, *Blood,* 73, 2202, 1989.
80. **Bianco, J.A., Pepe, M.S., Higano, C., Appelbaum, F.R., McDonald, G.B., and Singer, J.W.,** Prevalence of clinically relevant bacteremia after upper gastrointestinal endoscopy in bone marrow transplant recipients, *Am. J. Med.,* 89, 134, 1990.
81. **Appelbaum, F.R., Petersen, F.B., Buckner, C.D., Badger, C., Sandmaier, B., Storb, R., and Thomas, E.D.,** New preparative regimens prior to marrow transplantation for acute nonlymphoblastic leukemia, in *Bone Marrow Transplantation: Current Controversies,* Gale, R.P. and Champlin, R., Eds., Alan R. Liss, New York, 1989, 107.
82. **Uphoff, D.E.,** Alteration of homograft reaction by A-methopterin in lethally irradiated mice treated with homologous marrow, *Proc. Soc. Exp. Biol. Med.,* 99, 651, 1958.

83. **Thomas, E.D., Kasakura, S., Cavins, J.A., and Ferrebee, J.W.,** Marrow transplants in lethally irradiated dogs: the effect of methotrexate on survival of the host and the homograft, *Transplantation,* 1, 571, 1963.
84. **van Bekkum, D.W., Wagemaker, G., and Vriesendorp, H.M.,** Mechanisms and avoidance of graft-versus-host disease, *Transplant. Proc.,* 11, 189, 1979.
85. **Storb, R., Kolb, H.J., Deeg, H.J., Weiden, P.L., Appelbaum, F., Graham, T.C., and Thomas, E.D.,** Prevention of graft-versus-host disease by immunosuppressive agents after transplantation of DLA-nonidentical canine marrow, *Bone Marrow Transplant.,* 1, 167, 1986.
86. **Storb, R., Rudolph, R.H., Kolb, H.J., Graham, T.C., Mickelson, E., Erickson, V., Lerner, K.G., Kolb, H., and Thomas, E.D.,** Marrow grafts between DL-A-matched canine littermates, *Transplantation,* 15, 92, 1973.
87. **Storb, R., Rudolph, R.H., and Thomas, E.D.,** Marrow grafts between canine siblings matched by serotyping and mixed leukocyte culture, *J. Clin. Invest.,* 50, 1272, 1971.
88. **Deeg, H.J., Storb, R., Weiden, P.L., Graham, T., Atkinson, K., and Thomas, E.D.,** Cyclosporin-A: effect on marrow engraftment and graft-versus-host disease in dogs, *Transplant. Proc.,* 13, 402, 1981.
89. **Deeg, H.J., Storb, R., Weiden, P.L., Raff, R.F., Sale, G.E., Atkinson, K., Graham, T.C., and Thomas, E.D.,** Cyclosporin A and methotrexate in canine marrow transplantation: engraftment, graft-versus-host disease, and induction of tolerance, *Transplantation,* 34, 30, 1982.
90. **Deeg, H.J., Storb, R., Appelbaum, F.R., Kennedy, M.S., Graham, T.C., and Thomas, E.D.,** Combined immunosuppression with cyclosporine and methotrexate in dogs given bone marrow grafts from DLA-haploidentical littermates, *Transplantation,* 37, 62, 1984.
91. **Storb, R., Deeg, H.J., Whitehead, J., Appelbaum, F., Beatty, P., Bensinger, W., Buckner, C.D., Clift, R., Doney, K., Farewell, V., Hansen, J., Hill, R., Lum, L., Martin, P., McGuffin, R., Sanders, J., Stewart, P., Sullivan, K., Witherspoon, R., Yee, G., and Thomas, E.D.,** Methotrexate and cyclosporine compared with cyclosporine alone for prophylaxis of acute graft versus host disease after marrow transplantation for leukemia, *N. Engl. J. Med.,* 314, 729, 1986.
92. **Deeg, H.J., Spitzer, T.R., Cottler-Fox, M., Cahill, R., and Pickle, L.W.,** Conditioning-related toxicity and acute graft-versus-host disease in patients given methotrexate/cyclosporine prophylaxis, *Bone Marrow Transplant.,* 7, 193, 1991.
93. **Prümmer, O., Raghavachar, A., and Fliedner, T.M.,** Recovery of immune functions in dogs after total body irradiation and transplantation of autologous blood or bone marrow cells, *Exp. Hematol.,* 13, 891, 1985.
94. **Vriesendorp, H.M., Klapwyk, W.M., Heidt, P.J., Hogeweg, B., Zurcher, C., and van Bekkum, D.W.,** Factors controlling the engraftment of transplanted dog bone marrow cells, *Tissue Antigens,* 20, 63, 1982.
95. **Lösslein, L.K., Kolb, H.J., Porzsolt, S., Schäffer, E., Scholz, S., Meissner, H., Holler, E., Wilmanns, W., and Thierfelder, S.,** Hyperfractionation of total-body irradiation and engraftment of marrow from DLA-haploidentical littermates, *Transplant. Proc.,* 19, 2707, 1987.
96. **Deeg, H.J., Prentice, R., Fritz, T.E., Sale, G.E., Lombard, L.S., Thomas, E.D., and Storb, R.,** Increased incidence of malignant tumors in dogs after total body irradiation and marrow transplantation, *Int. J. Radiat. Oncol. Biol. Phys.,* 9, 1505, 1983.
97. **Storb, R., Epstein, R.B., Ragde, H., and Thomas, E.D.,** Marrow grafts by combined marrow and leukocyte infusions in unrelated dogs selected by histocompatibility typing, *Transplantation,* 6, 587, 1968.

98. **Deeg, H.J., Storb, R., Szer, J., Appelbaum, F.R., Hackman, R.C., and Thomas, E.D.,** Facilitation of engraftment of DLA-nonidentical marrow by treatment of the recipient with monoclonal anti-Ia antibody, *Transplant. Proc.,* 17, 493, 1985.
99. **Deeg, H.J., Storb, R., Raff, R.F., Weiden, P.L., DeRose, S., and Thomas, E.D.,** Marrow grafts between phenotypically DLA-identical and haploidentical unrelated dogs: additional antigens controlling engraftment are not detected by cell-mediated lympholysis, *Transplantation,* 33, 17, 1982.
100. **Atkinson, K., Shulman, H.M., Deeg, H.J., Weiden, P.L., Graham, T.C., Thomas, E.D., and Storb, R.,** Acute and chronic graft-versus-host disease in dogs given hemopoietic grafts from DLA-nonidentical littermates: two distinct syndromes, *Am. J. Pathol.,* 108, 196, 1982.
101. **Zurcher, C., van Kessel, A.C.M., and Vriesendorp, H.M.,** Graft-versus-host reactions in dogs after total body irradiation and bone marrow transplantation, in *Annual 1977,* Radiobiological Institute TNO, Rijswijk, 1977, 278.
102. **Kolb, H., Sale, G.E., Lerner, K.G., Storb, R., and Thomas, E.D.,** Pathology of acute graft-versus-host disease in the dog. An autopsy study of ninety-five dogs, *Am. J. Pathol.,* 96, 581, 1979.
103. **Deeg, H.J., Storb, R., Raff, R.F., Weiden, P.L., Atkinson, K., and Thomas, E.D.,** Long-term survival of dogs given marrow grafts from unrelated DLA-nonidentical donors and treated with Cyclosporin A and Methotrexate — Is tolerance mediated by suppressor cells?, in *Experimental Hematology Today,* Baum, S.J., Ledney, G.D., and Thierfelder, S., Eds., S. Karger, Basel, 1982, 127.
104. **Deeg, H.J., Raff, R.F., Severns, E., Appelbaum, F.R., Thomas, E.D., and Storb, R.,** Suppressor and cytotoxic cells in DLA nonidentical canine radiation chimeras given cyclosporine and methotrexate as prophylaxis for graft-versus-host disease, *Transplant. Proc.,* 15, 3042, 1983.
105. **Shulman, H.M., Luk, K., Deeg, H.J., Shuman, W.B., and Storb, R.,** Induction of hepatic veno-occlusive disease in dogs, *Am. J. Pathol.,* 126, 114, 1987.
106. **Hellstrom, I., Hellstrom, K.E., Storb, R., and Thomas, E.D.,** Colony inhibition of fibroblasts from chimeric dogs mediated by the dogs' own lymphocytes and specifically abrogated by their serum, *Proc. Natl. Acad. Sci. U.S.A.,* 66, 65, 1970.
107. **Tsoi, M.S., Storb, R., Weiden, P.L., Schroeder, M.L., and Thomas, E.D.,** Canine marrow transplantation: do serum blocking factors maintain stable graft-vs.-host tolerance?, *Transplant. Proc.,* 7, 841, 1975.
108. **Atkinson, K., Storb, R., Weiden, P.L., Deeg, H.J., Gerhard-Miller, L., and Thomas, E.D.,** *In vitro* tests correlating with presence or absence of graft-vs-host disease in DLA nonidentical canine radiation chimeras: evidence that clonal abortion maintains stable graft-host tolerance, *J. Immunol.,* 124, 1808, 1980.
109. **Weiden, P.L., Storb, R., Tsoi, M.-S., Graham, T.C., Lerner, K.G., and Thomas, E.D.,** Infusion of donor lymphocytes into stable canine radiation chimeras: implications for mechanism of transplantation tolerance, *J. Immunol.,* 116, 1212, 1976.
110. **Deeg, H.J., Severns, E., Raff, R.F., Sale, G.E., and Storb, R.,** Specific tolerance and immunocompetence in haploidentical but not in completely allogeneic canine chimeras treated with methotrexate and cyclosporine, *Transplantation,* 44, 621, 1987.
111. **Lennon, T.P., Yee, G.C., Kennedy, M.S., Torok-Storb, B., Bernstein, S.A., and Deeg, H.J.,** Monitoring of cyclosporine therapy with in vitro biological assays, *Transplantation,* 44, 799, 1987.
112. **Tutschka, P.J., Beschorner, W.E., Allison, A.C., Burns, W.H., and Santos, G.W.,** Use of cyclosporin A in allogeneic bone marrow transplantation in the rat, *Nature,* 280, 148, 1979.

113. **Tutschka, P.J., Hess, A.D., Beschorner, W.E., and Santos, G.W.,** Suppressor cells in transplantation tolerance, *Transplantation,* 33, 510, 1982.
114. **Deeg, H.J., Atkinson, K., Weiden, P.L., and Storb, R.,** Mechanisms of tolerance in canine radiation chimeras, *Transplant. Proc.,* 19, 75, 1987.
115. **Atkinson, K., Storb, R., Weiden, P.L., Deeg, H.J., Kopecky, K.J., Graham, T.C., and Thomas, E.D.,** Studies on transplantation tolerance in canine radiation chimeras, in *Biology of Bone Marrow Transplantation,* Gale, R.P. and Fox, C.F., Eds., Academic Press, New York, 1980, 271.
116. **Weiden, P., Robinett, B., Graham, T.C., Adamson, J.W., and Storb, R.,** Canine cyclic neutropenia, a stem cell defect, *J. Clin. Invest.,* 53, 950, 1974.
117. **Weiden, P.L., Storb, R., Graham, T.C., and Schroeder, M.L.,** Severe hereditary haemolytic anaemia in dogs treated by marrow Transplantation, *Br. J. Haematol.,* 33, 357, 1976.
118. **Storb, R., Marchioro, T.L., Graham, T.C., Willemin, M., Hougie, C., and Thomas, E.D.,** Canine hemophilia and hemopoietic grafting, *Blood,* 40, 234, 1972.
119. **Weiden, P.L., Hackman, R.C., Deeg, H.J., Graham, T.C., Thomas, E.D., and Storb, R.,** Long-term survival and reversal of iron overload after marrow transplantation in dogs with congenital hemolytic anemia, *Blood,* 57, 66, 1981.
120. **Weiden, P.L., Storb, R., Deeg, H.J., Graham, T.C., and Thomas, E.D.,** Prolonged disease-free survival in dogs with lymphoma after total-body irradiation and autologous marrow transplantation consolidation of combination chemotherapy induced remissions, *Blood,* 54, 1039, 1979.
121. **Deeg, H.J., Shulman, H.M., Albrechtsen, D., Graham, T.C., Storb, R., and Koppang, N.,** Batten's disease: failure of allogeneic bone marrow transplantation to arrest disease progression in a canine model, *Clin. Genet.,* 37, 264, 1990.
122. **O'Brien, J.S., Storb, R., Raff, R.F., Harding, J., Appelbaum, F., Morimoto, S., Kishimoto, Y., Graham, T., Ahern-Rindell, A., and O'Brien, S.L.,** Bone marrow transplantation in canine GM1 gangliosidosis, *Clin. Genet.,* 38, 274, 1990.
123. **Breider, M.A., Shull, R.M., and Constantopoulos, G.,** Long-term effects of bone marrow transplantation in dogs with mucopolysaccharidosis I, *Am. J. Pathol.,* 134, 677, 1989.
124. **Taylor, R.M., Stewart, G.J., and Farrow, B.R.H.,** Comparison of the effect of total body and total lymphoid irradiation on bone marrow engraftment in MHC-matched dogs, *Transplant. Proc.,* 21, 3820, 1989.
125. **Wulff, J.C., Durkopp, N., Aprile, J., Tsoi, M.-S., Springmeyer, S.C., Deeg, H.J., and Storb, R.,** Two monoclonal antibodies (DLy-1 and DLy-6) directed against canine lymphocytes, *Exp. Hematol.,* 10, 609, 1982.
126. **Szer, J., Deeg, H.J., Severns, E., and Storb, R.,** DLA-D-specific suppressor cells characterized by monoclonal antibodies, *Transplantation,* 39, 187, 1985.
127. **Krawiec, D.R. and Muscoplat, C.C.,** Development and characterization of a monoclonal antibody (Aby 6C6) that distinguishes medullary from cortical thymocytes, *Am. J. Vet. Res.,* 45, 499, 1984.
128. **McKenzie, J.L. and Fabre, J.W.,** Studies with a monoclonal antibody on the distribution of THY-1 in the lymphoid and extracellular connective tissues of the dog, *Transplantation,* 31, 275, 1981.
129. **Krawiec, D.R. and Muscoplat, C.C.,** Development and characterization of a hybridoma-derived antibody (Aby 1A1) with specificity to canine thymocytes and peripheral T lymphocytes, *Am. J. Vet. Res.,* 45, 491, 1984.
130. **Wulff, J.C., Deeg, H.-J., and Storb, R.,** A monoclonal antibody (DT-2) recognizing canine T lymphocytes, *Transplantation,* 33, 616, 1982.

131. **Ladiges, W.C., Durkopp, N., Urban, C., Wulff, J.C., and Storb, R.,** Monoclonal antibodies to canine cell surface antigens: a comparison of screening assays, *Hybridoma,* 3, 387, 1984.
132. **Carreno, M., Esqueriazi, V., Fuller, L., Milgram, M., Alejandro, R., Pardo, V., and Miller, J.,** A monoclonal antibody affecting canine T cell reactions, *Fed. Proc.,* 45(Abstr.), 501, 1986.
133. **Beatty, P.G., Ledbetter, J.A., Martin, P.J., Price, T.H., and Hansen, J.A.,** Definition of a common leukocyte cell-surface antigen (Lp95-150) associated with diverse cell-mediated immune functions, *J. Immunol.,* 131, 2913, 1983.
134. **Rieber, P., Lohmeyer, J., Schendel, D.J., and Riethmuller, G.,** Human T cell differentiation antigens characterizing a cytotoxic/suppressor T cell subset, *Hybridoma,* 1, 59, 1981.
135. **Pesando, J.M., Hoffman, P., and Conrad, T.,** Malignant human B cells express two populations of p24 surface antigens, *J. Immunol.,* 136, 2709, 1986.
136. **Sandmaier, B.M., Schuening, F.G., Bianco, J.A., Rosenman, S.J., Bernstein, I., Goehle, S., Storb, R., and Appelbaum, F.R.,** Biochemical characterization of a unique canine myeloid antigen, *Leukemia,* 5, 125, 1991.
137. **Bernstein, I.D., Andrews, R.G., Cohen, S.F., and McMaster, B.E.,** Normal and malignant human myelocytic and monocytic cells identified by monoclonal antibodies, *J. Immunol.,* 128, 876, 1982.
138. **Deeg, H.J., Graham, T.C., Gerhard-Miller, L., Appelbaum, F.R., Schuening, F., and Storb, R.,** Prevention of transfusion-induced graft-versus-host disease in dogs by ultraviolet irradiation, *Blood,* 74, 2592, 1989.
139. **Storb, R., Bean, M., Appelbaum, F., Schuening, F., Graham, T., and Raff, R.,** Treatment of marrow donor blood products with gamma-irradiation prevents transfusion-induced sensitization to DLA-identical marrow grafts, *Transplant. Proc.,* 23, 1697, 1991.

Chapter 20

Bone Marrow Transplantation in the Pig

Philip C. Guzzetta, Craig V. Smith, Kazuaki Nakajima, Arnold Mixon, and David H Sachs

CONTENTS

I. INTRODUCTION

The ability to induce transplantation tolerance by the creation of a lymphohematopoietic chimera was proven by the landmark studies of Billingham, Brent and Medawar in 1953 on tolerance induction in fetal mice.[1] Extensive studies of bone marrow transplantation (BMT) in rodents have revealed the relative ease with which chimerism can be induced across various major histocompatibility complex (MHC) barriers utilizing a variety of preparative regimens to render the recipient susceptible to engraftment of donor marrow. On the other hand, BMT across MHC barriers in man and other large animals[2] is severely limited by problems of engraftment and development of graft versus host disease (GVHD). In this laboratory, we have utilized a herd of partially inbred miniature swine[3] to evaluate BMT as a possible means for inducing tolerance across selective MHC barriers in a large animal model.

Initial studies of BMT in miniature swine indicated that the dose of total body irradiation (TBI) that was sufficient to cause lethal bone marrow suppression was 900 cGy in a single dose.[4] Doses of TBI in excess of 1100 cGy in a single dose were associated with an unacceptably high rate of gastrointestinal toxicity. The first rescues of lethally irradiated animals were performed with autologous BMT, in which it was shown that 0.9 $\times$ 10^8/kg or more bone marrow cells were needed for reliable engraftment.[4]

0-8493-3629-5/94/$0.00+$.50

The use of 900 cGy at 10 cGy/min of TBI allowed engraftment of autologous, MHC matched, Class I matched and Parent(P)→F1 BMT. Greater MHC disparity, including Class II matched, Class I mismatched BMT, did not permit engraftment with 900 cGy of TBI.[5]

Additional studies[12] revealed that consistent engraftment of F1→P, F1→F1, and P1→P2 BMT required 1300 cGy of TBI at 20 cGy/min split into two equal doses to minimize radiation toxicity. An equally effective preparative regimen was 500 cGy + 50 mg/kg cyclophosphamide the day prior to BMT and 650 cGy the day of BMT. Attempts to replace TBI with Busulfan (4 or 6 mg/kg for 4 days) plus cyclophosphamide (60 mg for 2 days or 50 mg for 4 days) were unsuccessful in the F1→P combination because of insufficient myeloablation and severe systemic toxicities.

As expected, GVHD was severe following with P→F1, F1→F1, and P1→P2 BMT. GVHD was mild, but did occur, following MHC matched and F1→P BMT presumably due to minor antigen disparities that are known to exist within the herd. Reduction of GVHD with consistent engraftment could be accomplished in the P→F1 model given 900 cGy TBI by T-cell depletion of the donor inoculum with complement and anti-porcine-CD4 and anti-porcine-CD8 prior to infusion of at least 2.0×10^8 marrow cells/kg.[6] T-cell depletion of donor marrow for F1→P or P1→P2 BMT invariably led to failure of engraftment.

Successful engraftment and diminished GVHD following P1→P2 BMT in rodents utilizing a mixed-marrow inoculum consisting of allogeneic marrow plus T-cell depleted autologous marrow,[7] stimulated a similar approach in the pig. The preparative protocol in the pig was 650 cGy × 2 at 20 cGy/min TBI on days –1 and 0 with administration of 7.5×10^8 allogeneic marrow cells/kg plus 1.0×10^8 T-cell depleted autologous marrow cells/kg on day 0 after TBI. Although there were some long-term survivors (one to 154 days), all animals died of GVHD complications, infections, or radiation toxicity.[8]

Acceptable engraftment rates and survival were obtained in F1→P or F1→F1 using 650 × 2 cGy TBI or 500 + 650 cGy TBI + 50 mg/kg cyclophosphamide. After some of these animals had attained prolonged survival and regained immunocompetence as measured by mixed lymphocyte reaction (MLR) and cell mediated lympholysis (CML), they were subjected to kidney transplants from donors MHC matched to the BMT donor but mismatched to the BMT recipient; they were found to be tolerant to the kidney.[9,10] These studies confirm the premise that engraftment of MHC-disparate bone marrow will lead to long-term tolerance of a vascularized organ MHC matched to the bone marrow donor.

II. TECHNIQUE OF BMT IN THE PIG

A. CARE PRIOR TO BMT

The MHC of all animals is controlled by strict pedigreed breeding and is confirmed by a cytotoxicity assay on the lymphocytes, utilizing rabbit complement and pig alloantisera detecting the three known MHC haplotypes in the herd. Additional confirmation is obtained by pre BMT MLR, which also serves as a baseline study of *in vitro* immunoreactivity for individual animals.

Animals used as bone marrow donors and recipients are 3 to 4 months of age and weigh 15 to 30 kg. The recipient animal arrives from the farm 16 days prior to the date for BMT (day 0), is weighed, receives a Betadine scrub, and is housed in a clean run. Food is withheld the day after arrival and day –14 the animal is taken to the operating room. There, the animal is sedated with i.m. Xylazine (2 mg/kg), Butorphanol Tartrate (0.5 mg/kg), and Ketamine Hydrochloride (22 mg/kg), intubated, and anesthetized with Halothane, at a level permitting spontaneous respiration. Keflin (1 g i.v.) is given before a skin incision is made, and a gastrostomy tube (GT) (size 18 F Foley catheter) is placed. Blood is taken

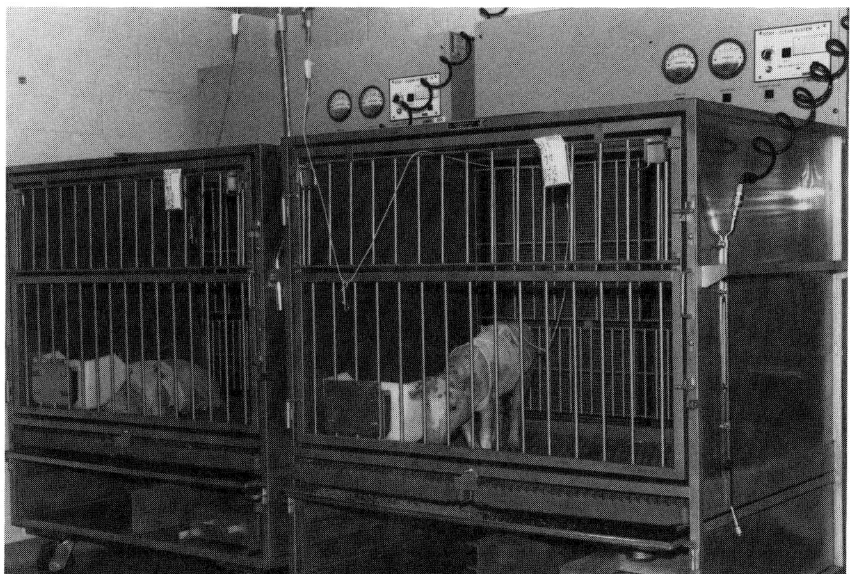

Figure 1 Laminar flow cages utilized for BMT recipients.

for baseline CBC, electrolytes, BUN, and serum creatinine while the animal is anesthetized. Following the operation, the animal is segregated from other animals. The animal is fed the day following surgery and receives Cephatabs (cephadroxil) (200 mg bid) through the gastrostomy for 2 days.

On day –7, the animal is moved into a laminar flow unit (Figure 1) which is scrubbed with a topical bactericidal agent daily. Persons handling the animal wear clean but not sterile gloves. A bowel prep is begun on day –7 and stopped on day –1, which consists of Tylocin 10 cc per GT q.d., Noroxin (norfloxacin) 200 mg per GT b.i.d., Ditrim (trimethoprin-sulfadiazine) 1.0 cc per GT q.d. and Nystastin 250,000 U per GT b.i.d. The recipient is kept NPO after day –1. On day –2, the donor animal arrives from the farm, receives a betadine scrub, and is placed in a clean run.

B. BONE MARROW HARVEST

On day –1, the donor and recipient are taken to the operating room, sedated, intubated, and anesthetized with the same technique as described for GT placement. The animals are prepped, scrubbed, and draped in sterile fashion.

1. Recipient

The external jugular vein is exposed and a silastic central venous catheter is placed so that its tip resides in the superior vena cava. The catheter is tunnelled through the skin in the dorsal neck and the catheter cuff is placed deep within the subcutaneous tissue. Keflin 1 g is then given i.v. If autologous marrow is needed, it is removed from the proximal humerae and tibiae, by cutting down over the bones, incising the periosteum, making a square window in the bone cortex with an osteotome, and scooping the marrow out using a curette. Care must be taken not to enter the joint space, since such entry may increase the animal's discomfort and limit its mobility after BMT, sometimes to the point of requiring euthanasia. The harvest sites are packed with methylmethacrylate, the bone cortex window replaced, and the wounds closed with absorbable suture.[4] An elastic netting is fitted for the upper torso to protect the venous catheter and the animal is allowed to awaken. This technique of marrow excision rather than aspiration, as is done in man, is necessary because the pig's marrow cavity is like a sponge rather than the more fluid marrow space in man.

2. Donor

The external jugular vein is exposed and cannulated with a segment of sterile plastic tubing with a large bore and the animal is exsanguinated. Marrow is harvested with curettes from the humerae, tibia, and distal femurs. If necessary, the vertebrae, which are very rich in cells in pigs as they are in man[11] are taken through a transabdominal approach. The bone fragments are minced with scissors and placed into a 1-l plastic container with 500 cc RPMI 1640 (without glutamine) and 1 mg% DNase. The number of cells that can be obtained by this method is approximately 1×10^{10} for a 20-kg animal, with about 30% more obtained when the vertebra are harvested. We have found that as the animals get older and bigger, their marrow becomes more fibrous and less cellular; thus the best donors weigh 20 kg or less. A split thickness skin graft is taken using the Brown dermatome at 0.020 in. to be prepared in the laboratory for controlled rate freezing in pig serum and DMSO by protocol.

C. MYELOABLATIVE THERAPY

Most animals are prepared for BMT with TBI and are given the first dose within hours following the central line placement. Our current TBI protocol for MHC mismatched BMT is either 650 cGy on day –1 and 650 cGy on day 0 or 50 mg/kg cyclophosphamide i.v. bolus followed by 500 cGy on day –1 and 650 cGy on day 0. When cyclophosphamide is used, the animal also receives 1000 ml Ringer's lactate i.v. to obviate the problem of hemorrhagic cystitis. The TBI is administered at 20 cGy/min and we have used a Cobalt irradiator or a linear accelerator with equal success. All radiation doses are calculated to the midline of the animal. The animals are placed in a transport cage with plastic siding which prevents them from turning during the TBI to ensure accurate radiation dosing.

Radiation toxicity remains a significant problem in this model, particularly when manifested by pulmonary problems. Less toxicity with equal immunoablative effect may be possible with more fractionation of the TBI dose, a possibility that is being investigated.

D. BONE MARROW PREPARATION

The marrow fragments are taken to the laboratory and stirred for 1 hour at room temperature. The marrow is prepared as described in Table 1 and is infused on day 0, shortly after the last dose of radiation. Before the marrow is administered, it is filtered a final time, recounted, and infused slowly by syringe via the central venous catheter. When the mixed marrow protocol is followed, the T-cell depleted autologous marrow is given first followed immediately by the allogeneic marrow.

E. POST BMT CARE

1. Platelet and Whole Blood Therapy

Because bleeding from thrombocytopenia is a common problem in the time prior to engraftment, the platelet count is followed closely in the first 2 weeks post BMT. When the platelet count is less than 50,000 or if bleeding occurs (usually evidenced by subcutaneous ecchymosis or gastrointestinal [GI] bleeding) with a higher platelet count but the animal has not yet engrafted, platelet transfusions are given. In animals requiring platelet transfusions, one U of platelets per day are given starting on day 5 or 6 and continuing for up to 10 days. In some animals 2 U per day may be necessary. The platelets are obtained from large (generally greater than 45 kg) blood donor animals who have had a central venous catheter placed for the purpose of blood donation. The protocol for blood

Table 1 **Bone Marrow Preparation**

1. In a sterile hood, filter bone marrow through nylon mesh stretched over 1000-ml beakers. Rinse the original marrow container with Hanks Balanced Salt Solution (without phenol red) and add to the beaker.
2. Pour filtrate into 50-ml conicals and spin at 1200 rpm for 10 min.
3. Slowly pour off supernatant, add 20 ml ACK lysing buffer to each pellet, vortex, and let sit for 5 min. Resuspend with Hanks and spin at 1200 rpm for 10 min.
4. Prepare 100 ml of bone marrow storage media.

RPMI 1640 (without glutamine)	88.5 ml
DNase (1.0 mg/ml)	5.0 ml
Serum from donor	5.0 ml
Penicillin (0.06 mg/ml)/Streptomycin (0.135 mg/ml)	0.5 ml
Gentamicin (50 mg/ml)	0.1 ml
L Glutamine (2 m*M*)	1.0 ml

   Filter (45 μ) medium into sterile 200-ml flask
5. After spinning the bone marrow, pour off supernatant and combine all conicals into sterile 275-ml culture flask. Use the bone marrow storage media to achieve a final volume of 100 ml of resuspended bone marrow cells.
6. Count the number of cells and store overnight at 4°C.
7. On day 0, refilter and recount, prior to infusion.

donation is shown in Table 2. All blood products given are MHC matched for the BMT donor and irradiated with 3000 cGy. Packed red cells are given when the hemoglobin is less than 8 g/dl or there has been an acute blood loss.

2. Feeding

Animals may vomit for a few days following TBI and are generally anorectic for up to 10 days. They are offered food and have *ad lib* access to water during that time, but require i.v. fluid support. Intravenous hyperalimentation is started on day 1 and is continued until the animal is tolerating an oral diet without severe diarrhea, usually

Table 2 **Blood and Platelet Donation Protocol**

Donor animals must be at least 45 kg in weight and MHC matched with the bone marrow donor. Under general anesthesia, a Hickman catheter is placed in the external jugular vein. The next day, animals are bled in the following manner,

1. The blood is withdrawn into a 60-ml syringe attached to a three-way stopcock, with one port attached to a citrate blood collecting bag. Before withdrawing any blood, 10 cc of citrate is brought into the syringe to prevent clotting.
2. 400 ml of blood is placed into each of two bags and 1500 ml of 5% dextrose lactated ringers (D5/RL) is given to the animal if no RBCs are being returned. The blood is placed in a blood bag, centrifuged, and the platelets, plasma, and RBCs are separated.
3. Usually only the platelets are given to the BMT recipient, thus the RBCs and plasma can be returned to the blood donor when blood is drawn the next time. The animals may be bled as frequently as three times per week. Following removal of the blood, the RBCs and plasma are warmed and returned to the blood donor with additional D5/RL to equal 1500 ml. Returning this blood allows the animals to tolerate bleeding for several weeks.
4. Once all fluids have been given, 2 g of Keflin are given i.v., the catheter is flushed with 10 units/ml heparin in saline, and the catheter is capped.

Table 3 **Intravenous Hyperalimentation Protocol Post BMT**

Day	No. bottles	D50 (ml)	10% AA (ml)	20% Lipid (ml)	KCl (mEq)	Vitamins (ml) B	C	K
1	1	200	250	150	20	1	1	0
2	1	300	350	250	20	1	1	0
3	1	300	350	250	20	1	1	0
4	1	300	350	250	20	1	1	1[a]
5	1	300	350	250	20	1	1	0
6—14	2	300	350	250	20	1	1	0
14—21	1	300	350	250	20	1	1	0
21+[b]								

[a] Vitamin K is given once a week; [b] Once the animal is able to tolerate oral or gastrostomy feedings and maintain its weight, it is rapidly tapered from i.v. hyperalimentation. Occasional animals require i.v. nutritional support beyond 21 days.

between day 14 and 21. The protocol for i.v. hyperalimentation is shown in Table 3. After day 7, the animals are encouraged to eat and if diarrhea is not a problem, the diet may be supplemented with p.o. or per GT elemental feedings such as Isocal. Every effort is made to return the animals to enteral feeding as rapidly as possible and often the diet must be individualized (some animals favor fruits while others prefer dog food!).

3. Antibiotics

Ditrim (trimethoprim-sulfadiazine), Noroxin (norfloxacin), and nystatin are given per GT from pre BMT, as mentioned previously, to 6 weeks post BMT. Broad spectrum anti-pseudomonal intravenous antibiotics Piracil (piperacillin) 1 g b.i.d. and Cefobid (cefoperzone sodium) 1 g b.i.d. are given days 1 to 10, or until engraftment has occurred. We have noted less problems with psuedomonas infections, particularly pneumonia, since the protocol has included these medications.

4. Central Venous Catheter Care

Because of the multiple blood samples necessary for clinical care and *in vitro* assays in addition to the need for intravenous medications and nutrition, it is essential to keep the central venous catheter functional and aseptic. Our preferred catheter is the Hickman because the wall is thicker than the Broviac and tearing or puncture of this catheter is unusual. Although catheter sepsis does occasionally occur, it is uncommon.

Strict attention to aseptic technique is essential. Clean gloves must be worn whenever manipulating the catheter. The catheter hub is wiped with alcohol prior to removing the cap for blood drawing or hyperalimentation administration. When phlebotomy is performed, 3 ml of blood is withdrawn and discarded. After samples have been drawn, the catheter hub is again wiped with alcohol and a new sterile catheter cap or intravenous line for hyperalimentation is connected. When an intravenous infusion is completed, the hub is again wiped with alcohol and a sterile cap applied. Following blood drawing or intravenous fluid infusion, the catheters are flushed with 5 ml of saline with 10 U heparin/ml and capped.

5. Vital Signs Monitoring

Temperature (via an ear thermometer) and respiratory rate are recorded q.d. for day 1 to 14 and thereafter only as needed. Attention to the respiratory rate is particularly important 21 to 50 days after BMT because the risk of radiation pneumonitis is greatest during that time. Body weight is recorded weekly. The animals are evaluated by the investigators twice daily until they are off all medications and have passed the risk period for GVHD and radiation pneumonitis, about 6 weeks.

6. Expected Course Following BMT

On day 0 and 1, the animals commonly have decreased levels of activity and may be anorectic. Thereafter, they appear more normal until symptoms of GI radiation toxicity such as diarrhea develop, usually by day 5. Engraftment (defined as WBC > 1000/ml^3), usually occurs between day 6 and 13 and only rarely later than day 16. Radiation control animals not given marrow or animals that do not engraft usually die before day 18 despite intensive supportive measures. The cause of death in aplastic animals is usually a combination of sepsis and hemorrhage with the GI tract and the lung most severely affected.

Animals receiving autologous, MHC matched or F1→P BMT have a survival rate in excess of 85% with TBI of 650 cGy × 2. Death in the other 15% is caused by failure to engraft, infection, or radiation pneumonitis. GVHD, which will be discussed later, occurs in some MHC matched and F1→P BMTs due to minor antigen disparity.

F1→F1 BMT have been performed using both unmodified allogeneic marrow or the mixed-marrow protocol (T-cell depleted autologous marrow plus non-T-cell depleted allogeneic marrow) with a survival rate of approximately 60% and no significant difference in survival between the animals receiving the two marrow inocula. TBI for the F1→F1 BMT was either with 650 cGy × 2 or 500 cGy + 50 mg/kg cyclophosphamide + 650 cGy. Death in the F1→F1 animals was commonly due to GVHD, infection, or radiation pneumonitis.

P1→P2 (complete MHC mismatched) BMT eventually resulted in a fatal outcome. TBI was 650 cGy × 2. When unmodified allogeneic marrow was used alone, most animals died within 20 days, usually from severe GVHD and infection. When the mixed-marrow protocol was used, 41% died before 21 days, 41% died between 21 and 34 days, 12% survived 35 to 58 days, and one animal (6%) survived 154 days.[12] GVHD and infection were still the most common cause of death, although the animals surviving more than 1 month had radiation pneumonitis as a contributing factor in their demise. In those F1→P and F1→F1 animals that survive long term, their general health is very good. They have normal GI function and recurrent infections are unusual. They do not grow to a normal size, attaining only about 60% of the normal body weight at maturity. Their activity is normal. At least one of the female BMT pigs has become pregnant. They do have chronic skin changes, with some loss of pigmentation in spotted animals and dry skin, probably due to TBI.

7. Radiation Toxicities

Toxicity of the preparative regimen of TBI, with or without cyclophosphamide, is a significant component of the morbidity and mortality of this model. As mentioned in the background information, the total TBI dose was determined by testing various MHC combinations with various TBI doses. It was shown that TBI of 1300 cGy split into two equal doses was necessary for consistent engraftment of F1→P BMT with an acceptable rate of radiation toxicity. Although most of the animals have diarrhea for a few weeks following BMT, they regain normal GI function and as long as their early post BMT nutritional needs are met by intravenous feeding they recover well.

Pulmonary problems are the usual cause of death in animals that have engrafted. The combination of GVHD, endogenous immunosuppression, infection, and radiation pneumonitis makes the pulmonary system the most susceptible to a fatal complication. In seven animals with radiation pneumonitis, we have been able to successfully treat the condition in four with i.v. prednisolone 1 mg/kg q.d. for 1 week then 0.5 mg/kg q.d. for 1 week. All animals treated for radiation pneumonitis are also given broad spectrum i.v. antibiotics.

Radiation pneumonitis generally occurred between day 21 and 50 and was evidenced by dyspnea, tachypnea, and a diffuse perihilar infiltrate on chest X-ray. When dyspnea was severe, the animals were placed in an oxygen box. Animals which responded favorably to the prednisolone treatment, did so within several days, and there were no recurrences of pulmonary problems following discontinuation of the prednisolone.

8. GVHD

GVHD remains a potent barrier to BMT in miniature swine, just as it does in man. With the exception of autologous BMT, all animals are at risk to develop GVHD. In the MHC matched or F1→P BMT, GVHD due to minor antigen disparity is mild and self limited. In P→F1, the GVHD is severe and usually lethal unless the allogeneic marrow is T cell depleted.[6] In F1→F1, the GVHD is severe but the majority of animals will spontaneously improve. In P1→P2, GVHD is a major cause of death.

The manifestations of GVHD in this model are similar to those in man. Development of a erythematous, macular rash coincides with engraftment as the WBC rises to greater than 800/ml^3 between day 5 and 14. The rash usually begins around the ears and face. If the rash is localized and mild, it is called Grade I. If the rash is generalized, but not exfoliative, it is called Grade II. If the skin is sloughed, leaving the dermis exposed, it is called Grade III.[6] Histologic evaluation of this rash has confirmed these changes to be consistent with GVHD of the skin. Grades I and II usually recede, with flaking of the epidermis and healing within 2 weeks. Grade III may heal, but is usually seen only in those animals with lethal GVHD. A few animals have had more than one episode of skin rash following BMT, but usually, once the rash has cleared it does not return.

The GI manifestations of GVHD may be masked by the bowel prep that all recipients receive and the acute radiation toxicity caused by TBI. Animals dying following BMT often have so much hemorrhage in the intestine that the presence of GVHD is difficult to detect. Animals with autologous BMT have diarrhea for 5 to 10 days, then rapidly return to normal GI function. Animals developing a Grade I or II GVHD rash also have diarrhea, but it is for a longer period of time, up to 1 month, suggesting that at least part of the GI dysfunction is due to GVHD. Hematochezia may occur during the time of thrombocytopenia, but is rare in those animals with spontaneous resolution of GVHD and survival. Anorexia occurs during the time that the GVHD rash is at its peak, although the use of i.v. hyperalimentation may be a contributory factor.

Hyperbilirubinemia is a common occurrence in animals with a severe GVHD rash and uncommon in animals without GVHD rash post BMT. Liver biopsies have shown lymphocytic infiltration but not the venoocclusive disease seen in human GVHD. A confounding factor in the liver dysfunction is that all animals have limited oral intake and are on i.v. hyperalimentation when jaundice appears, usually in the second week post BMT. Various manipulations of the hyperalimentation formula have not altered the incidence of jaundice. The liver dysfunction is usually self limited when the rash is Grade I or II, although some animals have died primarily of hepatic disease.

During active GVHD, the recipient is known to be at significant risk for infection due to T-cell dysfunction. The lung is very susceptible to infection during GVHD and often is the cause of death in lethal GVHD in this model. Once the animal is engrafted and beyond the stage of active GVHD rash, at about 1 month, the risk of serious infections dramatically decreases. Some animals develop recurrent skin infections or late pneumonias, but the vast majority of long-term survivors have no problems with infections, despite the fact most require 3 months or longer to regain *in vitro* T-cell responsiveness.

9. *In Vitro* Assays

In vitro assays are performed to confirm engraftment with allogeneic bone marrow and to determine the immunocompetence of the animal at various times following BMT. All animals have blood drawn for MHC typing and MLR pre BMT. The MHC typing is performed by Flow Cytometry (FCM) and is repeated at 2 weeks post BMT to confirm allogeneic engraftment. At 3 weeks, FCM is performed to evaluate recovery of the various T-cell subsets and MLR and PHA stimulation are performed to test *in vitro* T-cell responsiveness. At 4 weeks post BMT, FCM for MHC typing and for T cell subsets, and PHA, MLR, and CML are performed. These tests are repeated monthly, for 6 months, except for the typing, which is repeated only at 3 months to confirm persistence of allogeneic engraftment.

III. APPLICATION OF PIG BMT FOR EXPERIMENTAL STUDIES

The primary focus of this laboratory with pig BMT has been the induction of tolerance by creation of a lymphohematopoietic chimera in a large animal model across a known MHC barrier. An important component of this goal has been to generate chimeras without the use of long-term immunosuppressive medications so that immunocompetence is maintained to third party antigens, while recipients specifically lose reactivity to donor MHC. This goal has clearly been accomplished for F1→P and F1→F1 BMT.[10] As of yet, the P1→P2 BMT have not been successful long term, but innovative methods of GVHD prevention may permit this barrier to be crossed in the near future. The similarities in anatomy and physiology of man and pig make this model very attractive for use in experimental solid organ transplants, as discussed elsewhere in this book. This model thus allows careful immunologic study of both bone marrow and solid organ transplantation in the same animal.

The potential clinical implications of this model are numerous. Methods of large animal myeloablation, GVHD prevention and treatment, and testing of new products (e.g., growth factors) to encourage more rapid or successful engraftment of allogeneic marrow can be tested with this preclinical model. The exciting prospects of utilizing miniature swine for the introduction of genetically altered information are currently being investigated. Finally, the size of these animals closely approximates that of man and many of the supportive measures required after BMT are identical to those needed in patients, making these animals particularly useful in answering clinical questions of patient care after BMT.

ACKNOWLEDGMENTS

This work was supported in part by NIH grant RO1 CA55553.

REFERENCES

1. **Billingham, R.E., Brent, L., and Medawar, P.B.,** Actively acquired tolerance of foreign cells, *Nature,* 172, 603, 1953.
2. **Deeg, H.J., Severns, E., Raff, R.F., Sale, G.E., and Storb, R.,** Specific tolerance and immunocompetence in haploidentical, but not completely allogeneic, canine chimeras treated with methotrexate and cyclosporine, *Transplantation,* 44, 621, 1987.
3. **Sachs, D.H., Leight, G., Cone, J., Schwarz, S., Stuart L., and Rosenberg, S.,** Transplantation in miniature swine. I. Fixation of the major histocompatibility complex, *Transplantation,* 22, 559, 1976.
4. **Pennington, L.R., Sakamoto, K., Poplitz-Bergez, F.A., Pescovitz, M.D., McDonough, M.A., MacVittie, T.J., Gress, R.E., and Sachs, D.H.,** Bone marrow transplantation in miniature swine. I. Development of the model, *Transplantation,* 45, 21, 1988.

5. **Poplitz-Bergez, F.A., Sakamoto, K., Pennington, L.R., Pescovitz, M.D., McDonough, M.A., MacVittie, T.J., Gress, R.E., and Sachs, D.H.,** Bone marrow transplantation in miniature swine. II. Effect of selective genetic differences on marrow engraftment and recipient survival, *Transplantation,* 45, 27, 1988.
6. **Sakamoto, K., Sachs, D.H., Shimada, S., Popitz-Bergez, F.A., Pennington, L.R., Pescovitz, M.D., McDonough M.A., MacVittie, T.J., Katz, S.I., and Gress, R.E.,** Bone marrow transplantation in miniature swine. III. Graft-versus-host disease and the effect of T cell depletion of marrow, *Transplantation,* 45, 869, 1988.
7. **Ildstad, S.T. and Sachs, D.H.,** Reconstitution with syngeneic plus allogeneic or xenogeneic bone marrow leads to specific acceptance of allografts or xenografts, *Nature,* 307, 168, 1984.
8. **Suzuki, T., Sundt, T.M., Kortz, E.O., Mixon, A., Eckhaus, M.A., Gress, R.E., Spitzer, T.R., and Sachs, D.H.,** Bone marrow transplantation across an MHC barrier in miniature swine, *Transplant. Proc.,* 21, 3076, 1989.
9. **Sundt, T.M., Suzuki, T., Kortz, E.O., Eckhaus, M.A., Gress, R.E., and Sachs, D.H.,** Induction of specific transplant tolerance in a large animal model by bone marrow transplantation, *ACS Surg. Forum,* 39, 365, 1988.
10. **Guzzetta, P.C., Sundt, T.M., Suzuki, T., Mixon, A., Rosengard, B.R., and Sachs, D.H.,** Induction of kidney transplantation tolerance across MHC barriers by bone marrow transplantation in miniature swine, *Transplantation,* 51, 862, 1991.
11. **Sharp, T.G., Sachs, D.H., Matthews, J.G., Maples, J., Woody, J.N., and Rosenberg, S.A.,** Harvest of human bone marrow directly from bone, *J. Immunol. Meth.,* 69, 187, 1984.
12. **Smith, C.V., Suzuki, T., Guzzetta, P.C., Sundt, T.M., Mixon, A., Spitzer, T., Eckhaus, M.A., and Sachs, D.H.,** Bone marrow transplantation in miniature swine: IV. Development of myeloablatine regimes that allow engraftment across major histocompatibility barriers, *Transplantation,* 56, 541, 1993.

Section VII

Xenotransplantation

Chapter 21

The Use of Xenografts in Experimental Transplantation

Donald V. Cramer and Leonard Makowka

CONTENTS

I. INTRODUCTION

Transplantation of vascularized organs has become an accepted, effective treatment regimen for a variety of end-stage organ diseases. The success of transplantation has led to an increase in the frequency of procedures performed for all organs. The number of transplants has increased to such a level that the number of procedures that can be performed is primarily limited by the availability of donor organs. For the last 6 years, for example, the number of cadaveric renal transplants has been relatively stable at approximately 9500 per year, a number that closely reflects the total number of available donor organs. While the rate of transplantation of other vascularized organs has been growing rapidly, limitations in the number of donor organs will impose a similar plateau for transplantation of these organs in the near future. The rate of liver transplantation, for example, has grown approximately 20% each year and the number of transplants performed in 1990[1,2] represents approximately 60% of the total number of organs that would be available under the most favorable conditions. It is clear that the future application of organ transplantation at any substantial level of expansion will depend upon a significant increase in the number of available organs.

The use of xenografts for organ transplantation offers several distinct and clear advantages over cadaveric organs for organ transplantation. The first and most important is the potential availability of a predictable and ready supply of donor organs. The current shortage of donor organs is the most important limitation on the size of the pool of recipients that may benefit from organ transplantation. An increase in the availability of suitable organs will result in a large increase in the number of transplants performed. The availability of donor organs would also be accompanied by the opportunity of being able

0-8493-3629-5/94/$0.00+$.50

Table 1 **Cardiac Xenograft Rejection in Small Laboratory Species**

Donor	Recipient	N	Mean survival time
Hyperacute			
Guinea pig	Rat	5	14.8 min
Rat	Guinea pig	5	87.6 min
Mouse	Guinea pig	5	8.4 min
Hamster	Guinea pig	5	34.0 min
Accelerated			
Hamster	Rat	5	3.9 days
Mouse	Rat	5	2.7 days
Rabbit	Rat	5	2.0 days

Modified from References 4 and 15.

to perform the transplant as an elective procedure. The timing of the transplantation procedure is essential in obtaining a good outcome for the patient and the early application of a transplant has the effect of improving overall graft survival. A third important advantage of xenografting would be the opportunity to match the size of the graft with small adults and children. Despite the development of innovative surgical techniques, there remains a difficulty in obtaining suitably sized organs for this group of patients.

II. IMMUNOLOGY OF XENOGRAFT REJECTION

At the present time, the application of xenografts for the treatment of end-stage organ disease is limited by our ability to understand and control the xenograft reaction. The exchange of vascularized organs between two different species can frequently be accompanied by the immediate and catastrophic loss of the graft. A large number of species react to the foreign graft by mounting a very rapid (hyperacute) rejection of the graft (Tables 1 and 2). In experimental models and selected clinical cases, this rapid graft loss has been extremely difficult to control and has been a major factor in the reluctance to apply this technology in the treatment of patients.

A. CLASSIFICATION OF XENOGRAFT REACTIONS

The exchange of tissues between two species results in an immune-mediated rejection reaction that is more intense than that observed for the transplantation of tissue between members of the same species (allografts). The severity of the xenograft rejection reaction has traditionally been considered to reflect the genetic disparity between the donor and recipient species.[3] The more distant the two species, the more likely that the donor and recipient species will exhibit natural xenoantibodies that react with the tissues of the donor graft and precipitate a violent rejection reaction. Xenografts exchanged between species with high levels of natural antibody are considered to be "discordant" reactions. Discordant reactions are hyperacute reactions that occur within minutes after revascularization of the transplanted organ. They are primarily the result of an antibody-mediated vascular injury and disseminated coagulopathy within the small vessels of the graft. In some species combinations, (such as several species of rodents[4] and humans/rabbits[5]), however, hyperacute rejection of the graft may occur in the absence of anti-donor antibody, probably as the result of the direct activation by tissues of the graft of the alternate complement pathway.

Table 2 **Common Donor and Recipient Xenograft Combinations in Large Animal Species**

Donor	Recipient	Organ	Type of reaction
Pig	Dog	Kidney	Hyperacute (10 min)
	Rhesus monkey	Heart	Hyperacute
Hare	Rabbit	Kidney	Acute (6 day)
Rabbit	Cat	Kidney	Hyperacute (10 min)
	Pig	Kidney	Hyperacute (1 hour)
Pig	Rabbit	Kidney	Hyperacute (1 hour)
	Baboon	Heart	Hyperacute (1–8 hour)
		Liver	Accelerated (6 hour–3.5 day)
		Kidney	Hyperacute
Vervet monkey	Baboon	Heart	Acute (10 day)
Cynomolgus	Baboon	Heart	Acute (7 day)
Fox	Dog	Heart	Acute (8 day)
		Kidney	Acute (5–6 day)
Sheep	Goat	Kidney	Acute (11 day)
Calf	Goat	Heart	Acute (5–7 day)
Lamb	Goat	Heart	Acute (6 day)
Cat	Dog	Kidney	Hyperacute (1/2 hour)

Modified from References 3 and 27.

Xenografts exchanged between closely-related species may not exhibit high levels of preformed anti-donor antibodies and the xenograft is rejected in an aggressive, first-set rejection of the organ.[6] These "concordant" reactions occur with an accelerated pace that results in rapid graft loss within a few days. As described below, recent experimental evidence suggests that the rejection of the majority of xenografts is mediated by anti-donor antibody and that the tempo of the reaction reflects the level of antibody present in the recipient at the time of the organ transplant. The distinction between the two patterns of rejection implied by traditional classification schemes is based upon genetic disparity alone may therefore not be appropriate.[4]

B. HYPERACUTE XENOGRAFT REJECTION

Hyperacute rejection consists of an immediate, diffuse intravascular coagulopathy. The explosive and violent inflammatory and coagulative response can be the result of the binding of naturally occurring xenoantibodies of the recipient to the vascular endothelium of the donor graft, followed by the activation of the classical inflammatory cascades, the release of soluble mediators and the initiation of disseminated intravascular coagulation.[3,7] The final result is thrombosis of small vessels, ischemic necrosis of the tissue, and rapid loss of the graft. The antibody that mediates this reaction is primarily an IgM antibody that is present in the serum of normal recipients, probably in response to exposure to environmental antigens or intestinal organisms. In experimental animals, natural xenoantibodies are produced after birth and appear to be the result of exposure to environmental antigens such as food and microorganisms.[8]

The pathological features of the hyperacute rejection of discordant xenografts are poorly documented but are consistent with an antibody-mediated injury to vascular endothelium. The primary changes that have been described include endothelial cell swelling, necrosis, interstitial edema, platelet and fibrin thrombi, and hemorrhage.[9-11] Although the exact nature of the antibodies and target antigens that mediate the reaction

Table 3 **Rejection of Cardiac Xenografts by Sensitized Rat Recipients**

Challenge Graft	N	MST
LEW Recipients of Hamster Hearts		
Hamster	5	4.70 ± 2.9 min
Mouse	5	4.25 ± 0.5 min
Rabbit	3	4.00 ± 1.2 min
ACI	5	5.00 ± 2.0 day
LEW Recipients of Mouse Hearts		
Mouse	3	8.00 ± 2.0 min
Hamster	2	20.00 ± 10.0 min

is not clearly understood, the changes are compatible with antibody binding to the endothelium with activation of complement and other inflammatory cascades.

Hyperacute rejection is also seen between selected species combinations in the absence of preformed antibody.[5,12,13] As described below for several combinations of rodents, the mechanism by which this reaction is mediated is different than that observed in species with preformed antidonor antibody. In these models, the hyperacute rejection may represent a direct activation of the alternative complement pathway by antigens within the graft, rather than the traditional activation of complement by the binding of natural antibody to the donor graft.

C. ACCELERATED XENOGRAFT REJECTION

Some xenograft reactions are characterized by an accelerated rejection that occurs within a few days. In general, this type of reaction is seen between species that do not exhibit significant levels of naturally occurring xenoantibodies. It is clear from studies in rodents, however, that even in the species combinations with accelerated rejection, the primary damage to the graft is due to anti-donor antibody and that the delay in graft rejection is due to a lag in antibody production by the recipient (Table 3). Once the recipient has been sensitized to donor tissue, subsequent challenge grafts are rejected hyperacutely.[14,15] Similar studies in the fox-to-dog combination illustrates the role that humoral responses play in the rejection of xenografts between closely related species.[16] Untreated dogs reject fox kidneys in approximately 5 to 6 days posttransplantation, primarily due to humoral immune responses. Antibodies to fox tissues can be detected at approximately 4 days posttransplantation and treatment with CsA has little effect on preventing the reaction.

III. EXPERIMENTAL MODELS

The recognition that the basic pathogenetic mechanism involved in the immune rejection of xenografts is very similar, even for widely diverse donor/recipient combinations, provides greater confidence that the data derived from experimental studies is relevant for application to clinical xenografting in humans. In this section, we intend to briefly describe the most commonly used xenograft models, the pertinent features of the individual models for their application to clinical xenografting, and the species that represent the most likely candidates for use as donor for xenografts in patients.

A. RODENTS

Animals within the general order of *Rodentia* represent one of the most important and frequently used models for experimental studies of xenograft transplantation. The traditional advantages of using rodents for research, including cost, availability, ease of maintenance, and genetic uniformity, also apply to their use for xenografting. In addition, the variety of different species available provides for the choice of donor/recipient combinations that span a wide range in intensity of the xenograft reaction. The species most commonly used within this group include rats, mice, hamsters, guinea pigs, and rabbits (Order Lagomorpha).

The rat as a recipient for xenografts from other rodent donors has several distinct advantages when compared to other potential donor/recipient combinations. The rat has traditionally served as the most important small experimental animal for studying the rejection of vascularized organ allografts. The microsurgical techniques for the transplantation of organs such as kidney, liver, heart, pancreas, and small bowel are well established for the rat in many laboratories. These techniques are generally applicable, with minor modifications to account for size differences, to donor organs from other species.

The use of rats as recipients of xenografts also offers the opportunity to study a wide range of xenograft reactions, including reagents that are available for a detailed immunopathological characterization of the rejection reaction. Several species, including the mouse, hamster, and new-born rabbit, can serve as xenograft donors for rats. Heterotopic cardiac xenografts from these donors stimulate rejection of the graft by the rat recipients in 2 to 4 days posttransplantation (PTX) (Table 1). Of these species, the Syrian Golden hamster has been the most widely used as an organ donor. The heterotopic heart and orthotopic liver allograft procedures in the rat (Chapters 5 and 12) form the basis for the surgical techniques utilized for these organs in the hamster-to-rat model. After transplantation, the hamster heart grafts are rejected in 3 to 4 days and the liver grafts in approximately 7 days.

The rat normally exhibits low but demonstrable levels of IgM antibody that bind to hamster xenografts, primarily to the vascular endothelium. After transplantation and rejection of the graft, the levels of rat anti-hamster xenoantibodies rise sharply. The recipient is sensitized to the graft and a second cardiac graft is rejected in a hyperacute fashion.[15] At rejection, the lesions present in the heart xenograft are the same as those seen for hyperacute rejection and consist of antibody-mediated vascular injury with endothelial damage, edema, and hemorrhage. Small amounts of plasma from rat recipients at the time of rejection can be used to passively transfer the ability for naive recipients to mount a hyperacute rejection of heart xenografts from hamsters, other rodents (mice and guinea pigs) and newborn rabbits (Table 3).

It is not clear whether the sharp rise in anti-donor antibody seen in the concordant rodent models represents polyspecific natural antibody production or the induction of a primary T lymphocyte-dependent antibody response. It is possible that the response represents the stimulation of a preexisting population of lymphocytes, perhaps $CD5^+$ B lymphocytes, that are responsible for producing polyspecific IgM natural antibodies that cross-react with the graft.[17] Alternatively, the response may represent a traditional primary immune response in which the early antibodies produced are IgM and are stimulated by exposure to the heart graft antigens. The rejection of hamster cardiac xenografts is associated with the appearance of a large number of splenic lymphocytes in the rat that exhibit features of $CD5^+$ B lymphocytes. It remains to be established whether the antibodies produced have the broad reaction patterns of natural antibodies. The positive reaction patterns of these antibodies against a variety of different rodent species suggest that this may be true.[15] Alternatively, the rejection of the hamster hearts is associated with the emergence of antibodies that recognize new xenoantigens, suggesting that the response may be a combination of natural and induced immune responses.[18] While the details of the antibody response remain to be established, it is likely that these concordant combinations are appropriate models for comparison to clinical non-human primate-to-human xenografts.

Rodents also exhibit hyperacute rejection of xenografts (Table 1). Four examples of hyperacute rejection are seen for cardiac grafts exchanged between rats, guinea pigs, and hamsters.[15] The guinea pig-to-rat cardiac xenograft model has been most carefully studied and the role that natural antibody plays in this reaction is still controversial.[20] It would appear that the rat has natural antibodies that bind to guinea pig tissues, particularly lymphocytes and endothelial cells, but the role that these antibodies play in the hyperacute rejection of heart grafts has not been well-defined. Extensive removal of the antibodies prior to transplantation does not influence hyperacute rejection of the graft.[21] Rejection of the graft in the rat is accompanied by evidence of direct activation of the alternative, rather than classical complement pathways.[22] In this model, it is possible that the hyperacute rejection may not depend upon antibody binding to activate complement and that direct endothelial activation of the alternative complement pathway may be the cause of the reaction.

Our preliminary examination of the rat-to-guinea pig, hamster-to-guinea pig, and mouse-to-guinea pig combinations has suggested that the same mechanism of rejection is seen in all of these models. None of these species exhibit cytotoxic natural antibodies against each other, and the hyperacute rejection of the graft is characterized by a lack of antibody binding to the donor graft endothelium. In each of these models, there is immunohistochemical evidence for the activation of complement without the binding of recipient antidonor antibodies to graft endothelium at the time of rejection.[15] These results suggest that these species may not be appropriate models for studying the type of hyperacute rejection mediated by preformed antibody. This may be an important limitation for the rodent models as this is the type of xenograft reaction that would be anticipated for some potential xenograft donors, such as the pig, for clinical transplantation in humans. It remains to be established that this mechanism of hyperacute rejection is not important in humans, however, as the reaction of human blood to rabbit heart may be mediated by the alternative complement pathway.[5]

B. LARGE ANIMALS

A variety of larger species have been used for xenografting. The most commonly used combinations of donor and recipient species have been included in Table 2. The pig-to-dog combination has been used most frequently to examine the general problems associated with hyperacute rejection. Kidney grafts exchanged between these two species are associated with a hyperacute rejection within 20 min. Perfusion of pig kidneys with citrated dog blood in an *ex vivo* perfusion system can mimic the tempo and pathological features of the *in vivo* model, allowing for a detailed examination of the hematological factors required for the rejection process to occur.[23] The use of separate components of whole blood, including plasma, erythrocytes, platelets, and neutrophils is not sufficient as a single agent to produce hyperacute rejection. All of these components are required together, demonstrating the interdependence of the humoral (antibody, complement) and cellular (RBC, platelets, PMNs) components of whole blood on the completion of the reaction.

C. CLINICAL APPLICATION OF XENOGRAFTING IN HUMANS

The potential for the use of xenografts in humans has focused attention on the species that might be used for donors for this type of transplantation.[24] The selection of a suitable donor depends upon a variety of physical, physiological, and ethical issues. Two groups of animals, because of their size and physiological similarities, represent the most likely candidates as human xenograft donors and thereby the most important experimental models. They include non-human primates, particularly baboons and rhesus monkeys, and pigs.

The use of adult rhesus monkeys or baboons as donors has the distinct advantage because of their genetic similarity to man. This similarity may play an important role in reducing the aggressiveness of the host rejection response and improve the long-term ability of the graft to function in the host. The close genetic relationships of humans and other primates have formed the basis for previous attempts to use these species as donor for xenografts to humans. The first clinical xenografts using modern medical techniques were conducted by Reemtsma and his colleagues at Columbia.[25] Twelve kidney transplants from chimpanzees to humans were performed with variable success. Most of the grafts were rejected in a few weeks with one graft surviving 9 months. These cases clearly illustrated the ability of the transplanted organs to function normally in a xenogeneic host. They also demonstrated the ability of immunosuppressive therapy, although at that time limited in effectiveness, to prolong kidney xenograft survival.

The majority of the human xenografts that have been attempted have been conducted at a time when the surgical techniques were undergoing development and the therapeutic agents to prevent rejection were ineffective. The improvements in surgical techniques and the introduction of new, powerful immunosuppressive drugs has renewed interest in this type of transplantation. Recently, Starzl and colleagues transplanted a baboon liver to a male patient. The liver functioned normally for 2 months.[26] The patient died from sepsis and the liver once again displayed minimal evidence of cellular rejection, suggesting that this type of xenograft may be successful given sufficient experience in the medical management of these patients.

Despite the obvious advantages of using non-human primates for xenotransplantation, the disadvantages associated with the use of these animals precludes any widespread application of these species as xenograft donors. These disadvantages include ethical concerns of the use of these intelligent animals, the problems of shared susceptibility to serious diseases such as tuberculosis, hepatitis, retroviruses, and herpes B virus, and a limited number of animals to use as donors. All primates have low reproductive rates and the numbers of non-human donors of sufficient size available to serve as donors would not provide a solution to the current problem of the shortage of donor organs. It is likely that the future use of non-human primates will be for special, limited application, perhaps as donors for pediatric patients.

The most likely alternative to the use of non-human primates as xenograft donors to humans is the pig. The pig is anatomically and functionally very similar to humans[27] and could serve as a satisfactory donor for a variety of organs. The advantages associated with the use of pigs as xenograft donors are high reproductive and growth rates, factors which make it practical to develop sufficient numbers of grafts of suitable size for any potential application, the ethical acceptability of the pig as a donor, and a growing body of experimental data addressing the potential for this type of xenograft to succeed.

The primary disadvantage to the use of the pig as a xenograft donor is the potential for a hyperacute rejection of the graft due to the presence of preformed antibody. Normal humans have low to modest levels of circulating antibody that reacts to pig tissues. The antibody is primarily IgM, binds to pig aortic endothelium and lymphocytes, and is cytotoxic for both cell types.[28,29]

Although the expectation that pig-to-human xenografts will be rejected hyperacutely, there is data available from *ex vivo* perfusion studies to suggest that the pig liver may function effectively for limited periods of time in patients. The liver is generally less susceptible to hyperacute rejection,[30] perhaps due to a dual blood supply. One of the most reasonable early applications of pig xenografts for humans would be the use of these grafts for the temporary support of patients with fulminant liver failure. The pig liver has been used for short-term *ex vivo* support in approximately 100 patients to provide for metabolic detoxification due to acute liver failure.[24] This procedure provides for temporary and slight improvements in liver function tests. One of the most dramatic examples was the support of a single patient over a period of 11 weeks, using a combination of *ex vivo* perfusion of 16 liver xenografts from several different species.[31] These grafts included ten pigs, three baboons, and other miscellaneous species, including a calf. The patient responded to the perfusion as exemplified by the reversal of hepatic coma on at least eight separate occasions. The perfused livers produced bile and clotting factors that were functional. More recently, Makowka et al., performed an orthotopic pig liver xenograft to a patient with fulminent liver graft failure as a temporary bridge to transplantation with a human organ. This organ displayed functional activity but exhibited signs of hyperacute rejection 3 hours posttransplantation.[49]

D. EXPERIMENTAL MODELS FOR HUMAN XENOGRAFTING

The experiments described above demonstrate the potential for the use of the pig liver as a temporary xenograft for humans, particularly when combined with *ex vivo* absorption of natural antibodies and the use of immunosuppressive drugs to prevent new antibody formation. The placement of a pig xenograft for more prolonged periods will require substantial improvements in our understanding of the xenograft reaction, the effect exposure of the graft to a xenogeneic environment on organ function, and the development of improved immunosuppressive therapy. These improvements will involve the use of selected experimental models that most closely resemble the potential donor species and their human recipients.

There have been a variety of models used to examine the exchange of xenografts between primate species. The current status of xenografting with non-human primates suggests that the baboon will remain as the most frequent model for representing humans as potential xenograft recipients.[32] This is because of the size, genetic and anatomic similarity to humans, and current availability. Baboons and chimpanzees have been used as donors of hearts, livers, and kidneys for human xenotransplantation.[3] The baboon is also an important species to attempt to mimic the nature of the rejection that may result from the use of various species as potential donors in a clinical setting.[33] Examples of this include closely-related primate-to-primate transplants, including baboon to rhesus monkeys,[34] cynomolgus to rhesus monkeys,[35] vervet monkey to baboon,[33,36] and cynomolgus monkey to baboon[37] xenografts. In general, these species exhibit graft rejection that is extended over periods of several days, even a few weeks.

The potential for the use of the pig as a xenograft donor has led to experiments using xenografts exchanged between pigs and non-human primates, including baboons[38,39] and rhesus monkeys.[35] In both of these models, cardiac xenografts were hyperacutely rejected within minutes due to the deposition of immunoglobulin, complement, fibrin, and the formation of platelet thrombi in the donor graft. Preformed antidonor antibodies were reported to be present in the recipients and removal of the antibody prior to the transplant procedure may have been responsible for the prolonged survival of pig to baboon[40] and pig to rhesus[41] cardiac xenografts.

IV. FUTURE DIRECTIONS

Significant improvements in surgical techniques and immunosuppressive drugs have resulted in dramatic improvements in the survival of most vascularized allografts. These technical achievements have a direct application to the use of other species as donors for clinical transplantation. It is reasonable to expect that these advances will result in significant improvements in success rates for xenografts when compared to attempts made under less favorable conditions. It is also clear, however, that even the most recent attempt to perform baboon or pig liver xenografts to humans was not associated with any substantial improvement in graft survival, despite the use of the most modern technologies. This illustrates the necessity of making significant advances in our understanding of the xenograft reaction and our ability to mitigate aggressive rejection of the donor graft. We expect that the xenograft models described above will provide the experimental framework for the development of new knowledge in three general areas of future research; improved immunosuppressive control of the xenograft reaction, the induction of tolerance to donor tissues in the recipient, and procedures to modify donor xenograft antigen expression to avoid graft rejection.

Table 4 **The Survival of Hamster to Rat Cardiac Xenografts Following Treatment with a Combination of Brequinar Sodium and Cyclosporin A**

Experimental group	N	Median Survival (day)	P-value
Untreated controls	12	4.0 ± 0.4	—
BQR	12	5.5 ± 5.5	<0.001
CsA 15 mg/kg/day	8	5.0 ± 0.9	<0.03
BQR 3 mg/kg/day to Day 14; then 3 × week + CsA 10 mg/kg/day	7	7 × >100 day	<<0.001

Modified from Reference 45.

A. IMPROVED IMMUNOSUPPRESSION

The use of new immunosuppressive drugs, in combination with traditional immunosuppressive regimens, promises to provide an important improvement in our ability to prevent the xenograft reaction. Current immunosuppressive protocols that depend upon CsA, and the functionally related compound, FK 506, are highly effective for preventing allograft rejection. These compounds interfere with T-lymphocyte-mediated immune responses, the primary immunopathological responses responsible for rejection of allografts. These two compounds have little effect on the rejection of xenografts, perhaps due to a lack of significant disruption of B-lymphocyte-mediated antibody production. Several new immunosuppressive drugs, particularly Rapamycin and Brequinar sodium, display important effects on antibody production and act to significantly prolong xenograft survival.[42,43] The activity of these two drugs is particularly effective when used in combination with either CsA or FK 506. The use of either CsA or FK 506 in combination with BQR is highly synergistic and capable of inducing permanent survival of xenografts in rodents[44-46] (Table 4). These results are extremely encouraging as the removal of preformed antibody, followed by effective prevention of new antibody formation, provide the experimental framework for designing new polytherapeutic regimens to allow for permanent xenograft survival in larger species and humans.

B. INDUCTION OF TOLERANCE

The ability to induce immunological tolerance to donor tissues has the potential of providing for permanent survival of xenografts. Experimental studies currently being performed in rodents suggest that the creation of bone marrow chimerism between two different donor/recipient species induces tolerance to donor xenografts.[47,48] The establishment of mouse bone-marrow cells in rats allows for the permanent survival of donor skin xenografts while maintaining the ability of the rat recipients to reject third-party grafts. These procedures have not been successfully extended to vascularized grafts in larger species but the potential for this type of tolerance induction is an important area of investigation.

C. GRAFT MODIFICATION

The most effective method to prevent the recognition and rejection of xenografts may be modifications in the expression of the donor antigens that act as targets for the host xenograft immune response. While our current understanding of the nature of xenoantigens is not sufficiently detailed to address this issue directly, the technology to prevent antigen expression or to induce the expression of new antigens is currently available. This type of genetic manipulation could prevent some species from being recognized as the donors of foreign grafts by potential recipients. This represents an extremely interesting and active field of investigation, particularly in the creation of transgenic donors, such as the pig, for use in clinical transplantation. This development has the potential to significantly alter the current approaches to clinical transplantation.

REFERENCES

1. **Takasu, S., Sakagami, K., Morisaki, F., et al.,** Immunosuppressive mechanism of 15-deoxyspergualin on sinusoidal lining cells in swine liver transplantation: suppression of MHC class II antigens and interleukin-1 production, *J. Surg. Res.,* 51, 165, 1991.
2. **Cramer, D.V., Chapman, F.A., Jaffee, B.D., et al.,** The prolongation of concordant hamster-to-rat cardiac xenografts by Brequinar sodium, *Transplantation,* 54, 403, 1992.
3. **Auchincloss, H., Jr.,** Xenogeneic transplantation: a review, *Transplantation,* 46, 1, 1988.
4. **Cramer, D.V., Wu, G.-D., and Makowka, L.,** The pathogenesis of xenograft reactions in rodents and other small laboratory species, *Transplant. Proc.,* in press.
5. **Forty, J., Cary, N., White, D.J.G., and Wallwork, J.,** Hyperacute rejection of rabbit hearts by human blood is mediated by the alternative pathway of complement, *Transplant. Proc.,* 24, 488, 1992.
6. **Calne, R.Y.,** Organ transplantation between widely disparate species, *Transplant. Proc.,* 2, 550, 1970.
7. **Platt, J.L. and Bach, F.H.,** Mechanism of tissue injury in hyperacute xenograft rejection, in *Xenotransplantation. The Transplantation of Organs and Tissues Between Species,* Cooper, D.K.C., Kemp, E., Reemtsma, K., and White, D.J.G., Eds., Springer-Verlag, Berlin, 1991, 69.
8. **Hammer, C. and Hingerle, M.,** Development of preformed natural antibodies in gnotobiotic dogs and pigs, impact of food antigens on antibody specificity, *Transplant. Proc.,* 24, 707, 1992.
9. **Makowka, L., Chapman, F.A., Cramer, D.V., et al.,** The role of inflammatory reactions in xenotransplantation, in *Xenograft 25,* Hardy, M.A., Ed., Elsevier, New York, 1989, 159.
10. **Larsen, S. and Starklint, H.,** Histopathology of kidney xenograft rejection, in *Xenotransplantation. The Transplantation of Organs and Tissue Between Species,* Cooper, D.K.C., Kemp, E., Reemtsma, K., and White, D.J.G., Eds., Springer-Verlag, Berlin, 1991, 181.
11. **Rose, A.G.,** Histopathology of cardiac xenograft rejection, in *Xenotransplantation. The Transplantation of Organs and Tissues Between Species,* Cooper, D.K.C., Kemp, E., Reemtsma, K., and White, D.J.G., Eds., Springer-Verlag, Berlin, 1991, 231.
12. **Johnston, P.S., Wang, M.-W., Kim, S.M.L., Wright, L.J., and White, D.J.G.,** Discordant xenograft rejection in an antibody-free model, *Transplantation,* 54, 573, 1992.
13. **Wu, G.D., Cramer, D.V., Chapman, F.A., et al.,** A comparative immunopathologic study of hyperacute rejection in naive and sensitized rodent cardiac xenograft models, *Transplant. Proc.,* 25, 464, 1993.
14. **Corry, R.J. and Kelley, S.E.,** Immunological enhancement of primarily vascularized rat heart xenotransplants in mice, *Transplantation,* 18, 503, 1974.
15. **Wu, G.D., Cramer, D.V., Chapman, F.A., et al.,** Cardiac hyperacute rejection in concordant xenograft combinations induced by immunization, *Transplant. Proc.,* 24, 691, 1992.

16. **Hammer, C.,** Experimental xenotransplantation between closely related nonprimate species, in *Xenotransplantation. The Transplantation of Organ and Tissues Between Species,* Cooper, D.K.C., Kemp, E., Reemtsma, K., and White, D.J.G., Eds., Springer-Verlag, Berlin, 1991, 339.
17. **Jaffee, B.D., Jones, E.A., Loveless, S.E., and Chen, S.F.,** The unique immunosuppressive activity of Brequinar sodium, *Transplant. Proc.,* in press.
18. **Wu, G.D., Cramer, D.V., Cosenza, C., Tuso, P., Eiras-Hreha, G., Wang, H.K., and Makowka, L.,** Evidence for the role of preformed antibody in mediating cardiac rejection in the hamster-to-rat xenograft model, *Surg. Forum,* in press.
19. **Leventhal, J.R., Dalmasso, A.P., Cromwell, J.W., et al.,** Prolongation of cardiac xenograft survival by depletion of complement, *Transplantation,* 55, 857, 1993.
20. **Paul, L.C., Green, B., Davidoff, A., and Benediktsson, H.,** Xenoantibodies in the rat against guinea pig tissues, *Transplant Int.,* 3, 199, 1990.
21. **Leventhal, J.R., Flores, H.C., Gruber, S.A., et al.,** Evidence that 15-deoxyspergualin inhibits natural antibody production but fails to prevent hyperacute rejection in a discordant xenograft model, *Transplantation,* 54, 26, 1992.
22. **Miyagawa, S., Hirose, H., Shirakura, R., et al.,** The mechanism of discordant xenograft rejection, *Transplantation,* 46, 825, 1988.
23. **Bryan, B.A., Henry, M.L., Han, L.K., Sedmak, D.D., and Ferguson, R.M.,** Effects of formed elements of xenograft rejection in an *ex vivo* organ perfusion model, in *Xenotransplantation. The Transplantation of Organs and Tissues Between Species,* Cooper, D.K.C., Kemp, E., Reemtsma, K., and White, D.J.G., Eds., Springer-Verlag, Berlin, 1991, 405.
24. **Cramer, D.V., Sher, L., and Makowka, L.,** Liver xenotransplantation: clinical experience and future considerations, in *Xenotransplantation. The Transplantation of Organs and Tissues Between Species,* Cooper, D.K.C., Kemp, E., Reemtsma, K., and White, D.J.G., Eds., Springer-Verlag, Berlin, 1991, 559.
25. **Reemtsma, K.,** Renal heterotransplantation from nonhuman primates to man, *Ann. N.Y. Acad. Sci.,* 162, 412, 1969.
26. **Starzl, T.E., Fung, J., Tzakis, A., et al.,** Baboon-to-human liver transplantation, *Lancet,* 341, 65, 1993.
27. **Cooper, D.K.C., Ye, Y., Rolf, L.L., Jr., and Zuhdi, N.,** The pig as potential organ donor for man, in *Xenotransplantation. The Transplantation of Organs and Tissues Between Species,* Cooper, D.K.C., Kemp, E., Reemtsma, K., and White, D.J.G., Eds., Springer-Verlag, Berlin, 1991, 481.
28. **Tuso, P.J., Cramer, D.V., Yasunaga, C., Wu, G.D., Cosenza, C., and Makowka, L.,** Xenoperfusion of pig livers with human blood removes natural antibody to pig vascular endothelium, *Transplantation,* in press.
29. **Tuso, P.J., Cramer, D.V., Middleton, Y.D., et al.,** Pig aortic endothelial cell antigens recognized by human IgM natural antibodies, *Transplantation,* in press.
30. **Starzl, T.E. and Demetris, A.J.,** Liver transplantation: a 31-year perspective, *Curr. Prob. Surg.,* 27, 49, 1990.
31. **Abouna, G.M., Serrou, B., Boehmig, H.J., Amemiya, H., and Martineau, G.,** Long-term hepatic support by intermittent multi-species liver perfusions, *Lancet,* 1, 391, 1970.
32. **Cooper, D.K.C., Ye, Y., and Niekrasz, M.,** Heart transplantation in primates, in *Handbook of Animal Models in Transplantation Research,* Cramer, D.V., Podesta, L., and Makowka, L., Eds., CRC Press, Boca Raton, FL, 1992, in press.
33. **Cooper, D.K.C., Human, P.A., Rose, A.G., et al.,** The role of ABO blood group compatibility in heart transplantation between closely related animals species. An experimental study using the vervet monkey to baboon cardiac xenograft model, *J. Thorac. Cardiovasc. Surg.,* 97, 447, 1989.

34. **Marquet, R.L., Vav Es, A.A., Heystek, G.A., van Leersum, R.H., and Balner, H.,** Prolongation of baboon to rhesus monkey kidney xenograft survival by pretransplant rhesus blood transfusion, *Transplantation,* 25, 165, 1978.
35. **Calne, R.Y., Davis, D.R., Pena, J.R., et al.,** Hepatic allografts and xenografts in primates, *Lancet,* 1, 103, 1970.
36. **Cooper, D.K.C. and Rose, A.G.,** Experience with experimental xenografting in primates, in *Xenograft 25,* Hardy, M.A., Ed., Elsevier, Amsterdam, 1989, 95.
37. **Sadeghi, A.M., Robbins, R.C., Smith, C.R., Kurlansky, P.A., Michler, R.E., and Rose, E.A.,** Cardiac xenotransplantation in primates, *J. Thorac. Cardiovasc. Surg.,* 93, 809, 1987.
38. **Calne, R.Y., White, H.J.O., Herbertson, B.M., et al.,** Pig-to-Baboon liver xenografts, *Lancet,* 1, 1176, 1968.
39. **Lexer, G., Cooper, D.K.C., Rose, A.G., Wicomb, W.N., Keraan, M., and Du Toit, E.,** Hyperacute rejection in a discordant (pig to baboon) cardiac xenograft model, *J. Heart Transplant.,* 5, 411, 1986.
40. **Cooper, D.K.C., Human, P.A., Lexer, G., et al.,** Effects of cyclosporine and antibody adsorption on pig cardiac xenograft survival in the baboon, *J. Heart Transplant.,* 7, 238, 1988.
41. **Fischel, R.J., Bolman, R.M.,III, Platt, J.L., Najarian, J.S., Bach, F.H., and Matas, A.J.,** Removal of IgM antiendothelial antibodies results in prolonged cardiac xenograft survival, *Transplant. Proc.,* 24, 488, 1992.
42. **Wood, R.P., Katz, S.M., and Kahan, B.D.,** New immunosuppressive agents, *Transplant. Sci.,* 1, 34, 1991.
43. **Makowka, L. and Cramer, D.V.,** Brequinar sodium: a new immunosuppressive drug for transplantation, *Transplant. Sci.,* 2, 50, 1992.
44. **Yasunaga, C., Cramer, D.V., Cosenza, C.A., et al.,** The effect of Brequinar sodium on in vivo antibody production, *Transplant. Proc.,* 25 (Suppl. 2), 40, 1993.
45. **Cosenza, C.A., Cramer, D.V., Tuso, P.J., Chapman, F.A., Wang, H.K., and Makowka, L.,** Combination therapy with Brequinar sodium and cyclosporin A synergistically prolongs haster-to-rat cardiac xenograft survival, *J. Heart Lung Transplant.,* in press.
46. **Murase, N., Starzl, T.E., Demetris, A.J., et al.,** Hamster to rat heart and liver xenotransplantion with FK 506 plus antiproliferative drugs, *Transplantation,* in press.
47. **Ildstad, S.T. and Sachs, D.H.,** Reconstitution with syngeneic plus allogeneic or xenogeneic bone marrow leads to specific acceptance of allografts or xenografts, *Nature,* 307, 168, 1984.
48. **Slavin, S., Strober, S., Fuks, Z., and Kaplan, H.S.,** Induction of specific tissue transplantation tolerance using fractionated total lymphoid irradiation in adult mice: long-term survival of allogeneic bone marrow and skin grafts, *J. Exp. Med.,* 146, 34, 1977.
49. **Makowaka, L.,** unpublished data.

Section VIII

Immunosuppressive Agents

Chapter 22

The Use of Immunosuppressive Agents in Animal Models of Experimental Transplantation

Carlos A. Cosenza, Donald V. Cramer, and Leonard Makowka

CONTENTS

I. INTRODUCTION

The application of organ transplantation as a therapeutic procedure is limited by the rejection of the foreign graft by the host. The immunologic basis of this response was first demonstrated in 1943 by Gibson and Medawar.[1] Since then a great deal of effort has been dedicated to the search for effective and safe methods of suppressing the immune system to prevent graft rejection. A critical issue that must be addressed in the performance of these experiments is the choice of animal models for the transplant experiments and the relevance that these models have for the extrapolation of the data for potential clinical application.

Experiments performed in 1959 by Schwartz and Damashek first demonstrated the potential of the antimetabolite, **6-mercaptopurine,** for the prolongation of canine kidney allograft survival due to suppression of the host's immune response. This drug could not be introduced into the clinical practice, however, because of severe hepatic toxicity and

0-8493-3629-5/94/$0.00+$.50

myelosuppression.[2] Subsequent to these early experiments, Hitchings and Elion synthesized a series of mercaptopurine derivatives, searching for a compound suitable for parenteral administration and with an improved ratio of immunosuppression to bone marrow toxicity.[3] The result of this work was the development of the antimetabolite, **azathioprine,** an S-imidazole derivative of 6-mercaptopurine, whose mode of action is to block DNA and RNA synthesis. It does not prevent activation of T cells by allograft antigens but inhibits their subsequent proliferation by disruption of purine biosynthesis.[4] Azathioprine also decreases the level of circulating leukocytes by inhibiting the proliferation of promyelocytes within the bone marrow. The myelosuppressive effects of azathioprine administration are most consistently seen as a depression of the total white blood cell count in the peripheral blood. Additional adverse effects of this drug include infections, impairment of wound healing, and gastrointestinal complications.[5]

Although other antimetabolites were subsequently evaluated for their potential immunosuppressive action, none demonstrated an improvement in efficacy for preventing graft rejection when compared to azathioprine. Azathioprine inhibits the primary allograft response, but has little effect upon the secondary response, and is, therefore, not useful for reversing ongoing rejection. It has been most commonly used in combination with steroids and was the primary immunosuppressive agent in transplantation for almost 20 years.

The introduction in the 1980s of the fungal metabolite **cyclosporine** by Borel and associates produced a dramatic improvement in pharmacologic immunosuppression.[6] The survival rates of all types of organ transplants increased substantially, and the use of this immunosuppressive agent has made organ transplantation an effective therapy for many life-threatening diseases. Cyclosporin A (CsA), when used in high doses for prolonged periods of time, however, exhibits important renal and hepatic toxic side effects in experimental animals and human allograft recipients.[7-9]

The success of CsA has stimulated the extended screening of natural products of fungi for immunosuppressive activity and in 1984, lead to the discovery of a new immunosuppressive agent, **FK 506.**[10] This compound has the structure of a macrolide and has shown a potency *in vitro* at least tenfold greater than CsA.[11] This substance is currently being tested in clinical trials and early reports suggest that this drug can be an effective alternative to the use of CsA.[11]

At the end of the 1980s additional new compounds with promising immunosuppressive effects in experimental animals have been introduced including: **Rapamycin,** a natural macrolide produced by a fungus,[12] **15-deoxyspergualin,** an anti-tumor agent produced by a bacterium,[13] and **mycophenolate mofetil (RS-61443),** a morpholinoethylester derived of mycophenolic acid.[14] Finally, a novel compound, **Brequinar sodium,** that is an inhibitor of the *de novo* pyrimidine synthesis, has recently been shown to be an effective primary immunosuppressive agent for *in vitro* and *in vivo* immune responses, including heart, liver and kidney allograft survival in rodents.[15]

II. IMMUNE RESPONSE AND GRAFT REJECTION

Acute rejection of allografts involves the participation of a variety of cells including cytotoxic T cells, helper T cells, antibody-forming B cells, macrophages, and natural killer cells (NK cells).[16] The allograft response is initiated by allospecific antigen receptors on the surface of a subpopulation of T lymphocytes that recognize donor allograft antigens. T-cell receptor (TCR) activation precipitates a series of intracellular events that result in the production of a variety of specific lymphokines.[17,18] Recently, a group of cytoplasmatic proteins with enzymatic activity (immunophilins) have been demonstrated to be part of the mechanism that mediates signal transmission pathway

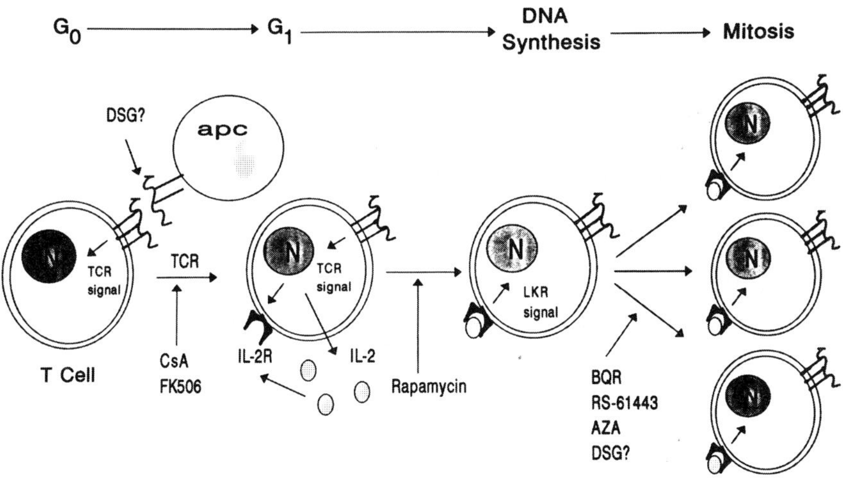

Figure 1 Possible sites of action of cyclosporin A (CsA), FK506, Rapamycin (RPM), mycophenolic acid (MPA), Brequinar sodium (BQR), and 15-deoxyspergualin (DSG) on T cell activation.

from the cytoplasm to the nucleus for T-cell activation. These cytosolic proteins exhibit cis-trans peptidyl-propyl isomerase activity and have been suggested to express a common receptor for structurally different immunosuppressants, including CsA, FK 506 and Rapamycin.[19-21] The activation of macrophages elicits a signal for the production of interleukin 1(IL-1) by transcription of specific messenger RNA in the nucleus of the cell.[19] IL-1 acts as an autocrine factor inducing the manufacture of interleukin 6 (IL-6) by macrophages. IL-6 acts to support T-cell activation. Activated T cells, in turn, secrete the T-cell growth factor, interleukin 2 (IL-2) and simultaneously express cell surface IL-2 receptors.

The release of IL-2 represents a crucial event in the rejection response as it stimulates the secretion of bioactive T cell proteins (lymphokines) that activate other T cells, B cells, and macrophages. IL-2-stimulated T cells produce activating factors for B cells, such as IL-4 and IL-5, that enable these cells to elaborate specific anti-graft antibodies.[22] Simultaneously, IL-2 causes helper cells to secrete factors that will permit the cytotoxic capacity of T cells to develop. IL-2 stimulates the release of gamma interferon (IFN-γ), which activates the cytolytic activity of macrophages. Additionally, the interaction of IL-2 with IL-2 receptors induces proliferation of both helper and cytotoxic T cells,[23] creating an expanding circle of rejection-associated events.

All steps involved in the rejection mechanisms are potential targets for immunosuppressive drugs (Figure 1). Cyclosporin A, FK 506, and Rapamycin selectively inhibit T-lymphocyte activation by inhibiting signal transduction for specific cytokines. Other drugs, such as 15-deoxyspergualin, seems to act on earlier events, primarily through impairment of monocyte and macrophage processing of antigen. Further, lymphocyte proliferation in response to activation requires a supply of nucleotides for nucleic acid synthesis. These reactions are inhibited by compounds (e.g., azathioprine, mizoribine) that compete as antimetabolites in the production of nucleotides. Two novel drugs that specifically inactivate a single enzymatic step in the *de novo* nucleotide synthetic pathways have been recently described. They include Brequinar sodium, a compound that blocks the action of dihydroorotate dehydrogenase, and mycophenolate mofetil (RS-61443), a synthetic derivative of mycophenolic acid, a compound that acts as a noncompetitive inhibitor of inosine monophosphate dehydrogenase.

Table 1 **Pharmacokinetic Parameters**

Parameter	CsA	FK 506
Absorption		
Time to peak	2–4 hour	5–4 hour
Extent of absorption	30%	25%
Distribution		
Blood/plasma ratio	1–2	> 4
Lipoprotein binding	High	Low
Elimination		
Metabolism	> 98%	> 98%
Urinary excretion	< 6%	< 2%
Metabolite excretion	Biliary	Biliary
Half-life in plasma	5–12 hour	4–14 hour
Plasmatic clearance	10 ml/min/kg	43 ml/min/kg

III. ANIMAL MODELS OF IMMUNE SUPPRESSION

During the last two decades, the use of experimental animal models of organ transplantation has been critical for the development and characterization of new immunosuppressive agents. In this section, the most important data regarding doses, immunosuppressive effectiveness, and toxicity of these substances is summarized to provide a guide to the experimental use and efficacy comparisons for these agents.

A. CYCLOSPORINE

Cyclosporine is a small cyclic peptide derived from the fungus, *Tolupocladium inflatum Gams.*[24] Its primary mode of action is to inhibit the production of IL-2 by blocking the activation of IL-2 genes within the T cells. Through this mechanism, T cells do not proliferate or release lymphokines, such as the macrophage-stimulating INF-γ and activating factors for B cells, thereby sharply reducing the events that are the result of T-cell activation.[25]

The molecular basis of CsA action is not completely understood. This agent does not affect the initial plasma membrane events of alloantigen recognition and activation. Rather, it has been recently postulated that an intracytoplasmatic binding protein, cyclophilin, with a molecular weight of 15 to 20 kDa, represents the specific receptor through which CsA exerts its immunosuppressive effects.[20,26] The antigen recognition by the TCR initiates a series of events, including lymphokine gene transcription, that lead to T-cell activation. Recently, it has been hypothesized that calcineurin is an essential enzyme in the T-cell signalling cascade.[27] Calcineurin, a Ca^{++} calmodulin-dependent phosphatase would activate some specific cytoplasmic transcription factors (such as NF-AT and NF-IL-2A), resulting in the subsequent activation of specific nuclear factors. These nuclear factors regulate the transcription of T-cell activation genes, such as the IL-2 gene. The complexes formed between CsA with its intracytoplasmic receptor, cyclophilin, and FK-506 with its cytoplasmic receptor, FK binding protein (FKBP), are potent inhibitors of this TCR-mediated signal transduction pathway by virtue of their blockage of calcineurin activity.[28]

The most important pharmacokinetic characteristics of CsA are summarized in Table 1. CsA has a significant level of distribution in red blood cells and this characteristic must be considered when analyzing of cyclosporine levels. CsA is converted by the hepatic cytochrome P-450 enzymatic complex into metabolites that retain variable amounts of biological activity.[29] CsA and its metabolites are primarily eliminated in bile (90%) and approximately 6% appears in the urine.

Table 2 **Experimental Transplantation Models Using Cyclosporine**

Species	Graft	Route	M.E.D.[a]	Ref.
Mouse	Skin	i.m.	25 mg/kg	31,32
Rat	Skin	i.m.	40 mg/kg	34
		s.c.	15 mg/kg	153
	Cardiac	i.m.	5 mg/kg	37
		Oral	10 mg/kg	59
	Kidney	i.m.	5 mg/kg	47
		Oral	10 mg/kg	
	Liver	Oral	15 mg/kg	49
	Small bowel	Oral	15 mg/kg	53
	Small bowel (GVH)	i.m.	25 mg/kg	54
	GVHR (b.m.)	Oral	15 mg/kg	154
	Pancreas	i.m.	5–2 mg/kg	155
		Oral	15–20 mg/kg	156,157
Rabbit	Skin	i.m.	10 mg/kg	35
	Kidney	Oral	25 mg/kg	46
Dog	Skin	Oral	30 mg/kg	33
	Cardiac	Oral	20 mg/kg	40
	Kidney	Oral	20 mg/kg	158
	Liver	Oral	20 mg/kg	50
	Lung	Oral	17 mg/kg	41
	Small bowel	i.m./oral	25 mg/kg	160,56
	GVHR (b.m.)	Oral	15–25 mg/kg	62
Pig	Cardiac	Oral	25 mg/kg	39
	Liver	Oral	20 mg/kg	52
	Small bowel	Oral	25–30 mg/kg	55,57
	Kidney	Oral	25 mg/kg	
Cynomolgus	Cardiac	i.m.	25 mg/kg	38,161
Rhesus	Kidney	Oral	25 mg/kg	162
Baboon	Pancreas	Oral	25–50 mg/kg	163

[a] M.E.D., Minimal effective dose.

Cosenza, C.A., et al., unpublished observation.

1. Prevention of Allograft Rejection

The effectiveness of CsA for preventing graft rejection depends upon the type of transplanted organ, the experimental animal model, the nature of the vascular supply to the tissue, the dosage of the drug, and the length of the treatment period. CsA has shown to be effective in preventing allograft rejection for a wide variety of grafts in small and large animal species[30,31] and the data presented in Table 2 summarizes the influence of each of these factors in graft survival. Skin grafts can be protected from rejection by CsA in mice,[32,33] rats,[34] rabbits,[35] and dogs,[33,36] although these allografts are acutely rejected following cessation of CsA therapy. Cardiac transplants between two inbred strains of rats were the first solid organ allografts used to examine the effectiveness of CsA as an immunosuppressive drug.[37] These and subsequent studies showed that CsA is a highly effective immunosuppressive agent and that a relatively short course of treatment was enough to prevent graft rejection for long periods of time. The timing of drug administration was shown to be particularly important, since CsA treatment started on day 4 was ineffective for preventing rejection. Similar results have been achieved in larger animals (pigs, dogs, and monkeys) transplanted with histoincompatible allografts.[38-40] The use of

CsA has allowed for long-term survival of lung transplants in rats and dogs.[41-43] The importance of timing of CsA administration has also been demonstrated in rats with lung allografts. A single dose of CsA given immediately after transplantation can delay rejection for 3 weeks. If a single dose was administered on day 3, most of the grafts survived long term.[44] The effect of CsA on survival of kidney grafts has been studied in rats, rabbits, pigs, dogs, and monkeys. Once again, graft acceptance can be induced by a short course of treatment in rats and rabbits.[45,46] CsA has proved to be effective in prolonging kidney graft survival in pigs, dogs and monkeys, but long-term survival of the grafts after withdrawal of CsA has been more difficult to demonstrate. While a small number of dogs survived after the drug was stopped, kidney grafts in the majority of animals were acutely rejected.[47] Long-term treatment in pigs (60 days) or monkeys (120 days) resulted in acceptance of the allograft after CsA was discontinued.[48] In rats, lower doses of the drug are required to induce tolerance to liver grafts than for other organs, including skin, heart, and kidney. The specificity of the tolerance induced by CsA to liver grafts appears to be time dependent. The initial period of tolerance is nonspecific and it is not until approximately 100 days of treatment that a donor-specific tolerance develops in the recipient.[49] Liver grafts in dogs treated with CsA initially function well, but cessation of the drug results in chronic rejection of the graft.[50,51] Pigs appear to respond more favorably to treatment interruption and long-term acceptance of liver grafts has been reported after administration of CsA to the recipient for 25 to 50 days.[52] The efficacy of CsA for preventing rejection of small bowel transplants and disrupting the development of graft-versus-host disease (GVHD) has been controversial. In rats, high doses of CsA prevented rejection and GVHD during the period of drug administration.[53,54] Similar doses of this drug have been shown to be effective in pigs and dogs in some reports,[55,56] and ineffective in others.[57]

In summary, CsA can be considered particularly effective when used as a primary immunosuppressive drug to prevent the rejection of vascularized organ grafts including heart, kidney, and liver. CsA has demonstrated limited effectiveness in abrogating the graft-versus-host reaction in bone marrow and small bowel transplantation, especially in large animal models, such as the dog and pig.[36,57] In general, CsA has been less effective at protecting secondary grafts and suppressing antibody production in sensitized animals.[58]

2. Tolerance Induction

The use of CsA to prevent graft rejection can be associated with the induction of a permanent state of acceptance (tolerance) of the donor graft. The ability of CsA to induce tolerance depends on the organ and species used. A prolonged state of specific acceptance can be achieved in rats and rabbits that have received organ grafts and short-term (7 days) CsA therapy.[31,59,60] The kinetics and specificity of CsA-mediated tolerance in these species have been subdivided into three different stages.[61] The first stage coincides with the cessation of CsA therapy at day 7. At this time, tolerance is not present and the placement of a second graft results in rejection, followed by loss of the first graft. Stage 2 emerges after approximately 14 days of CsA treatment. Following cessation of the drug, the placement of a second graft results in prolonged but not permanent graft survival (approximately 60 days). Finally, after 50 days of treatment (stage 3), a specific state of tolerance is present and all second donor-specific grafts survive indefinitely. A third-party graft is acutely rejected without affecting the survival of the original transplant.[59,60] In contrast to solid organ allografts, skin allografts are frequently rejected when the drug is discontinued despite prolonged treatment.[32,33] The mechanism by which CsA induces tolerance is largely unknown despite extensive experimental investigation. As suggested above, it would appear that one of the critical effects of this agent on the development of specific tolerance involves the sparing of CsA-sensitive suppressor T cells that produce soluble mediators responsible for graft survival, thereby profoundly impairing the ability of the host to respond to allografts.[61]

The induction of tolerance to the graft following CsA treatment is much more difficult to achieve using larger experimental species. In dogs and primates, acute graft rejection has been described for kidney and heart grafts approximately 14 days after the discontinuation of CsA therapy.[62] This may occur even in animals that have received more than 3 months of treatment. Pigs may be more sensitive to the effects of drug treatment and may develop CsA-mediated tolerance after treatment for 60 and 90 days for different models of organ transplantation (heart, liver, and small bowel).[39,52,55]

3. Toxic Side Effects

Prolonged treatment with CsA is associated with several toxic side effects. These side effects have been documented in experimental animals, especially when high therapeutic doses were used. These signs include the nonspecific symptoms of weight loss and lack of appetite, hepatotoxicity, nephrotoxicity, and central nervous system symptoms.[63]

The results of experimental studies of CsA-induced nephrotoxicity have been conflicting due to the lack of standardized methods, and the use of different protocols, species, and strains. While the administration of high doses of CsA in mouse and dog transplant models has not resulted in significant decreases in kidney function,[6,64] nephrotoxic damage in the rat has been demonstrated at therapeutic doses of CsA.[65] For this reason, the nephrotoxic side effect of CsA has been extensively studied in rat models. Human kidneys seems to be sensitive to CsA, and increase of creatinine levels in blood has been observed in patients when the whole blood levels of CsA exceed 250 ng/ml.[66] The nephrotoxic injury mediated by CsA can be associated with effects at various levels of the nephron: arteriole, glomerulus, and proximal tubule. This toxic effect with decrease in glomerular filtration rates may be due to: (1) damage of the proximal tubule cells, either due to an intracellular CsA deposition or alteration of cell lysosomes, (2) a glomerular toxicity as suggested by a progressive focal or total glomerular atrophy and concomitant fibroblastic proliferation, or (3) renal hemodynamic alterations with increased vascular resistance and increased renin activity.[67]

In order to avoid CsA-induced, toxic side effects, a number of therapies that include combinations of drugs with different modes of immunosuppressive action have been attempted. Steroids and/or azathioprine, in combination with CsA, have been extensively tested clinically and in several experimental models. The combination of subtherapeutic levels of CsA and azathioprine has been demonstrated to exert an enhancing effect on heterotopic heart allograft survival in rats and kidney transplantation in dogs.[68,69] The effectivity of CsA in combination with azathioprine or steroids has also been shown in clinical trials.[70,71]

The results of the combination of CsA and FK 506 initially reported in rats[72] and dogs[73] suggested a prolongation of the graft survival, when compared to single therapy. However, a subsequent study using subtherapeutic doses of both drugs in a heterotopic heart allograft model in the rat demonstrated that this combination results in an antagonistic effect.[74] This observation was confirmed by the data presented by Starzl et al., where CsA and FK 506 in combination, produced increased toxicity without evidence of significant prolongation of allograft survival in humans.[75]

Recently, the combination of cyclosporine and Rapamycin has been shown to exert a synergistic immunosuppressive effect for heart allografts in rats, using concentrations of the drugs that were individually ineffective.[76] This synergistic effect has been attributed to differences in the mechanism of action of these two drugs. Unlike FK 506, Rapamycin apparently exerts its immunosuppressive activity at a later stage in T-lymphocyte activation.[77]

B. FK 506

FK 506 is a macrolide antibiotic isolated from the fungus, *Streptomyces tsukubaensis.*[78] The drug has a mechanism of action that is similar to cyclosporine, with inhibition of the

Table 3 **Experimental Transplantation Models Using FK-506**

Species	Graft	Route	M.E.D.[a]	Ref.
Rat	Skin	i.m.	0.32 mg/kg	164
	Cardiac	Oral	1.0 mg/kg	84
		i.m.	1.28 mg/kg	72, 91
	Liver	i.m.	1.28 mg/kg	90,165
	Small bowel	i.m.	2.0 mg/kg	86
	Lung	i.m.	0.3 mg/kg	166
	Pancreas islet	s.c.	0.32 to 16 mg/kg	167
	GVH (b.m.)	i.m.	1.0 mg/kg	168
Dog	Kidney	i.m.	0.32 mg/kg	91
		Oral	1.0 mg/kg	91,169
	Liver	i.m.	1.0 mg/kg	73
	Lung	i.m.	0.1 mg/kg	170
Piglet	Liver	i.v.	0.03 mg/kg	171
Baboon	Kidney	i.m.	0.5 to 2 mg/kg	169,172
Cynomolgus	Liver	i.m.	1.0 mg/kg	95
	Pancreas	i.m.	1.0 mg/kg	173

[a] M.E.D., Minimal effective dose.

production of IL-2, gamma-INF, and IL-2 receptor expression. FK 506 does not inhibit the proliferation of activated T cells or B cells.[79,80] Like CsA, there is no evidence that FK 506 has an important effect on B-cell or macrophage function.

The molecular basis of the activity of FK-506 is not fully known. As described previously, both FK-506 and CsA appear to interfere with cytoplasmic signal transduction by binding with cytoplasmatic immunophilins and disrupting the expression of DNA.[23] A cytoplasmic protein with enzymatic (rotamase) activity has been recently described as a cytoplasmic receptor for FK 506.[81,82] The complex formed between FK 506 and its binding protein (FKBP) is capable of inhibiting the enzyme calcineurin, a key signalling protein on T-lymphocyte activation.[27]

The most important features of the pharmacokinetics and metabolism of FK 506 are summarized in Table 1. FK 506 displays pharmacological properties that are very similar to CsA. Both drugs are primarily eliminated by hepatic metabolism and less than 2% of the drug is excreted in the urine as the parent compound.[18] One difference for the two compounds is that FK 506 exhibits a 10- to 100-fold greater potency than that of cyclosporine.[78] When FK 506 has been administrated in patients in combination with CsA the usage of a common metabolic pathway results in enhancement of the drug-related toxic side effects.[83]

1. Prevention of Allograft Rejection

The immunosuppressive potency of FK 506 for prolonging the survival of rat skin and heart allografts was first tested in Japan by Ochiai et al.[84] Ochiai's work demonstrated that short courses of treatment with FK-506 (as few as 4 days) induces prolonged acceptance of heterotopic heart allografts.[85]

The use of FK 506 in additional experimental models has been summarized in Table 3. It is of interest to note that FK 506 has been shown to be particularly effective in prolonging small bowel transplantation in rats. In a strongly histoincompatible (LEW-to-ACI) small bowel model, Hoffman et al., compared the effectiveness of FK 506 and CsA. There was a significant prolongation of graft survival and prevention of GVHD in those animals treated with FK 506, demonstrating that this drug is more effective than CsA in preventing the rejection of the small bowel.[86] Recently, Murase et al. have shown that the particular efficacy of FK 506 in preventing GVHD in rat small bowel allograft is related to the strain combination used.[87] FK 506 has been also capable of reversing acute GVHD after bone-marrow transplantation in rats, and displayed a marked prolongation of disease-free intervals when compared to CsA-treated bone-marrow recipients.[88]

FK 506 has an unusual capacity of preventing rejection after only brief periods of treatment early in the rejection process. This effect has been demonstrated for heart and liver allografts in rats,[89,90] and kidney and liver grafts in dogs.[73,85] This unique ability of FK 506 to prevent rejection after short periods of treatment suggests that a disruption in early lymphocyte interaction is responsible for the subsequent induction of tolerance. The mechanisms of long-term graft acceptance have been not yet fully examined, but there is some evidence that the primary effect of FK 506, like CsA, spares suppressor T-cells involved in the reaction to donor antigens in the induction phase of graft acceptance.[91,92] In addition, FK-506 is capable of inhibiting established heart allograft rejection, a characteristic that is not exhibited by CsA. This distinction between the actions of the two drugs suggests the potential for complementary activity of FK 506 in selected patients that do not display prolonged graft survival following treatment with CsA.

2. Toxic Side Effects

The toxic effects associated with treatment with FK 506 remain poorly defined. The first reports of the toxic side effects of FK 506 in dogs[93] described a severe fibrinoid necrosis of medium-sized arteries, particularly in the gastrointestinal tract and heart. Conversely, a separate report concluded that the histopathological picture of vasculitis was also observed in the control animals.[73] These lesions were not described in rats,[94] or monkeys,[73,95] which suggest the existence of significant differences in species sensitivity for this agent. More recently, an arteriosclerotic lesion of the coronary arteries in both the transplanted and native heart of rabbits was described.[96] An FK 506-mediated nephrotoxicity has been reported in experimental animals[97] and clinical trials.[98] This side effect may be similar to that observed with CsA, and may be due to a vasoconstrictor effect of the two drugs.[99] A potential diabetogenic effect of FK 506 has been demonstrated in experimental and clinical reports.[73,100] Functional and pathological changes in the pancreas of rats and dogs have been described.[73,94,101] Finally, neurotoxic complications, similar to those observed with CsA, have been reported in patients receiving FK 506.[102]

C. RAPAMYCIN

Rapamycin is a lipophilic macrolide antibiotic that is produced by *Streptomyces hydroscopicus.* The drug is structurally related to FK 506 and was originally described as an antifungal agent. Rapamycin was subsequently shown to have significant immunosuppressive properties.[112] Unlike FK-506 and CsA, Rapamycin has little or no effect on inhibition of IL-2 production and/or IL-2 receptor expression. Instead, Rapamycin appears to suppress T-cell proliferation by impairing T-cell responses to growth-promoting lymphokines.[104] While both CsA and FK 506 inhibit stimulated lymphocytes from entering the cell division cycle at the G_0/G_1 interface, Rapamycin acts later and blocks progress through the cell division cycle during the G_1 phase.[105] Although the molecular mechanisms of its action is unknown, *in vitro* studies have demonstrated that Rapamycin acts as a reciprocal antagonist for FK 506, suggesting interaction of the two drugs with the same FKBP cytoplasmatic binding protein.[168] IL-2 stimulates the phosphorylation and activation of the cytoplasmic p70 S6 protein kinase complex that may represent the regulatory signalling pathway for the entry of T cells into S phase and subsequent cell proliferation. Recently, it has been shown that the complex formed between Rapamycin and the cytoplasmic receptor, FKBP, selectively blocks this p70 S6 kinase activation cascade.[107] Furthermore, Rapamycin strongly suppress *in vitro* PWM-stimulated IgG production. This effect is approximately 1000 times more potent than that of CsA.[108]

Table 4 **Experimental Transplantation Models Using Rapamycin**

Species	Graft	Route	M.E.D.[a]	Ref.
Mouse	Skin	i.p.	0.25 to 4 mg/kg	110
		Oral	2.5 to 40 mg/kg	110
	Heart	i.p.		
	GVH	i.p.	12 mg/kg	111
Rat	Heart	i.m.	0.5 to 50 mg/kg	112
	Heart	i.v.	0.08 to 0.8 mg/kg	
	Kidney	i.v.		113
	S. bowel	i.v.		113
	Pancreas	i.v.		174
Pig	Kidney	Oral	2 mg/kg	112
Dog	Kidney	Oral	0.25 to 3 mg/kg	112
Baboon	Kidney	Oral	2 mg/kg	114

[a] M.E.D., Minimal effective dose

1. Prevention of Allograft Rejection

The immunosuppressive effect of Rapamycin has been investigated in several animal models of organ transplantation (Table 4). This drug exhibits a marked insolubility in aqueous buffers. The potency and the efficacy of Rapamycin is therefore dependent on the method in which it is prepared for administration. A solution of Rapamycin in Cremophor EL/ethanol for oral administration or a suspension in carboxymethyl cellulose have been shown to have the highest immunosuppressive potency in experimental transplant models.[109]

A variety of allograft and xenograft rejection experiments have suggested that Rapamycin is more potent and effective than either FK 506 or CsA.[109] Eng et al.[110] reported that Rapamycin was highly effective in mice for prolonging skin-graft survival. This compound was over 100 times more potent than CsA. Similarly, Morris et al. used a nonvascularized mouse heart transplant model and demonstrated that Rapamycin (3 mg/kg i.p.) prolonged allograft survival to 146 days.[111] Heterotopic heart transplants in rats exhibited prolonged graft survival (82 ± 7 days) when recipients were treated with 10 mg/kg i.m. for 14 days. Specific tolerance to the graft donor developed following treatment with this regimen.[112] Greater efficacy was achieved by Stepkowski et al. testing the effectiveness of this drug for heart and kidney allograft in rats (WF to BUF). These investigations demonstrated that low doses of Rapamycin (0.08 to 0.8 mg/kg) significantly increased allograft survival, particularly when the drug was administered via continuous infusion in a polysorbate and polyethylene glycol solution.[113] As mentioned above, differences in formulation and route of administration of this drug may be responsible for the widely disparate organ graft survival results.

The efficacy of Rapamycin as an antirejection agent has also been tested in large animal models. Calne et al.[112] demonstrated that an oral dose of 2 mg/kg was sufficient to prolong the survival of orthotopic kidney transplants in pigs. Prolonged graft acceptance was seen after cessation of drug treatment in three of nine animals.

2. Toxic Side Effects

The information available describing the toxic side effects of Rapamycin is limited. Calne et al. demonstrated that treatment with Rapamycin at doses of 0.25 to 3 mg/kg p.o. following renal transplantation in mongrel dogs and baboons was associated with several toxic effects.[112,114] These side effects included acute necrotizing vasculitis of arterioles

Table 5 **Experimental Transplantation Models Using 15-Deoxyspergualin**

Species	Graft	Route	M.E.D.[a]	Ref.
Mice	Bone Marrow	i.p.	3 mg/kg	123
Rat	Heart	i.p.	2.5 mg/kg	121
	Limb	i.p.	2.5 mg/kg	159
	Kidney	i.p.	2.5 mg/kg	122
	Skin	i.p.	2.0 mg/kg	119
	Pancreas	i.p.	2.5 mg/kg	176
Dog	Kidney	i.v.	1.8 to 0.8 mg/kg	126
		i.v.	0.6 to 0.8 mg/kg	124
Baboon	Heart	i.v.	4 mg/kg	129
	Kidney	i.v.	4 mg/kg	
Cynomolgus	Kidney	s.c.	3 to 6 mg/kg	127
Heart		i.m.	2 to 7.5 mg/kg	128

[a] M.E.D., Minimum effective dose

and small arteries. Moreover, toxicity studies of Rapamycin in rats have demonstrated that this drug alone (1.5 mg/kg/day i.p.), or in combination with CsA (15 mg/kg/day orally) produces impairment of renal function. Rapamycin alone, or in combination with CsA, did not impair liver function. Additionally, focal myocardial necrosis has been observed in rats receiving Rapamycin.[115]

Recent studies of combination therapy using low doses of Rapamycin and CsA have shown a synergistic effect in rat heart and kidney allograft models, suggesting that a tenfold reduction in the dose of this drug could mitigate the important vascular effects observed with monotherapy in species sensitive to the toxic effects of the drug.[113]

D. 15-DEOXYSPERGUALIN

15-Deoxyspergualin (DSG) is a polyamine derivative of spergualin, a natural anticancer drug produced by *Bacillus laterosporus.* Although this compound has been shown to be an effective immunosuppressive agent, its mechanism of action is largely unknown. The mechanism by which DSG mediates its immunosuppressive effects, however, does appear to be different from that of other fungal or bacterial agents. A variety of *in vitro* immune functional assays has been used to assess the immunosuppressive action of DSG. Unfortunately, the data that these studies have produced are difficult to interpreted because: (1) DSG is highly unstable in tissue-culture medium; (2) toxic metabolites are produced in the presence of calf serum that are not present in *in vivo* studies; and (3) the *in vitro* concentrations of the drug that have been used are much greater than the maximum tolerated dose *in vivo.* Despite these limitations, it is clear that DSG inhibits lymphocyte proliferation in response to different mitogens (PHA > Con A > PWM).[116] Additionally, DSG effectively inhibits antibody production in mice after primary and secondary immunization.[117]

1. Prevention of Allograft Rejection

The immunosuppressive effect of 15-Deoxyspergualin (DSG) *in vivo* (Table 5) has been confirmed in several organ transplantation models . Dickneite et al. were the first to demonstrate that 15-DSG (at 2.5 mg/kg i.p. for 10 days) was effective in prolonging skin, heart, and limb allograft survival in rats.[118-120] These results were confirmed by Suzuki et al. for rat heart allografts using a 15-day treatment period.[121] Further, Walter et al.

obtained indefinitive kidney allograft survival in rats after ten doses of 2.5 mg/kg i.p.[122] Both a prophylactic and curative effect for GVHD was achieved by Nemoto et al. following bone-marrow transplantation in mice (C57BL/6 to CBA) treated with 3 mg/kg i.p. of DSG for 30 days. A combination of DSG and methotrexate yield better survival times than either drug administered alone.[123]

The experience with this drug in larger animals has primarily been accumulated by studies conducted in Japan. Amemiya et al. demonstrated that the optimal dose for prolongation of renal allograft survival in dogs was 0.6 to 0.8 mg/kg of DSG administered intravenously.[124] Severe alterations of the gastrointestinal tract, however, were described,[125] suggesting that therapy must be restricted to short periods of time postoperatively to avoid the appearance of toxic side effects. Fukao et al.[126,127] used decreasing doses of DSG (2.4 to 0.6 mg/kg) during a 7-day period to demonstrate the effectiveness of 15-DSG as rescue therapy for kidney transplantation in dogs.

Heart and kidney allograft studies conducted in primates treated with DSG have shown that this drug prolongs graft survival at high doses (7.5 mg/kg i.m.). These treatment levels, however, were associated with severe systemic toxicity.[128,129] The combination of low doses of CsA and low doses of DSG resulted in an enhanced effect for prolonging primate heart graft survival.

2. Toxic Side Effects

Toxic side effects of DSG have been described in rodents (hematologic alterations)[118,119] and larger animals (gastrointestinal disturbances, bone-marrow depression, and loss of weight).[124,127,128] These adverse effects are correlated with the dose and rate of drug administration. In addition to the toxic side effects, there are important limitations for the oral administration of DSG, and these problems may restrict the prolonged use of this drug in clinical trials. Amemiya et al. presented the first clinical use of this compound following administration by slow infusion as rescue therapy in kidney allograft recipients. The preliminary results of these studies indicated a high rate of remission (79%) following short-term treatment for 5 days. The primary adverse reactions observed in the studies consisted of reduction in WBC and platelet levels in 53 and 28% of the patients, respectively.[130]

E. MYCOPHENOLATE MOFETIL (RS-61443)

Mycophenolic acid was isolated early in this century from a corn mold, *Penicillin glaucum.* Although this compound displayed immunosuppressive effects, its use in transplantation was delayed until the development of a semisynthetic morpholinoethyl ester, mycophenolate mofetil (MPM;RS-61443), provided improved oral bioavailability.[131] Following gastrointestinal absorption, MPM is hydrolyzed to the active ingredient, mycophenolic acid.

Mycophenolic acid inhibits the *de novo* purine synthetic pathway by reversibly blocking two key enzymes, inosine monophosphate (IMP) dehydrogenase and guanosine monophosphate (GMP) synthetase.[14] *In vitro* studies have demonstrated that alloantigen or mitogen-activated lymphocytes have an increased activity of IMP dehydrogenase. This characteristic may explain the sensitivity of these cells to the disruption of this biosynthetic pathway by MPM.[132] MPM inhibits activated lymphocyte proliferation with no detectable effects on IL-1 and IL-2 production.[14] This compound selectively inhibits T and B lymphocyte proliferation to a degree greater than that seen with other cell types, including fibroblasts and endothelial cells.[132] MPM is also capable of inhibiting primary and secondary antibody responses to tetanus toxoid *in vitro.*[133]

1. Prevention of Allograft Rejection

The efficacy of MPM (RS-61443) has been tested in several different allograft models (Table 6). Prolonged heart allograft survival (> 100 days) has been reported in a BN to

Table 6 **Experimental Transplantation Models Using RS-61443**

Species	Graft	Route	M.E.D.[a]	Ref.
Mice	Heart	i.p.	40 mg/kg	131
	Pancreas islet	Oral	8 mg/kg	139
Rat	Heart	Oral	20 mg/kg	134
Dog	Kidney	Oral	40 mg/kg	135
	Liver	Oral	20 mg/kg	177
Cynomolgus	Heart	Oral	70 mg/kg	134

[a] M.E.D., Minimum effective dose.

LEW rat heterotopic cardiac model following treatment of the recipients with 20 mg/kg/day orally for 50 days.[134] Treatment with 40 mg/kg/day during the same period of time induced permanent donor-specific unresponsiveness.

Preliminary studies in large animals have also reported prolongation of graft survival. The drug is effective in prolonging kidney graft survival in dogs at a dose of 40 mg/kg/day. Gastrointestinal symptoms, including gastritis, diarrhea, and anorexia, however, were very common.[135] In the same study, a dose of 20 mg/kg/day, combined with 5 mg/kg/day of CsA and 0.1 mg/kg/day of methylprednisolone, resulted in prolonged graft survival (122.4 ± 38 days) without severe toxic effects.[135]

One interesting finding is that MPM is highly effective in preventing acute rejection and achieving prolonged heart graft survival in rats when drug administration is delayed until the fifth day following transplantation.[134] A similar efficacy in reversing acute rejection was demonstrated in a kidney allograft model in dogs. Administration of MPM at high doses (80 mg/kg) during 3 days significantly reversed acute rejection in more than 85% of the animals.[136] The preliminary results in a sensitized model in rats have failed to show that MPM is effective in preventing antibody production and subsequent allograft rejection.[137]

2. Toxic Side Effects

The toxic side effects reported with the use of MPM in dogs include weight loss, anorexia, and gastrointestinal signs following treatment with 40 mg/kg/day.[135] In the same study, a dose of 20 mg/kg/day in combination with CsA and steroids, was not associated with hematologic, hepatic, or nephrologic side effects. Clinical trials in psoriatic patients have suggested that mycophenolic acid is a safe drug in humans at levels of 20 to 60 mg/kg/day. Gastrointestinal symptoms were observed, however, during the first year of treatment. Viral infections (herpes zoster and simplex) were described in 10 to 25% of the patients during the treatment period.[138] Preliminary results from the first clinical trial in kidney transplant patients treated with different doses of MPM (100 to 3500 mg/day) have provided no evidence of organ toxicity or bone marrow depression.[139]

F. BREQUINAR SODIUM

Brequinar sodium is a novel antiproliferative compound that is a quinoline-carboxylic acid derivative selected for clinical evaluation because of its broad spectrum of antineoplastic activity *in vitro* and *in vivo*.[140] A series of *in vivo* and *in vitro* immune assays in mice and rats have demonstrated that Brequinar sodium (BQR) has potent immunosuppressive activity.[141,142]

BQR is a noncompetitive inhibitor of the enzyme dihydroorotate acid dehydrogenase(DHO-DH) in the *de novo* biosynthetic pathway of pyrimidine nucleotides. The enzyme DHO-DH catalyzes the oxidation of dihydroorotic acid to orotate.[143] BQR has

Table 7 **Experimental Transplantation Model Using Brequinar Sodium**

Species	Graft	Route	M.E.D.[a]	Ref.
Rat	Heart	Oral	6 mg/kg/3 × wk	15
	Kidney	Oral	6 mg/kg/3 × wk	15
	Liver	Oral	6 mg/kg/3 × wk	15
	Small bowel	Oral	5 mg/kg/3 × wk	175
Cynomolgus	Heart	Oral	2 mg/kg/3 × wk	142
Xenograft				
Hamster→	Heart	Oral	5 mg/kg/day	150
Rat	Liver	Oral	4 mg/kg/day	178

[a] M.E.D., Minimal effective dose.

been shown to be effective in suppressing the afferent/sensitization phase of the immune response. Additionally, BQR is a potent suppressor of an anamnestic contact sensitivity response.[141] BQR also effectively suppresses both T-cell dependent and T-cell independent antibody responses in mice.[144] BQR is a water soluble compound with high bioavailability when administered by mouth. This drug is rapidly absorbed from the gastrointestinal tract in all species tested and it exhibits a high level of protein binding in the plasma.[145] The breakdown products of BQR in the rat are excreted primarily in the urine (23%) and feces (66%). Its metabolic degradation has not been well characterized, but the presence of a large amount of the metabolites in the feces suggests significant hepatic metabolism of the drug.

1. Prevention of Allograft Rejection

The efficacy of BQR as an immunosuppressant *in vivo* in different transplant models has been recently evaluated in our laboratory.[142] This compound was tested for its ability to prevent the rejection of rat heart, kidney, and liver allograft at different doses (6, 12, and 24 mg/kg) administered three times weekly for a period of 4 weeks after surgery. The drug was highly effective in preventing allograft rejection in all of the models tested. In the case of the liver and kidney allograft recipients, the majority of the graft recipients developed a donor-specific tolerance that resulted in survival of the grafts for more than 200 days following treatment with BQR.[15,146] In contrast, cardiac grafts were rejected approximately 2 weeks after cessation of drug treatment, even if treatment was continued for 90 days. These data demonstrate that cardiac grafts are more resistent to the induction of stable tolerance.

When used in rat liver allograft recipients, BQR exhibits a high level of effectiveness in reversing ongoing rejection. The administration of 12 mg/kg BQR for a short period of time (days 6, 7, or 8) following transplantation, disrupts the rejection process and induces a permanent acceptance of the graft. Treatment with the same dosage schedule earlier or later in the rejection process is ineffective in inducing tolerance.[147]

One of the most important characteristics of the immunosuppressive activity of BQR is its synergistic interaction with other immunosuppressive drugs. The efficacy of subtherapeutic doses of BQR and CsA used in combination were studied in a rat heart allograft model. The combination of 1.5 to 3.0 mg/kg of BQR, given orally three times a week for 30 days, and 2.5 to 0.75 mg/kg of CsA for 14 days displayed a significant synergistic effect on the prolongation of graft survival.[148]

The ability of BQR to prevent antibody production make this drug particularly useful in special situations, such as presensitized patients and xenotransplantation, in which humoral immune responses play an important role. In a sensitized cardiac allograft model in rats, the administration of BQR 36 mg/kg three times weekly was effective in preventing both IgM and IgG production and prolonging heart graft survival.[149] The ability of BQR to inhibit antibody production is responsible for the most effective prolongation of concordant hamster-to-rat heart xenografts following treatment with a single drug.[150] Furthermore, the use in combination of subtherapeutic doses of BQR (3 mg/kg) and CsA (10 mg/kg) in the same model induces indefinitive xenograft survival (> 120 days).[151]

2. Toxic Side Effects

The toxic side effects of BQR result from the inhibition of pyrimidine synthesis and subsequent disruption of cell proliferation, particularly in the bone marrow. The adverse side effects we have observed in rodents, pigs, and cynomolgus monkeys consisted primarily of the antiproliferative effect of BQR on the gastrointestinal tract and bone marrow. The most common drug-related side effects were anorexia, leukopenia, diarrhea, and, in some cases, pancytopenia and hemorrhage when BQR was administered at higher dose levels.

The preliminary experimental data with BQR have demonstrated that this compound is a highly effective immunosuppressive drug, capable of preventing a wide variety of immune responses. This effective immunosuppressive activity, combined with the synergistic interaction of BQR with CsA and other immunosuppressive drugs, its ability to prevent antibody production, good bioavailability, and quantitative methods of monitoring drug blood levels indicate the potential for the use of BQR as a component of future regimens for preventing graft rejection in patients.[152]

IV. CONCLUSION

The immunosuppressive agents presently in use for clinical transplantation have been directly responsible for the increased number of patients who have been successfully treated with organ allografts. Allograft rejection still represents, however, one of the most frequent complications that limits the broader application of allografting as a therapeutic modality. Newer knowledge of the cellular and humoral components of the allograft rejection has provided insights on the characteristics important for the development of new immunosuppressive drugs. While early acute rejection is primarily mediated by T cells and their products, B cells and the antibodies they produce play an important role in the production of accelerated or hyperacute rejection of grafts in hypersensitized patients or following xenografting. Immunosuppressive agents that are capable of preventing both cellular and humoral immune responses will be important components of therapy in these forms of graft rejection. New drugs, such as 15-deoxyspergualin, Mycophenolate mofetil (RS-61443), and Brequinar sodium, have all been shown to be effective in inhibiting both T-cell and B-cell responses. Many of these agents display evidence of synergistic activities with the classically specific T-lymphocyte immunosuppressive agents, CsA and FK 506. In the future, significant experimental efforts will be directed at testing the usefulness of conventional immunosuppressive drugs when used in combination with these new immunosuppressive compounds. Identification of synergistic immunosuppressive regimens will allow maximum pharmacologic immunosuppression and a reduction in the occurrence of drug-related toxic side effects.

REFERENCES

1. **Gibson, T. and Medawar, P.B.,** The fate of skin homografts in man, *J. Anat.,* 77, 299, 1943.
2. **Schwartz, R. and Damashek, W.,** Drug-induced immunological tolerance, *Nature,* 183, 1682, 1959.

3. **Hitchings, G.H., and Elion, G.B.,** Chemical immunosuppression of the immune response, *Pharmacol. Rev.,* 15, 365-405, 1963.
4. **Bach, J.F. and Strom, T.B.,** *The Mode of Action of Immunosuppressive Agents,* 2nd ed., Elsevier, Amsterdam, 1986, 105-158.
5. **Starzl, T.E., Marchioro, R., and Waddell, W.,** The reversal of rejection in human renal homografts with subsequent development of homograft tolerance, *Surg. Gynecol. Obstet.,* 117, 385, 1963.
6. **Borel, J.F., Feurer, C., Magnee, C., and Staehelin, H.,** Effects of the new antilymphocyte peptide cyclosporin A in animals, *Immunology,* 32, 1017, 1977.
7. **Thomson, A.W., Whiting, P., Blair, J., Davidson, R.J., and Simpson, J.G.,** Pathological changes developing in the rat during a three week course of high-dose cyclosporin A and their reversal following drug withdrawal, *Transplantation,* 32, 271, 1982.
8. **Klintmalm, G.B., Iwatzuki, S., and Starzl, T.,** Nephrotoxicity of cyclosporin A in liver and kidney transplant patients, *Lancet,* 1, 470-471, 1981.
9. **Thiru, S., Calne, R.Y., and Nagington, J.,** Lymphoma in renal allograft patients treated with cyclosporin A as one of the immunosuppressive agents, *Transplant. Proc.,* 13, 359, 1981.
10. **Kino, T., Hatanaka, H., Hashimoto, M., et al.,** FK 506, a novel immunosuppressant isolated from a streptomyces. I. Fermentation, isolation, physico-chemical and biological characteristics, *J. Antibiot. (Tokyo),* 60, 1249-1255, 1987.
11. **Thomson, A.W.,** FK 506- How much potential?, *Immunol. Today,* 10, 6-9, 1989.
12. **Eng, C.P., Sehgal, S.N., and Vezna, C.,** Activity of Rapamycin (AY-22,989) against transplanted tumors, *J. Antibiot. (Tokyo),* 37, 1231-1237, 1984.
13. **Umeda, Y., Moriguchi, M., Kuroda, H. et al.,** Synthesis and antitumor activity of spergualin analogues. I. Chemical modification of 7guanidino-3-hydroxyacyl moiety, *J. Antibiot. (Tokyo),* 38, 836, 1985.
14. **Allison, A.C., Almquist, S.J., Muller, C.D. and Eugui, E.M.,** In vitro immunosuppressive effects of mycophenolic acid and an ester pro-drug, RS-61443, *Transplant. Proc.,* 23 (Suppl 2), 10-14, 1991.
15. **Cramer, D.V., Chapman, F.A., Jaffee, B.D., et al.,** The effect of a new immunosuppressive drug, brequinar sodium, on heart, liver, and kidney allograft rejection in the rat, *Transplantation,* 53, 303-308, 1992.
16. **Tilney, N.L. and Kupiec-Weglinski, J.W.,** The immunobiology of acute allograft rejection, in *Organ Transplantation, Current Clinical and Immunological Concepts,* Bailliere Tindall, Amsterdam, 1989, 19-37.
17. **Rudd, C.E.,** CD4, CD8 and the TCR-CD3 complex, a novel class of protein-tyrosine kinase receptor, *Immunol. Today,* 11, 400-405, 1990.
18. **Lechler, R.I., Lombardi, G., and Batchelor, J.R.,** The molecular basis of alloreactivity, *Immunol. Today,* 11, 83-88, 1990.
19. **Auron, P.E., Warner, S.J., and Webb, A.C.,** Studies of the molecular nature of human interleukin 1, *J. Immunol.,* 138, 1447-1456, 1987.
20. **Takahashi, N., Hayano, T., and Suzuki, M.,** Peptydil-propyl cis-trans cyclosporin A-binding protein cyclophilin, *Nature,* 337, 473-475, 1989.
21. **Bierer, B.E., Mattila, P.S., and Standaert, R.F.,** Two distinct signal transmission pathways in T lymphocytes are inhibited by complexes formed between an immunophilin and either FK 506 or Rapamycin, *Proc. Natl. Acad. Sci. U.S.A.,* 87, 9231-9235, 1990.
22. **Biron, C.A., Young, H.A., and Kasaian, M.T.,** Interleukin 2-induced proliferation of murine natural killer cells in vivo, *J. Exp. Med.,* 171, 173-188, 1990.
23. **Cantrell, P.A. and Smith, K.A.,** The interleukin-2 T cell system. A new cell growth model, *Science,* 224, 1312, 1984.

24. **Borel, J.F., Feurer, C., Gubler, H.U., and Staehelin, H.,** Biological effects of cyclosporin A, A new antilymphocytic agent, *Agents Actions,* 6, 468, 1976.
25. **Ryffel, B., Donatsch, P., Gotz, U., and Tschopp, M.,** Cyclosporine receptor on mouse lymphocytes, *Immunology,* 41,913, 1980.
26. **Schreiber, S.L.,** Chemistry and biology of the immunophilins and their immunosuppressive ligands, *Science,* 251, 283-287, 1991.
27. **Clipstone, N.A. and Crabtree, G.R.,** Identification of calcineurin as a key signalling enzyme in T-lymphocyte activation, *Nature,* 357, 695-697, 1992.
28. **O'Keefe, S.J., Tamura, J., Kincaid, R.L., Tocci, M.J., and O'Neill, E.A.,** FK-506 and CsA-sensitive activation of the interleukin-2 promoter by calcineurin, *Nature,* 357, 692-694, 1992.
29. **Kunzendorf, U., Brockmueller, J., Jochimsen, F., Roots, I., and Offermann, G.,** Immunosuppressive properties of cyclosporine metabolites, *Lancet,* 1, 734-735, 1989.
30. **Morris, P.J.,** Cyclosporin A, *Transplantation,* 32, 349-354, 1981.
31. **Green, C.J.,** Experimental transplantation and cyclosporine, *Transplantation,* 46(Suppl.), 3-10, 1988.
32. **Lems, S.P., Capel, P.J., and Koene, R.A.,** Prolongation of mouse skin allograft survival by cyclosporin A. Graft rejection after withdrawal of therapy, *Transplant. Proc.,* 12, 283, 1980.
33. **Borel, J.F. and Meszaros, J.,** Skin transplantation in mice and dogs, effect of cyclosporin A and dihydrocyclosporin C, *Transplantation,* 29, 161, 1980.
34. **White, D.J., Rolles, K., Ottawa, T., and Turrell, O.,** Cyclosporin A induced long term survival of fully incompatible skin and heart grafts in rats, *Transplant Proc.,* 12, 261, 1980.
35. **Jolley, W.B., Knierim, K., Ham, J., and Longerbeam, J.K.,** The effect of cyclosporine on simultaneous skin and pancreatic islet allograft in the rabbit, *Transplant. Proc.,* 15(Suppl. 1), 3011, 1983.
36. **Deeg, H.J., Storb, R., Gerhard-Miller, L., and Thomas, E.D.,** Cyclosporin A, a powerful immunosuppressant *in vivo* and *in vitro* in the dog, fails to induce tolerance, *Transplantation,* 29, 230-235, 1980.
37. **Kostakis, A.J., White, D.J., and Calne, R.Y.,** Prolongation of rat heart allograft survival by cyclosporin A, *IRSC Med. Sci.,* 5, 280, 1977.
38. **Jamieson, S.W., Burton, N.A., Beiber, C.P., Reitz, B.A., Oyer, P.E., Stinson, E.B., and Shumway, N.E.,** Cardiac allograft survival in primates treated with cyclosporin A, *Lancet,* 1, 545, 1979.
39. **Calne, R.Y., White, D.J., Rolles, K., Smith, D.P., and Herbertson, B.M.,** Prolonged survival of pig orthotopic heart graft treated with cyclosporin A, *Lancet,* 1, 1183-1185, 1978.
40. **Suzuki, S., Mizuochi, I., and Amemiya, H.,** Immunological study of cyclosporine in heterotopic transplantation of canine hearts, *Transplantation,* 39, 565-567, 1985.
41. **Veith, V.J., Norin, A.J., Montefusco, C.M., Pinsker, K.L., Kamholz, S.L., Gliedman, M.L., and Emeson, E.,** Cyclosporin A in experimental lung transplantation, *Transplantation,* 32, 474, 1981.
42. **Norin, A.J., Emeson, E.E., Kamholz, S.L., Pinsker, K.L., Montefusco, C.M., Matas, A.J., and Veith, F.J.,** Cyclosporin A as the initial immunosuppressive agent for canine lung transplantation, *Transplantation,* 34, 372-375, 1982.
43. **Reitz, B.A., Bieber, C.P., Raney, A.A., Pennock, J.L., Jamieson, S.W., Oyer, P.E., and Stinson, E.B.,** Orthotopic heart and combined heart and lung transplantation with cyclosporin-A immune suppression, *Transplant. Proc.,* 13, 393-396, 1981.
44. **Prop, J., Bartels, H.L., Petersen, A.H., Wildevuur, C.R., and Nieuwenhuis, P.,** A single injection of cyclosporin A reverses lung allograft rejection in the rat, *Transplant. Proc.,* 15, 511, 1983.

45. **Homan, W.P., Fabre, J.W., Williams, K.A., Millard, P.R., and Morris, P.J.,** Studies of the immunosuppressive properties of cyclosporine in rats receiving renal allograft, *Transplantation*, 29, 361-366, 1980.
46. **Green, C.J. and Allison, A.C.,** Extensive prolongation of rabbit kidney allograft survival after short-term cyclosporin A treatment, *Lancet,* 1, 1182-1183, 1978.
47. **Homan, W.P., French, M.E., Millard, P.R., and Morris, P.J.,** Studies of the effects of cyclosporin A upon renal allograft rejection in the dog, *Surgery,* 88, 168, 1980.
48. **Borleffs, J.C., Neuhaus, P., and Balner, H.,** Cyclosporin A as optimal immunosuppressant after kidney allografting in rhesus monkeys, *Heart Transplant.,* 2, 111, 1983.
49. **Engemann, R., Ulrichs, K., Thiede, A., Mueller-Ruchholtz, W., and Hamelmann, H.,** Graft tolerance after orthotopic liver transplantation in a primarily nontolerant rat strain combination following temporary cyclosporin A treatment, in *Microsurgical Models in Rats for Transplantation Research,* Thiede, A., Ed., Springer-Verlag, Berlin, 1985, 247.
50. **William, J.W., Peter, T.G., Haggitt, H.R., and van Voorst, S.,** Cyclosporine in transplantation of the liver in dogs, *Surg. Gynecol. Obstet.*, 156, 767, 1983.
51. **Todo, S., Porter, K.A., Kam, I., Lynch, S., Venkataramanan, R., DeWolf, A., and Starzl, T.E.,** Canine liver transplantation under Nva2-Cyclosporine versus cyclosporine, *Transplantation* 41, 296-300, 1986.
52. **Flye, M.W., Rodgers, G., Kacy, S., Mayschak, D.M., and Thorpe, L.,** Prevention of fatal rejection of SLA-mismatched orthotopic liver allograft in inbred miniature swine with cyclosporin A, *Transplant. Proc.,* 15, 1269-1271, 1983.
53. **Hatcher, P.A., Deaton, D.H., and Bollinger, R.R.,** Transplantation of the entire small bowel in inbred rats using cyclosporine, *Transplantation,* 43, 478-484, 1987.
54. **Saat, R.E., De Bruin, R.W., Heineman, E., Jeekel, J., and Marquet, R.L.,** The limited efficacy of cyclosporine in preventing rejection and graft-versus-host disease in orthotopic small bowel transplantation in rats, *Transplantation,* 50, 374-377, 1990.
55. **Grant, D., Duff, J., Zhong, R., Garcia, B., Lipohar, C., Keown, P., and Stiller, C.,** Successful intestinal transplantation in pigs treated with cyclosporine, *Transplantation,* 45, 279-284, 1988.
56. **Craddock, G.N., Nordgren, S.V., Reznick, R.K., Gilas, T., Lossing, A.G., Cohen, Z., Stiller, C.R., Cullen, J.B., and Langer, B.,** Small bowel transplantation in the dog using cyclosporine, *Transplantation,* 35, 284-288, 1983.
57. **Pritchard, T.J., Madara, J.L., Tapper, D., Wilmore, D.W., and Kirkman, R.L.,** Failure of cyclosporine to prevent small bowel allograft rejection in pigs, *J. Surg. Res.*, 38,553-558, 1985.
58. **Homan, W.P., Fabre, J.W., Millard, P.R., and Morris, P.J.,** Effect of cyclosporin A upon second-set rejection of rat renal allografts, *Transplantation,* 30, 354-357, 1980.
59. **Hall, B.M., Jelbart, M.E., Gurley, K.E., and Dorsch, S.E.,** Specific unresponsiveness in rats with prolonged cardiac allograft survival after treatment with CsA, *J. Exp. Med.*, 162, 1683-1694, 1985.
60. **Nagao, T., White, D.J., and Calne, R.Y.,** Kinetics of unresponsiveness induced by a short course of cyclosporin A, *Transplantation,* 33, 31, 1982.
61. **Kupiec-Weglinski, J.W., Filho, M.A., Strom, T.B., and Tilney, N.L.,** Sparing of suppressor cells, A critical action of cyclosporine, *Transplantation,* 38, 97-100, 1984.
62. **Deeg, H.J., Storb, R., Weiden, P.L., Graham, T., Atkinson, K., and Thomas, E.D.,** Cyclosporin A and methotrexate in canine marrow Transplantation, engraftment, graft-versus-host disease, and induction of tolerance, *Transplantation,* 34, 30-35, 1982.
63. **Whiting, P.H., Thomson, A.W., Blair, J.T., and Simpson, J.G.,** Experimental cyclosporin A nephrotoxicity, *Br. J. Exp. Pathol.*, 63, 88, 1982.

64. **Homan, W.P., French, M.E., and Morris, P.J.,** Effect of cyclosporin A upon the function of ischemically damaged renal autografts in the dog, *Transplantation,* 30, 228-230, 1980.
65. **Thomson, A.W., Whiting, P.H., Cameron, I.D., Lessels, S.E., and Simpson, J.G.,** A toxicological study in rats receiving immunotherapeutic doses of cyclosporin A, *Transplantation,* 31,121-124, 1981.
66. **Kahan, B.D.,** Cyclosporine nephrotoxicity, pathogenesis, prophylaxis, therapy and prognosis, *Am. J. Kidney Dis.*, 8, 323-331, 1986.
67. **Perico, N. and Remuzzi, G.,** Cyclosporine-induced renal dysfunction in experimental animals and humans, *Transplant Rev.*, 5, 63-79, 1991.
68. **Squifflet, J.P., Sutherland, D.E., Rynasiewicz, J.J., Field, J., Heil, J., and Najarian, J.S.,** Combined immunosuppressive therapy with cyclosporin A and azathioprine, *Transplantation,* 34, 315-318, 1982.
69. **Aeder, M.I., Sutherland, D.E., Lewis, W.I., and Najarian, J.S.,** Combination immunotherapy with low-dose cyclosporine and azathioprine in splenectomized canine recipients of renal allograft, *Transplant. Proc.,* 15(Suppl. 1), 2933-2937, 1983.
70. **Starzl, T.E., Klintmalm, G.B., Weil, R.III, Porter, K.A., Iwatsuki, S., Schroter, G.P., Fernandez-Bueno, C., and MacHugh, N.,** The use of cyclosporin A and prednisone in sixty-six cadaver kidney recipients, *Surg. Gynecol. Obstet.*, 153, 486-494, 1981.
71. **Simmons, R.L., Canafax, D.M., and Strand, M.,** Management and prevention of cyclosporine nephrotoxicity after renal Transplantation, use of low doses of cyclosporine, azathioprine and prednisone, *Transplant. Proc.,* 17(Suppl 1), 266-275, 1985.
72. **Murase, N., Todo, S., Lee, P.H., Fung, J.J., and Starzl, T.E.,** Heterotopic heart transplantation in the rat receiving FK-506 alone or with cyclosporine, *Transplant. Proc.,* 19, 71, 1987.
73. **Todo, S., Ueda, Y., Demetris, J.A., Imventarza, O., Nalesnik, M., Venkararaman, R., Makowka, L., and Starzl, T.E.,** Immunosuppression of canine, monkey and baboon allograft by FK 506, with special reference to synergism with other drugs and to tolerance induction, *Surgery,* 104, 239-249, 1988.
74. **Vathsala, A., Goto, S., Yoshimura, N., Stepkowski, S., Chou, T.C., and Kahan, B.D.,** The immunosuppresive antagonism of low doses of FK 506 and cyclosporine, *Transplantation,* 51, 232-239, 1991.
75. **Starzl, T., Todo, S., Fung, J., Demetris, A., Venkataramman, R., and Jain, A.,** FK 506 for liver, kidney and pancreas Transplantation, *Lancet,* 2, 1000-1004, 1989.
76. **Stepkowski, S.M. and Kahan, B.D.,** Rapamycin and cyclosporine synergistically prolong heart and kidney allograft survival, *Transplant. Proc.,* 23, 3262-3264, 1991.
77. **Kimball, P.M., Kerman, R.H., and Kahan, B.D.,** Production of synergistic but nonidentical mechanism of immunosuppression by rapamycin and cyclosporine, *Transplantation,* 51, 486-490, 1991.
78. **Kino, T., Hatanaka, H., and Miyata, S.,** FK 506, a novel immunosuppressant isolated from a Strewptomyces. II. Immunosuppressive effect of FK 506 in vitro, *J. Antibiot. (Tokyo)*, 40, 1256, 1987.
79. **Yoshimura, N., Matsui, S., Hamashima, T., and Oka, M.,** Effect of a new immunosuppressive agent, FK 506, on human lymphocyte responses in vitro. I. Inhibition of expression of alloantigen-activated suppressor cells, as well as induction of alloreactivity, *Transplantation,* 47, 351-356, 1989.
80. **Yoshimura, N., Matsui, S., Hamashima, T., and Oka, T.,** Effect of a new immunosuppressive agent, FK 506, on human lymphocyte responses in vitro. II. Inhibition of the production of IL-2 and gamma-INF, but not the B cell-stimulating factor 2, *Transplantation* 47, 356-359, 1989.

81. **Siekierka, J.J., Staruch, M.J., Hung, S.H., and Sigal, N.H.,** FK 506, a novel immunosuppressive agent, binds to a cytosolic protein which is distinct from the cyclosporin A-binding protein cyclophilin. *J. Immunol.,* 143, 1580-1583, 1989.
82. **Haeding, M.W., Galat, A., Uehling, D.E., and Schreiber, S.L.,** A receptor for the immunosuppressant FK 506 is a cis-trans peptidyl-prolyl isomerase, *Nature,* 341, 758-760, 1989.
83. **Omar, G., Ali Shah, I., Thomson, A.W., Whiting, P.H., and Burke, M.D.,** FK 506 inhibition of cyclosporine metabolism by human liver microsomes, *Transplant. Proc.,* 23, 934-935, 1991.
84. **Ochiai, T., Nakajima, K., Nagata, M., Suzuki, T., Asano, T., Uematsu, T., Goto, T., Hori, S., and Kenmochi, T.,** Effect of a new immunosuppressant agent, FK 506, on heterotopic cardiac allotransplantation in the rat, *Transplant. Proc.,* 19, 1284-1285, 1987.
85. **Ochiai, T., Nakajima, K., Nagata, M., Hori, S., Asano, T., and Isono, K.,** Studies of the induction and maintenance of long-term graft acceptance by treatment with FK 506 in heterotopic cardiac allotransplantation in rats. *Transplantation,* 44, 734-738, 1987.
86. **Hoffman, A.L., Makowka, L., Banner, B., Cai, X., Cramer, D.V., Pacualone, A., Todo, S., and Starzl, T.E.,** The use of FK 506 for small intestine allotransplantation. Inhibition of acute rejection and prevention of fatal GVHD, *Transplantation,* 49, 483-490, 1990.
87. **Murase, N., Demetris, A.J., Woo, J., Tanabe, M., Furuya, T., Todo, S., and Starzl, T.E.,** Graft-versus-host disease after Brown Norway-to-Lewis and Lewis-to-Brown Norway rat intestinal transplantation under FK506, *Transplantation,* 55, 1-7, 1993.
88. **Markus, P.M., Cai, X., Ming, W., Demetris, A.J., Fung, J.J., and Starzl, T.E.,** FK 506 reverses acute graft-versus-host disease after allogeneic bone marrow transplantation in rats, *Surgery,* 110, 357-364, 1991.
89. **Murase, N., Kim, D., Todo, S., Cramer, D.V., Fung, J., and Starzl, T.E.,** FK 506 suppression of heart and liver allograft rejection. II. The induction of graft acceptance in rats, *Transplantation,* 50, 739-744, 1990.
90. **Murase, N., Kim, D., Todo, S., Cramer, D.V., Fung, J., and Starzl, T.E.,** Suppression of allograft rejection with FK 506. I. Prolonged cardiac and liver survival in rats following short-course therapy, *Transplantation,* 50,186-189, 1990.
91. **Ochiai, T., Nagata, M., Nakajima, K., Suzuki, T., Sakamoto, K., Enomoto, K., Gunji, Y., Urmatsu, T., Goto, T., Hori, S., Kenmochi, T., Nakagouri, T., and Asano, T.,** Studies of the effects of FK 506 on renal allografting in the beagle dog, *Transplantation,* 44, 729-733, 1987.
92. **Ueda, Y., Todo, S., Eiras, G., Furukawa, H., Imventarza, O., Wu, Y.M., Oks, A., Zeevi, A., Oguma, S., and Starzl, T.E.,** Induction of graft acceptance after dog kidney or liver Transplantation, *Transplant. Proc.,* 22(Suppl. 1), 80-82, 1990.
93. **Thiru, S., Collier, D.J., and Calne, R.Y.,** Pathologic studies in canine and baboon renal allograft recipients immunosuppressed with FK 506, *Transplant. Proc.,* 19(Suppl. 6), 98-99, 1987.
94. **Haba, T., Hachisuka, T., Ohtosuka, S., Tanaka, Y., Satoh, E., Hayashi, S., and Takagi, H.,** Pathological and immunohistochemical examination in the rat treated with FK 506, *Transplant. Proc.,* 23, 2229-2232, 1991.
95. **Monden, M., Gotoh, M., Kanai, T., Valdivia, L.A., Umeshita, K., Endoh, W., Nakano, Y., Kawai, M., Ohzato, H., Ukei, T., Dono, K., and Tono, T.,** A potent immunosuppressive effect of FK 506 in orthotopic liver transplantation in primates, *Transplant. Proc.,* 22(Suppl. 1), 66-71, 1990.
96. **Shibata, T., Ogawa, N., Kaneko, I., Hokazono, K., and Omoto, R.,** Does FK 506 accelerate the development of coronary artery disease in transplanted heart as well as native heart?, *15th Int. Congr. Transplant. Soc.,* 24(Abstr.), 1992.

97. **Yamada, K., Sugisaki, Y., Akimoto, M., and Yamanaka, N.,** Short-term FK 506-induced morphological changes in rat kidneys, *Transplant. Proc.,* 23, 3130-3132, 1991.
98. **Fung, J.J., Alessiani, M., Abu-Elmagd, K., Todo, S., Shapiro, R., Tsakis, A., Van Thiel, D., Armitage, J., Jain, A., McCauley, J., Selby, R., and Starzl, T.E.,** Adverse effects associated with the use of FK 506, *Transplant. Proc.,* 23, 3105-3108, 1991.
99. **Moutabarrik, A., Ishibashi, M., Fukunaga, M., Kameoka, H., Takano, Y., Kokado, Y., Takahra, S., Jiang, H., Sonoda, T., and Okuyama, A.,** FK 506 mechanism of nephrotoxicity, stimulatory effect on endothelin secretion by cultured kidney cells and tubular cell toxicity *in vitro*, *Transplant. Proc.,* 23, 3133-3135, 1991.
100. **Fung, J., Abu-Elmagd, K., Todo, S., Shapiro, R., Tzakis, A., Jordan, M., Armitage, J., Jain, A., Bronster, O., Stieber, A., Kormos, R., Selby, R., Gordon, R., and Starzl, T.E.,** FK 506 in clinical organ Transplantation, *Clin. Transplant.,* 5, 517-522, 1991.
101. **Tze, W.J., Tai, J., Murase, N., Tzakis, A., and Starzl, T.E.,** Effect of FK 506 on glucose metabolism and insulin secretion in normal rats, *Transplant. Proc.,* 23, 3158-3160, 1991.
102. **Eidelman, B.H., Abu-Elmagd, K., Wilson, J., Fung, J.J., Alessiani, M., Jain, A., Takaya, S., Todo, S., Tzakis, A., Van Thiel, D., Shannon, W., and Starzl, T.E.,** Neurologic complications of FK 506, *Transplant. Proc.,* 23, 3175-3178, 1991.
103. **Matel, R.R., Klicius, J., and Galet, S.,** Inhibition of immune response by Rapamycin, a new antifungal antibiotic, *Can. J. Physiol. Pharmacol.,* 55, 48, 1977.
104. **Dumont, F.J., Staruch, M.J., Koprak, S.L., Melino, M.R., and Sigal, N.H.,** The immunosuppressive macrolides FK 506 and Rapamycin act as reciprocal antagonists in murine T cells, *J. Immunol.,* 144, 251-258, 1990.
105. **Metcalfe, S.M. and Richards, F.M.,** Cyclosporine, FK506, and Rapamycin. Some effects on early activation events in serum-free, mitogen-stimulated mouse spleen cells, *Transplantation,* 49, 798-802, 1990.
106. **Dumont, F.J., Melino, M.R., Staruch, M.J., and Koprak, S.L.,** The immunosuppressive macrolides FK506 and Rapamycin act as reciprocal antagonists in murine T cells, *J. Immunol.,* 144, 1418-1424, 1990.
107. **Kuo, C.J., Chung, J., Fiorentino, D.F., Flanagan, W.M., Blenis, J., and Crabtree, G.R.,** Rapamycin selectively inhibits interleukin-2 activation of p70 S6 kinase, *Nature,* 358, 70-73, 1992.
108. **Luo, H., Chen, H., Daloze, P., Chang, J., and Wu, J.,** Rapamycin suppress in vitro immunoglobulin production by human lymphocytes, *Transplant. Proc.,* 23, 2236-2238, 1991.
109. **Kahan, B.D., Chang, J.Y., and Seghal, S.N.,** Preclinical evaluation of a new potent immunosuppressive agent, Rapamycin, *Transplantation,* 52, 185-191, 1991.
110. **Eng, C.P., Gullo-Brown, J., Chang, J., and Seghal, S.N.,** Inhibition of skin graft rejection in mice by Rapamycin, A novel immunosuppressive macrolide, *Transplant. Proc.,* 23, 868-869, 1991.
111. **Morris, R.E., Meiser, B.M., Wu, J., Shorthouse, R., and Wang, J.,** Use of Rapamycin for the suppression of alloimmune reactions in vivo, schedule dependence, tolerance induction, synergy with cyclosporine and FK 506, and effect on host-versus-graft and graft-versus-host reactions, *Transplant. Proc.,* 23, 521, 1991.
112. **Calne, R.Y., Lim, S., Samaan, A., Collier, D.J., Pollard, S.G., and White, D.J.,** Rapamycin for immunosuppression in organ allografting, *Lancet,* 2, 227, 1989.
113. **Stepkowski, S.M., Chen, H., Daloze, P., and Kahan, B.D.,** Rapamycin, a potent immunosuppressive drug for vascularized heart, kidney and small bowel transplantation in the rat, *Transplantation,* 51, 22-26, 1991.
114. **Collier, D.S., Calne, R.Y., Pollard, S.G., Friend, P.J., and Thiru, J.,** Rapamycin in experimental renal allograft in primates, *Transplant. Proc.,* 23, 2246-2247, 1991.

115. **Whiting, P.H., Woo, J., Adam, B.J., Hasam, N.U., Davidson, R.J., and Thomson, A.W.,** Toxicity of rapamycin — a comparative and combination study with cyclosporine at immunotherapeutic dosage in rat, *Transplantation,* 52, 203-208, 1991.
116. **Kerr, P.G. and Atkins, R.S.,** The effects of deoxyspergualin in lymphocytes and monocytes in vivo and in vitro, *Transplantation,* 48, 1048-1052, 1989.
117. **Tepper, M.A., Petty, B., Bursuker, I., Pasternak, R.D., Cleaveland, J., Spitalny, G.L., and Schacter, B.,** Inhibition of antibody production by the immunosuppressive agent, 15-deoxyspergualin, *Transplant. Proc.,* 23,328-331, 1991.
118. **Dickneite, G., Schorlemmer, H.U., and Walter, P.,** The influence of 15-deoxyspergualin on experimental transplantation and its immunopharmacological mode of action, *Boehring Inst. Mitt.,* 80, 93, 1986.
119. **Dickneite, G., Schorlemmer, H.U., Walter, P., and Sedlacek, H.H.,** Graft survival in experimental transplantation could be prolonged by the action of the antitumoral drug 15-deoxyspergualin, *Transplant. Proc.,* 18, 1295-1296, 1986.
120. **Walter, P., Thies, J., Harbauer, G., Dickneite, G., Sedlacek, H.H., and Vonnahme, F.,** Allogeneic heart transplantation in the rat with a new antitumoral drug, 15-deoxyspergualin, *Transplant. Proc.,* 18, 1293-1294, 1986.
121. **Suzuki, S., Kanashiro, M., and Amemiya, H.,** Effect of a new immunosuppressant, 15-deoxyspergualin, on heterotopic rat heart transplantation in combination with cyclosporine, *Transplantation,* 44, 483-487, 1987.
122. **Walter, P., Dickneite, G., Feifel, G., and Thies, J.,** Deoxyspergualin induces tolerance in allogeneic kidney Transplantation, *Transplant. Proc.,* 19, 3980-3981, 1987.
123. **Nemoto, K., Hayashi, M., Ito, J., Sugawara, Y., Mae, T., Fujii, H., Abe, F., Fujii, A., and Takeuchi, T.,** Deoxyspergualin in lethal murine graft-versus-host disease, *Transplantation,* 51, 712-715, 1991.
124. **Amemiya, H., Suzuki, S., Niiya, S., Fukao, K., Yamanaka, N., and Ito, J.,** A new immunosuppressive agent, 15-deoxyspergualin as a immunosuppressive in dog renal allografting, *Transplant. Proc.,* 21, 3468-3470, 1989.
125. **Koyama, I., Kadokura, M., Hoshino, T., and Omoto, R.,** Effective use of 15-deoxyspergualin in kidney Transplantation, *Transplant. Proc.,* 21, 1088-1089, 1989.
126. **Fukao, K., Otsuka, M., Iwasaki, H., Yuzawa, K., and Iwasaki, Y.,** Immunosuppressive effect of deoxyspergualin on acute renal allograft rejection in dogs, *Transplant. Proc.,* 21, 1090-1093, 1989.
127. **Fukao, K., Iwasaki, H., Yuzawa, K., Otsuka, M., Yu, Y., Sharma, N., Iwasaki, Y., Hori, T., Murayama, Y., Terao, K., and Yamashita, J.,** Immunosuppressive effect and toxicity of 15-deoxyspergualin in cynomolgus monkeys, *Transplant. Proc.,* 23, 556-558, 1991.
128. **Kapelanski, D.P., Perelman, M.J., Faber, L.A., Paez, D.E., Rose, E.F., and Behrendt, D.M.,** 15-deoxyspergualin and primate heart Transplantation, *Heart Transplant.,* 9, 668-674, 1990.
129. **Reichenspurner, H., Hildebrandt, A., Human, P.A., Boehm, D.H., Rose, A.G., Odell, J.A., Reichart, B., and Schorlemmer, H.U.,** 15-deoxyspergualin for induction of graft nonreactivity after cardiac and renal allotransplantation in primates, *Transplantation,* 50, 181-185, 1990.
130. **Amemiya, H., Suzuki, S., Ota, K., Takahashi, K., Sonoda, T., Ishibashi, M., Omoto, R., Koyama, I., Dohi, K., Fukuda, Y., and Fukao, K.,** A novel rescue drug, 15-deoxyspergualin. First clinical trials for recurrent graft rejection in renal recipients, *Transplantation,* 49, 337-343, 1990.
131. **Morris, R.E., Hoyt, E.G., Murphy, M.P., Eugui, E.M., and Allison, A.C.,** Mycophenolic acid morpholinoethylester (RS-61443) is a new immunosuppressant that prevents and halts heart allograft rejection by selective inhibition of T- and B-cell purine synthesis, *Transplant. Proc.,* 22, 1659-1662, 1990.

132. **Allison, A.C., Hovi, T., Watts, R.W., and Webster, A.D.,** the role of the "de novo" purine synthesis in lymphocyte transformation, *Ciba Found. Symp.*, 48, 207, 1977.
133. **Burlingham, W.J., Grailer, A.P., Hullet, D.A., and Sollinger, H.W.,** Inhibition of both MLC and in vitro Ig G memory response to tetanus toxoid by RS-61443, *Transplantation,* 51, 545-547, 1991.
134. **Morris, R.E., Wang, J., Blum, J.R., Flavin, T., Murphy, M.P., and Almquist, S.J.,** Immunosuppressive effects of the morpholinoethylester of MPA (RS-61443) in rat and nonhuman primate recipients of heart allograft, *Transplant. Proc.,* 23(Suppl. 2), 19-25, 1990.
135. **Platz, K.P., Sollinger, H.W., Hullet, D.A., Eckhoff, D.E., Eugui, E.M., and Allison, A.C.,** RS-61443, a new, potent immunosuppressive agent, *Transplantation,* 51, 27-31, 1991.
136. **Platz, K.P., Bechstein, W.O., Eckhoff, D.E., Suzuki, Y., and Sollinger, H.W.,** RS-61443 reverses acute allograft rejection in dogs, *Surgery,* 110, 736-741, 1991.
137. **Knechtle, S.J., Wang, J., Burlingham, W.J., Beeskau, M., Subramanian, R., and Sollinger, H.W.,** The influence of RS-61443 on antibody-mediated rejection, *Transplantation,* 53, 699-701, 1992.
138. **Marinari, R., Fleischmajer, R., Schagger, A.H., and Rosenthal, A.L.,** Mycophenolic acid in the treatment of psoriasis, *Arch. Dermatol.,* 113, 930, 1977.
139. **Sollinger, H.W., Eugui, E.M., and Allison, A.C.,** RS-61443, Mechanism of action, experimental and early clinical results, *Clin. Transplant.,* 5, 523-526, 1991.
140. **Peters, G.J., Nadal, J.C., Laurensee, E.J., de Kant, E., and Pinedo, H.M.,** Retention of in vivo antipyrimidine effects of Brequinar sodium in murine liver, bone marrow and colon cancer, *Biochem. Pharmacol.,* 39, 135-144, 1990.
141. **Jaffee, B.D., Jones, E., Cheng, J., Ray, M.R., Cramer, D.V., and Makowka, L.,** Brequinar sodium is a novel immunosuppressive agent with a unique mechanism of action, *Proc. 10th Ann. Meet. Am. Soc. Transplant Phys.,* 270(Abstr.), 1991.
142. **Makowka, L. and Cramer, D.V.,** Brequinar sodium, mode of action and effects on graft rejection, in *Immunosuppressive Drugs, Developments in Anti-Rejection Therapy,* Thomson, A.W., Starzl, T.E., and Kent, T.E., Eds., Edward Arnold, London, in press.
143. **de Kant, E., Pinedo, H.M., Laurensee, E., and Peters, G.J.,** The relation between inhibition of cell growth and of dihydroorotic acid dehydrogenase by Brequinar sodium, *Cancer Lett.,* 46, 123-127, 1989.
144. **Jaffee, B.D., Jones, E.A., Loveless, S.E., and Chen, S.F.,** The unique immunosuppressive activity of Brequinar sodium, *Transplant. Proc.,* 25 (Suppl. 2), 19-22, 1993.
145. **Schwartsmann, G., van der Vijgh, W.J., van Hennik, M.B., Klein, I., et al.,** Pharmacokinetics of Brequinar sodium (NSC 368390) in patients with solid tumors during phase I study, *Eur. J. Cancer Clin. Oncol.,* 25, 1675-1681, 1989.
146. **Cramer, D.V., Chapman, F.A., Jaffee, B.D., Eiras-Hreha, G., Yasunaga, C., Wu, G.D., and Makowka, L.,** The effect of a new immunosuppressive drug, Brequinar sodium, on concordant hamster-to-rat cardiac xenografts, *Transplant. Proc.,* 24, 720-721, 1992.
147. **Cramer, D.V., Knoop, M., Chapman, F.A., and Makowka, L.,** Short course therapy with Brequinar sodium prevents liver allograft rejection in rats, *Transplantation,* 54, 752-753,1992.
148. **Cosenza, C.A., Cramer, D.V., Hreha-Eiras, G., Cajulis, E., Wang, H., and Makowka, L.,** Brequinar sodium and cyclosporine are synergistic when used in combination to prevent allograft rejection in the rat, *Transplantation,* in press.
149. **Yasunaga, C., Cramer, D.V., Cosenza, C.A., Tuso, P.J., Chapman, F.A., Barnett, M., Wu, G., Putnam, B.A., and Makowka, L.,** The effect of Brequinar sodium on in vivo antibody production, *Transplant. Proc.,* in press.

150. **Cramer, D.V., Chapman, F.A., Jaffee, B.D., Zajac, I., Hreha-Eiras, G., Yasunaga, C., Wu, G.D., and Makowka, L.,** The prolongation of concordant hamster-to-rat cardiac xenografts by Brequinar sodium, *Transplantation,* 54, 403-408,1992.
151. **Cosenza, C.A., Tuso, P.J., Chapman, F.A., Middleton, Y.D., Cramer, D.V., Wu, G., and Makowka, L.,** Prolonged xenograft survival following combination therapy with brequinar sodium and cyclosporine, *Transplant. Proc.,* 25 (Suppl. 2), 59-60, 1993.
152. **Makowka, L. and Cramer, D.V.,** Brequinar sodium, a new immunosuppressive drug for Transplantation, *Transplant Sci.,* 2, 50-54, 1992.
153. **Guillen, F.J., Hancock, W.W., Towpik, E., Kupiec-Weglinski, J.W., Rickles, F.R., Tilney, N.L., and Murphy, G.F.,** Inhibition of rat skin rejection by cyclosporine. In situ characterization of the impaired local immune response, *Transplantation,* 41, 734-739, 1986.
154. **Tutschka, P.J., Beschorner, W.E., Allison, A.C., and Santos, G.W.,** Use of cyclosporin A in allogeneic bone marrow transplantation in the rat, *Nature,* 20, 284, 1979.
155. **Rynasiewicz, J.J., Sutherland, D.E., Kawahara, K., and Najarian, J.S.,** Cyclosporine prolongation of segmental pancreatic and islet allograft function in rats, *Transplant. Proc.,* 12, 270, 1980.
156. **Klempnauer, J., Wagner, E., Steiniger, B., Wonigeit, K., and Pichlmayr, R.,** Pancreas and kidney allograft rejection responds differently to cyclosporine immunosuppression, *Transplant. Proc.,* 15(Suppl. 1), 3001, 1983.
157. **Dugoni, W.E. and Bartlett, S.T.,** Evidence that cyclosporine prevents rejection and recurrent diabetes in pancreatic transplants in the BB rat, *Transplantation,* 49, 845-848, 1990.
158. **Kyriakides, G.K., Olson, L., Flaa, C., and Miller, J.,** Reversal of kidney and prevention of pancreas transplant rejection with cyclosporine in beagles, *Transplant. Proc.,* 15(Suppl. 1), 2950-2952, 1983.
159. **Walter, P., Menger, M.D., Thies, J., Wolf, B., and Dickneite, G.,** Prolongation of graft survival in allogeneic limb transplantation by 15-deoxyspergualin, *Transplant. Proc.,* 21, 3186, 1989.
160. **Diliz-Perez, H.S., McClure, J., Bedetti, C., Hong, H-Q., de Santibanez, E., Shaw, B.W., Van Thiel, D., Iwatsuki, S., and Starzl, T.E.,** Successful small bowel allotransplantation in dogs with cyclosporine and prednisone, *Transplantation,* 37, 126-129, 1984.
161. **Cosimi, A.B., Shield, C.F., and Peters, C.,** Prolongation of allograft survival by cyclosporin A, *Surg. Forum,* 30, 287, 1978.
162. **Neuhaus, P., Borleffs, J.C., Maequet, R.L., and Balner, H.,** Results of kidney transplantation in rhesus monkeys treated with cyclosporin A and standard immunosuppression, *Transplant. Proc.,* 14, 111-112, 1982.
163. **Du Toit, D.F., Heydenrych, J.J., Louw, G., Zuurmond, T., Laker, L., Els, D., and Woolfe-Coote, S.,** The effect of cyclosporine alone and in combination with steroids on experimental segmental pancreatic allograft in the baboon, *Transplant. Proc.,* 15(Suppl. 1), 2992-2995, 1983.
164. **Inamura, N., Nakahara, K., Kino, T., Goto, T., Aoki, H., Yamaguchi, I., Kohsaka, M., and Ochiai, T.,** Prolongation of skin allograft survival in rats by a novel immunosuppressive agent, FK 506, *Transplantation,* 45, 206-209, 1988.
165. **Tsuchimoto, S., Kusumoto, K., Nakajima, Y., Kakita, A., Uchino, J., Natori, T., and Aizawa, M.,** Orthotopic liver transplantation in rats receiving FK 506, *Transplant. Proc.,* 21, 1064-1065, 1989.
166. **Katayama, Y., Yada, I., Namikama, S., and Kusagawa, M.,** Immunosuppressive effects of FK 506 in rat lung Transplantation, *Transplant. Proc.,* 23, 3301, 1991.

167. **Yasunami, Y., Ryu, S., Kamei, T., and Konomi, K.,** Effects of a novel immunosuppressive agent, FK 506, on islet allograft survival in the rat, *Transplant. Proc.,* 21, 272, 1989.
168. **Markus, P.M., Cai, X., Selvaggi, G., Cooper, M., Harnaha, J., Fung, J.J., and Starzl, T.E.,** The effect of cyclosporine, rapamycin and FK 506 in the survival following bone marrow Transplantation, *Transplant. Proc.,* 23, 3232-3233, 1991.
169. **Collier, D.J., Calne, R.Y., Thiru, S., Friend, P.J., Lim, S., White, D.J., Kohno, H., and Levickis, J.,** FK 506 in experimental renal allograft, *Transplant. Proc.,* 19, 3975-3979, 1987.
170. **Yokoshime, H., Hirai, T., Inui, K., Hasegawa, S., Aoki, M., Wada, H., and Hitomi, S.,** Immunosuppressive effects of FK 506 in canine lung Transplantation, *Transplant. Proc.,* 23, 3302, 1991.
171. **Lautenschlager, I., Hoeckerstedt, K., Maekisalo, H., Orko, R., and Taskinen, E.,** Efficiency of FK 506 and CsA to prevent acute cellular rejection of pig liver allograft, *Transplant. Proc.,* 23, 2233-2235, 1991.
172. **Imventarza, O., Todo, S., Eiras, G., Ueda, Y., Furukawa, H., Wu, Y.M., Zhu, Y., Oks, A., Demetris, A.J., and Starzl, T.E.,** Renal transplantation in baboons under FK 506, *Transplant. Proc.,* 22(Suppl. 1),64-65, 1990.
173. **Ericzon, B.G., Kubota, K., Groth, C.G., Wijnen, R., Tiebosch, T., Buurman, W., and Koostra, G.,** Pancreaticoduodenal allotransplantation with FK 506 in the Cynomolgus monkey, *Transplant. Proc.,* 22(Suppl. 1), 72-73, 1990.
174. **Chen, H.F., Wu, J.P., Luo, H.Y., and Daloze, P.M.,** The immunosuppressive effect of Rapamycin on pancreaticoduodenal transplants in the rat, *Transplant. Proc.,* 23, 2239-2240, 1991.
175. **Fabian, M.A., Collins, B., Jaffee, B., and Bollinger, R.R.,** Prolonged survival of small intestinal allograft using brequinar sodium, a specific inhibitor of the pyrimidine biosynthetic pathway, *15th Int. Congr. Transplant. Soc.,* 393(Abstr.), 1992.
176. **Schubert, G., Stoffregen, C., Timmermann, W., Schang, T., and Thiede, A.,** Comparison of the new immunosuppressive agent 15-deoxyspergualin and cyclosporin A after highly allogeneic pancreas Transplantation, *Transplant. Proc.,* 19, 3978-3979, 1987.
177. **Bechstein, W.O., Schilling, M., Steele, D.M., Hullett, D.A., and Sollinger, H.W.,** RS-61443-cyclosporine combination therapy prolongs canine liver allograft survival, *15th Int. Congr. Transplant. Soc.,* 384(Abstr.), 1992.
178. **Murase, N., Starzl, T.E., Demetris, A.J., Valdivia, L., Tanabe, M., Cramer, D.V., and Makowka, L.,** Hamster to rat heart and liver xenotransplantation with the FK 506 plus antiproliferative drugs, *Transplantation,* 55, 701-708, 1993.

Index

INDEX

C

D

E

F

G

H

I

M

N

O

P

R

S

T

U

V

X